P9-DNQ-280

Manual of Ornithology

Yale University Press   New Haven and London

Manual of Ornithology
Avian Structure & Function

Noble S. Proctor
Southern Connecticut State University

Patrick J. Lynch
Yale University School of Medicine

Illustrated by Patrick J. Lynch

Foreword by Roger Tory Peterson

Copyright © 1993 by Yale University.

All rights reserved.
This book may not be reproduced, in whole
or in part, including illustrations, in any
form (beyond that copying permitted by
Sections 107 and 108 of the U.S. Copyright
Law and except by reviewers for the public
press), without written permission from the
publishers.

Designed by Patrick J. Lynch.
Set in Adobe Palatino, Michelangelo, and
Helvetica typefaces. Printed in the United
States of America by Edwards Brothers, Inc.,
Ann Arbor, Michigan.

*Library of Congress Cataloging-in-
Publication Data*
    Proctor, Noble S.
        Manual of ornithology : avian
structure and function / Noble S. Proctor,
Patrick J. Lynch ; illustrated by Patrick J.
Lynch ; foreword by Roger Tory Peterson.
            p.   cm.
        Includes bibliographical references and
index.
        ISBN 0-300-05746-6
        1. Birds—Anatomy—Laboratory
manuals. I. Lynch, Patrick J., M. S.   II. Title.
    QL697.P76   1993
    598.2 '4 '078—dc20
        92-17066

A catalogue record for this book is available
from the British Library.

The paper in this book meets the guidelines
for permanence and durability of the
Committee on Production Guidelines for
Book Longevity of the Council on Library
Resources.

10   9   8   7   6   5   4   3   2

To Carolyn, Adam, and Eric

To Cheryl, Kathryn, and Virginia

Contents

Foreword

Noble Proctor has watched birds obsessively all over the globe while concurrently for some twenty years he has taught ornithology at the university level. As senior author of this new manual, he bridges the gap between the classroom with its academic procedures and the field, where the binocular and the telescope come into play. With eyes that are nearly as sharp as those of a red-tailed hawk, Noble Proctor is almost embarrassing to most other field birders whose vision is a mere twenty-twenty.

Patrick Lynch began his interest in birds in Noble's class and coupled it with his wonderful ability as an artist and biological illustrator. The two men have combined their talents to produce a manual that shows to best advantage the teacher and the artist. The informative text by Proctor is illustrated by more than 235 new drawings by Lynch, far more than in any other such book. The artwork is accurate in every way, and rather than the usual two-dimensional presentation of biological illustrations, Lynch has added a third-dimensional aspect that makes such details as skeletal and internal structures more clearly understandable. For the first time, many complex structures, such as the nervous and the air sac systems, are shown in situ rather than as bland diagrams. As we slowly evolve from the age of dissection, here is a book that will help us to understand the internal structuring of birds without the need to dissect them in the laboratory. Another unique feature is the presentation of complex information in attractive charts that are easy to interpret. The information has been gathered and presented in a quick and easy-to-understand manner. This will be the perfect book for any ornithology course, acting as a lab manual as well as a general reference work. For the less intensive course it can serve as a basic text. Anatomists will find it useful and accurate, and any birder will find it a gold mine of facts. Inasmuch as birds are the only living creatures with feathers, the chapter on feathers is of special interest.

Sections on bird migration and weather patterns represent years of field experience by both authors. I could elaborate on a myriad of other innovations that Proctor and Lynch have brought to this manual; you will find visual information at every turn of the page. Without doubt, this presentation will make a significant impact on academia while at the same time being easy to use for the layperson. Too often within the world of academia is information presented in a static, albeit well-documented, format. Proctor and Lynch have broken this static mold, presenting the same high-quality packet of information in a more visually stimulating format. Every library and biology department, as well as every birder, should have a copy close at hand.

Roger Tory Peterson
East Lyme, Connecticut

Preface

This book had its origins more than fifteen years ago, when as teacher and student we commiserated over the lack of a good laboratory manual for ornithology courses. It is ironic that although birds may be the most heavily illustrated of all animal groups, in vertebrate biology courses their anatomy and natural history are often slighted. There are many excellent references on particular facets of avian anatomy, but these papers and chapters in books are scattered throughout the professional literature and are often only available to those students and professors lucky enough to have easy access to large research libraries. This manual is our attempt to collect under one cover the visual references necessary for an undergraduate laboratory course in ornithology. As the book progressed, we also realized that a good deal of the material we chose to include bordered on the information one might normally find in an ornithology textbook. Walking the tightrope between textbook and reference manual has been difficult. We did not want to limit the use of this book to formal college courses. We hope the book will address the entire spectrum of interest in bird anatomy, from the backyard birder looking for facts about a particular aspect of a bird's life to artists and naturalists who require detailed references to avian topography and internal anatomy. We hope there will be something in this text for everyone and that it will become a useful reference for all. When used as a laboratory manual, this book should supplement such standard ornithology texts as Gill's *Ornithology* or Welty and Baptista's *Life of Birds*.

For the instructor of the ornithology lab we have provided worksheets at the end of each chapter; these may be photocopied from the book and used directly or may serve as the basis for class projects and discussions. The structure of each chapter assumes that a variety of laboratory specimens are available as references for students. We have also assumed that students will be performing at least a few direct dissections of preserved or fresh birds. If preserved bird skins, skeletal references, and other laboratory specimens are not available, this text should be supplemented with additional books, photographs, videotapes, and other suitable references to the spectrum of avian anatomy and behavior. Although we do not advocate laboratory manuals as a substitute for dissection, we recognize that dissections are increasingly unpopular in undergraduate courses. For this reason, we have tried to make the figures in the internal anatomy chapters as detailed as possible. In choosing which species to illustrate, we have usually chosen genera of birds that are reasonably common throughout the United States, Canada, and Great Britain. We urge laboratory instructors to collect as wide a variety of specimens as possible to expose students to the full range of avian diversity. We have also developed a chart reference format and applied this concept throughout the book. All too often, students are overwhelmed by anatomic terminology, such as muscle origins, insertions, and functions. In this book, terminology references have been consolidated in tables scattered at appropriate points throughout the text. We hope that both students and their instructors will find this approach helpful.

Many people have given generously of their time and advice during the preparation of this volume. We are especially grateful to the following individuals for their comments, suggestions, assistance, and counsel: Jon Ahlquist, Grit Ardwin, Larry Balch, Benton Basham, David Bolinsky, Milan Bull, Chester Faunce, Davis Finch, Adam Fry, Frank Gallo, Dan Gibson, Stephen Gilbert, William Hamilton, Carl Jaffe, Peter Marra, William Martha, Les Mehrhoff, John Ostrom, Roger Tory Peterson, Virginia Peterson, Wayne Petersen, Linda Seigneur, Fred Sibley, Virginia Simon, Jeffrey Spendelow, Eleanor Stickney, Thede Tobish, Edward Tufte, and all the ornithology students over the years whose comments and suggestions improved this book in substantial ways. Each has contributed to this volume, but we alone are responsible for any errors of fact or emphasis that remain. Special thanks to Walter Bock, George Clark, and Steve Zack for their time and effort in reviewing the entire manuscript.

We thank our wives, Carolyn and Cheryl, and our children, Adam, Eric, and Kathryn, for their support and encouragement over the course of this project. At Yale University Press, we thank our editors, Edward Tripp, Jean Thomson Black, and Laura Dooley, and designer Jim Johnson for their help (and patience!) during the preparation of this book.

Noble S. Proctor
Branford, Connecticut

Patrick J. Lynch
Madison, Connecticut

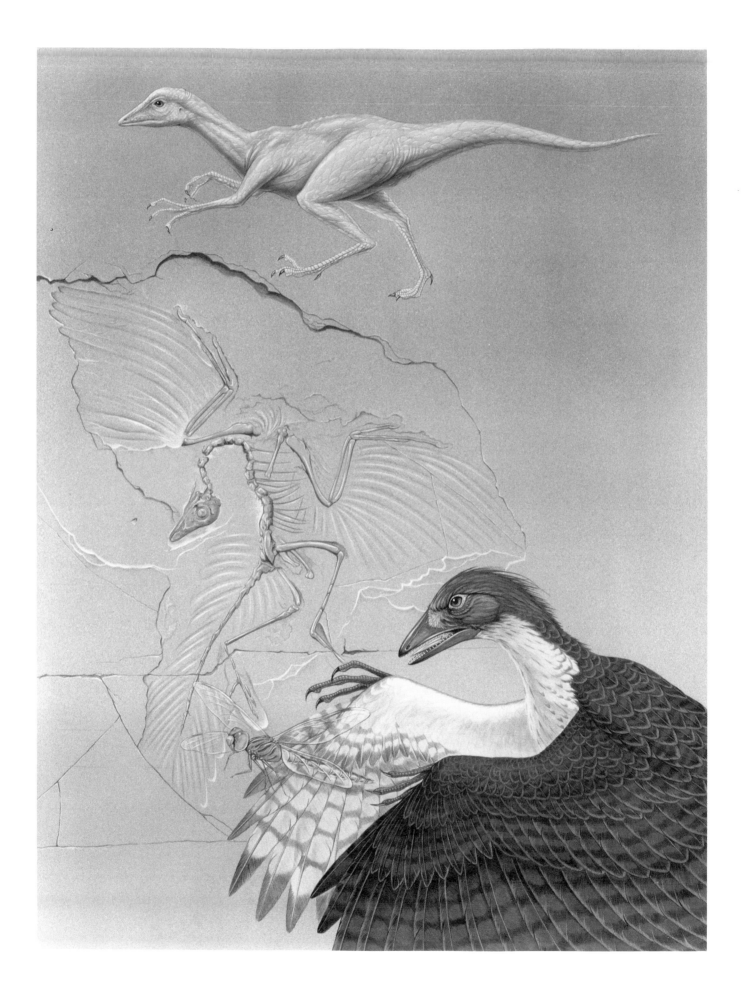

Introduction

1

Since the earliest humans first recognized birds as a separate life-form, birds have been inextricably woven into the fabric of human life. Humans were undoubtedly first interested in birds as a source of food. But human interaction with birds has evolved to the point at which legends, folklore, and myths concerning birds have become a part of every culture on earth, and birds or bird eggs remain an essential part of the diet of most people on the planet.

The first explicit human references to birds can be seen in Paleolithic cave paintings created nearly fifteen thousand years ago. Since then birds have appeared in human artwork and visual symbolism and as metaphoric references in the literature of most cultures. We are all familiar with the depiction of birds of prey as symbols of war or power and as messengers of the gods. In contrast, the dove is almost universally seen as a symbol for peace or love (except to the Japanese, who in ancient times considered the dove a symbol for war!). Doves appear often in biblical passages and Christian visual symbolism, where they are associated with the love of Jesus Christ or the Virgin Mary. Other bird families have played major roles in the legends and folklore of many cultures. To Native Americans of the Pacific Northwest, the Common Raven (*Corvus corax*) was the symbol of a wide variety of human interactions from betrayal to guidance and was even seen as a symbol of death. Avian symbols remain a part of modern life around the world. In Europe, storks have been symbols of childbirth since the Middle Ages. Eagles have symbolized power, strength, and faithfulness since Roman times, and the Bald Eagle (*Haliaeetus leucocephalus*) appears on most of the official symbols of the United States of America. Benjamin Franklin was a vocal advocate of the Wild Turkey (*Meleagris gallopavo*) as an appropriate emblem for the fledgling republic but was outvoted by supporters of the Bald Eagle. An observant naturalist, Franklin saw the Wild Turkey as a wary and intelligent bird with majestic display plumage, but today the word *turkey* is synonymous with foolishness or stupidity. In most European cultures, owls are seen as symbols of intelligence and mystery, and they feature prominently in many of our most popular children's tales and movies. Owls can also symbolize death or the dread of dark places, and the Barn Owl (*Tyto alba*) is probably the source of most ghost myths that surround the abandoned buildings Barn Owls favor as nest sites.

In music the songs of birds have long been a source of inspiration, and many familiar classical compositions refer to bird songs or calls. With its clear, diagnostic "cuck-oo" call the Common Cuckoo (*Cuculus canorus*) is the species most often imitated in musical references. In the eighteenth century Luigi Boccherini wrote perhaps the most complex of all musical pieces using bird songs. His work *The Aviary* incorporates many readily identifiable bird songs.

People have always noted the movements and behavior of birds to learn more about the world around them. In their drive to explore beyond their South Sea island homes, the Polynesians recognized that Asiatic Golden Plovers (*Pluvialis fulva*) arrived each autumn to spend the winter in Polynesia and disappeared northward for the summer months. They guessed that the birds must be traveling to some other substantial landmass to the north, and so they set sail, following the migrating plovers to find their summer breeding grounds. And find it they did—the Hawaiian Islands. Many Central African cultures have developed a unique relationship with the Black-throated Honeyguide (*Indicator indicator*). As its name suggests, the honeyguide leads human hunters to the nests of wild honeybees, and the villagers then reward the bird with the wax combs that are plundered from the bees' nest. In one of the most famous uses of birds as a source of information, the coal miners of nineteenth-century England used caged canaries (*Serinus canaria*) to gauge the health of the air within the coal mine. If the miner's canary remained alert and healthy, the air in the mine shaft was probably not contaminated with deadly methane gas.

The annual movements of birds continue to fascinate many people, and the spring and fall migrations are two of the most evocative signs of the changing seasons. The arrival of Cliff Swallows (*Hirundo pyrrhonota*) at California's San Juan Capistrano Mission each spring is well publicized, and the little Ohio town of Hinkley has gained notoriety for celebrating the annual spring arrival of their "buzzards" (actually Turkey Vultures, *Cathartes aura*). In fact, most bird species are just as regular in their migratory patterns. No one speaks of "the Phoebes of New Haven, Connecticut," or the "Sharp-tailed Sparrows of Brigantine, New Jersey," yet these birds arrive as regularly at their nesting grounds as do the swallows of Capistrano or the vultures of Hinkley. To see this yourself, set up a bird feeder or nesting box in your yard, and note the date your common backyard birds arrive in spring. From year to year these dates will be remarkably consistent, typically varying by only a few days.

Think for a moment about how much birds are part of our lives. Because of the low fat content of chicken meat, the poultry industry is booming, and what other animal has a day dedicated to it the way we have institutionalized the roast turkey at Thanksgiving? Wild birds and their eggs are a significant elements in the diets of many hunter-gatherer peoples in Africa, Asia, and South America, and millions of people in developed nations eat chicken eggs for breakfast every day. In the caves of Borneo many a life has been risked to collect the tiny nests of the Edible Nest Swiftlet (*Collocalia fuciphaga*). The tiny swiftlet nests, made entirely of the swiftlet's hardened saliva, are scraped from precarious locations on cave walls and ceilings and then sold for use in bird's nest soup and other delicacies. Trained falcons, goshawks, and eagles are still used for sport hunting, particularly in the Arab countries of the Middle East and North Africa. Feathers are an important insulating material in the clothing industry—eiderdown is unequaled as a barrier to winter winds, and goose down fills many fine pillows, comforters, and mattresses. A small but lucrative industry centers around "ranching" Ostriches (*Struthio camelus*) for their decorative plumes and lean meat, and Ostrich skin is even used as an exotic but highly durable substitute for cow leather in making boots and other articles of clothing.

As humans have roamed the planet they have taken their local birds with them, not just as household pets but also by deliberately releasing them into the wild to populate the area with "familiar" birds from home. What better way to preserve a sense of home than by bringing along a bird that reminds the traveler of a far-away homeland. Unfortunately, when exotic bird species are introduced to new areas they often drive out indigenous bird species, and ecological disaster results. The global spread of such nuisance birds as the House Sparrow (*Passer domesticus*), the Rock Dove, or Common Pigeon (*Columba livia*), and the European Starling (*Sturnus vulgaris*) is due largely to human carelessness and sentimentality, as well as the sheer adaptability of these three species to live in and around humans. The avian pet industry has developed rapidly in the twentieth century, and for too many species of colorful tropical birds the results have been devastating. Many species of parrots are doubly threatened by loss of habitat due to the worldwide destruction of tropical rain forests and through hunting by collectors for the pet trade. Each year untold numbers of tropical finches, parrots, and other colorful birds are collected from the wild only to die from exposure, hunger, or careless handling before reaching pet stores in developed countries.

The pleasure that birds provide people is difficult to measure, but the multimillion-dollar sales of bird seed and feeders are a good indirect measure of that interest. The popularity of feeding wild birds has actually changed the ranges of several common songbirds. Many observers credit backyard feeders for the northward expansion of such common species as the Cardinal (*Cardinalis cardinalis*) and the Tufted Titmouse (*Parus bicolor*), which often rely on feeders to carry them through the cold winter months in the northeastern United States.

Birdwatching, and the economic, environmental, and even political activity that surrounds it, has become an important part of the lives of millions around the world. Because of their growing numbers birdwatchers have become an international economic force, one that affects the welfare of natural areas around the globe. In 1934, Roger Tory Peterson's *Field Guide to the Birds* removed many of the mysteries of bird identification and literally changed the face of natural history and field biology in the United States. With Peterson's book the public finally had a simple way to make accurate identifications of birds in the field, and interest in birds (and many other forms of wildlife) has grown ever since. Now in its fourth edition, Peterson's guide ranks as one of the ten best-selling books of all time and will likely be regarded by future historians as a milestone in the development of the environmental and conservation movements of the late twentieth century.

Interest in wildlife has also created **ecotourism,** a new term for travelers who roam the globe to observe and photograph wild animals in their natural habitats. This multimillion-dollar business could become an important factor in preserving the wildlife and natural ecosystems of third-world

nations. Developing countries have often used or sold their natural resources without regard to the growing economic value of undisturbed
habitat and thriving wildlife populations. Once such a nation realizes the
importance of its wildlife and how much income can be generated from
ecotourism, the protection of native species often becomes a much higher
priority. Peterson's original *Field Guide*, Rachel Carson's *Silent Spring*, and
the work of such groups as the National Audubon Society and National
Wildlife Foundation helped create a strong environmental awareness in the
populations of the United States and other developed countries. This
ecological consciousness has spread to developing nations, which are
increasingly aware of the economic and social value of protecting and
maintaining their wilderness areas.

Newcomers to birdwatching usually focus simply on identifying the wild
birds they encounter. This soon grows into an appreciation of various
aspects of bird behavior and a better understanding of the habitats within
which birds live. Identifying bird songs, becoming familiar with the plant
communities of your area, and observing the changes of wild bird populations through the seasons of the year are all facets of avian natural history
that enrich the experience of watching birds. Each discovery or observation
leads the birder toward a more complete or "ecological" appreciation of the
way birds interact with the environment. From this start many birders often
begin to participate in bird-banding programs or in more formal ornithological studies done by local, state, or federal wildlife authorities or
university faculty members. In fact, most large studies of bird populations
in the United States would be impossible without the enthusiastic partici-
pation of thousands of "amateur" birders, some of whom know much more
about the local birds than do the professionals.

Why do birds receive more attention than other forms of wildlife? One
reason is that birds are highly visible and vocal residents or visitors in most
habitats on earth, from the most remote ocean islands to the highest
Himalayan peaks. Birds have even been seen in the high latitudes of the
Arctic and Antarctic, near the North and South poles. Chances are that
there are at least several birds within a hundred feet of you right now,
whether you are reading this book in a city office building or a rural
farmhouse. This *accessibility* of birds cannot be overestimated as a reason
for their popularity. You might, for example, have an intense interest in the
local mammals of your area, but in most of North America you would be
lucky to see more than five or six species of mammals with any regularity,
even if you actively look for them every day. Most mammals are secretive
nocturnal animals and are therefore difficult to find and watch recreationally.
The wide assortment of shapes, sizes, colors, and behaviors of wild birds
also make them an inherently fascinating group of animals to observe.
Without heroic effort it is possible to see at least four hundred of the more
than eight hundred species found in North America. Anyone with a field
guide and a pair of binoculars can begin to enjoy birdwatching with little
or no study or training.

Finally, the fact that birds can do something we cannot has long held our
interest. Our stiff-winged mechanical approach to flight may be an efficient
time condenser on vacation trips, but to fly like a bird has been a human
dream at least since the myth of Icarus in ancient Greece. Familiarity,

beauty, exquisite songs, and the ability to fly all contribute to the age-old human fascination with birds and their world. The natural history and biology of birds can be enjoyed on many levels, from casually observing the birds that gather around the backyard feeder to formal training and research in the science of ornithology.

We have developed this book for a multifaceted audience: ornithology students, birdwatchers, bird banders, professional naturalists, and all others who study avian anatomy and biology. For the undergraduate student the book will act as a laboratory manual, to be used along with an ornithology textbook, giving the student a visually oriented lab reference to the anatomy and dissection of birds. For most other readers this book will be a reference volume, a collection of many aspects of avian structure and anatomy under a single cover. This book will have served its purpose if it leads you to a fuller understanding and enjoyment of the birds around you.

1 ANATOMIC TERMINOLOGY

For effective communication, vertebrate anatomists have developed a specialized vocabulary of terms to describe the regions of the body, the relative locations of structures, and the major planes of section used to illustrate vertebrate anatomy. To ensure consistency in the application of these terms, the long axis of a bird's body is always assumed to be horizontal, as it would be if the bird were standing normally or flying. Thus the bird's back is assumed to be upward or dorsal, and the belly or abdomen is likewise assumed to be downward or ventral. The most commonly encountered anatomic terms of orientation are summarized in the tables and figure below.

Planes of Section

Cross-sections of the body are an old and valuable convention for illustrating the relative location and structure of organs within the body. Anatomists use three major planes of section, each perpendicular to the other two: sagittal, transverse, and frontal. These planes may be used as imaginary reference points to describe the location of one structure relative to another, to demonstrate anatomic structures in actual dissection cuts through the body, or to illustrate anatomic structures in artwork.

Sagittal Plane

The sagittal plane runs lengthwise and vertically through the bird's body, dividing the body into right and left portions. If the plane lies directly on the midline of the body and divides the body into equal right and left halves, the plane is called the median or midsagittal plane. If a sagittal section lies lateral to the midline of the body, it is sometimes called a parasagittal plane.

Transverse Plane

In birds and quadrupeds the transverse plane is a vertical plane that divides the body into anterior and posterior portions. Sections that cut through a limb perpendicular to the long axis of the limb are also called transverse sections.

Frontal Plane

In birds the frontal plane (sometimes called the coronal plane) is a horizontal plane that divides the bird's body into upper (dorsal) and lower (ventral) portions.

Common Anatomic Terms — Relative Positions and Directions

Dorsal

Ventral

The terms *dorsal* and *ventral* describe locations in the top and bottom portions of a bird's body, as if it were split in two by a horizontal plane. The dorsal direction (*dorsally*) is toward the back or top side; the ventral direction (*ventrally*) is toward the lower or abdominal side.

Superior

Inferior

In birds the long axis of the body is horizontal, making the terms *superior* and *inferior* synonymous with *dorsal* and *ventral*. Both pairs refer to the upper (dorsal) and lower (ventral) portions of the body.

Anterior

Posterior

These terms refer to locations in the front and back halves of a bird's body. *Anteriorly* (sometimes *rostrally* or *cranially*, see also *cephalad*) means directed toward the head; *posteriorly* (sometimes *terminally*, see also *caudad*) means directed toward the tail.

Cephalad

Caudad

These terms refer to directions between the front and back halves of a bird's body. *Cephalad* (sometimes *anteriorly*, *cranially*, or *rostrally*) means toward the head; *caudad* (sometimes *terminally* or *posteriorly*) means toward the tail.

Proximal

Distal

These are terms of location relative to the distance from the body's midline or point of attachment to the body. *Proximal* structures are closer to the midline or primary point of attachment; *distal* structures are farther from the midline or point of attachment. Your hand is distal to your shoulder, whereas your thigh is proximal to your knee.

Lateral

Medial

These terms of location assume an imaginary line that splits the body into equal right and left halves. *Medial* structures are closer to the midline of the body; *lateral* structures are farther from the midline. The mouth is medial to the ears; the shoulder is lateral to the sternum. *Ipsilateral* structures are on the same side of the body, whereas *contralateral* structures are on opposite sides.

Adductor

Abductor

Adductors are muscles that draw toward the midline of the body or appendage; *abductors* draw away from the center of the body or appendage.

Extensor

Flexor

Extensors are muscles that extend a part of the body away from the midline, extend an appendage, or extend part of an appendage; *flexors* pull parts of the body toward the midline, pull appendages in toward the body, or pull parts of an appendage toward the body.

Pronator

Supinator

Pronators are muscles that rotate wing bones forward and ventrally (leading edge points downward); *supinators* rotate wing bones backward and dorsally (leading edge points upward). These muscles and the *pronation* and *supination* actions they produce are essential for flapping flight.

Reference: Lucas and Stettenheim, 1972.

Common Anatomic Terminology

This figure illustrates many of the terms that describe the relative position of structures within the bird's body, and also shows two of the major planes of section used to describe the internal anatomy of birds. Note that many of the terms may be synonymous with each other; the opposing terms medial and lateral often mean the same thing as proximal and distal. Because the long axis of a bird's body is normally horizontal, the terms anterior and posterior are synonymous with cephalad (meaning toward the head) and caudad (meaning toward the tail). When in flight birds can also be described with the same three-dimensional terms used to describe other aircraft; thus the roll, yaw, and pitch axes are illustrated here.

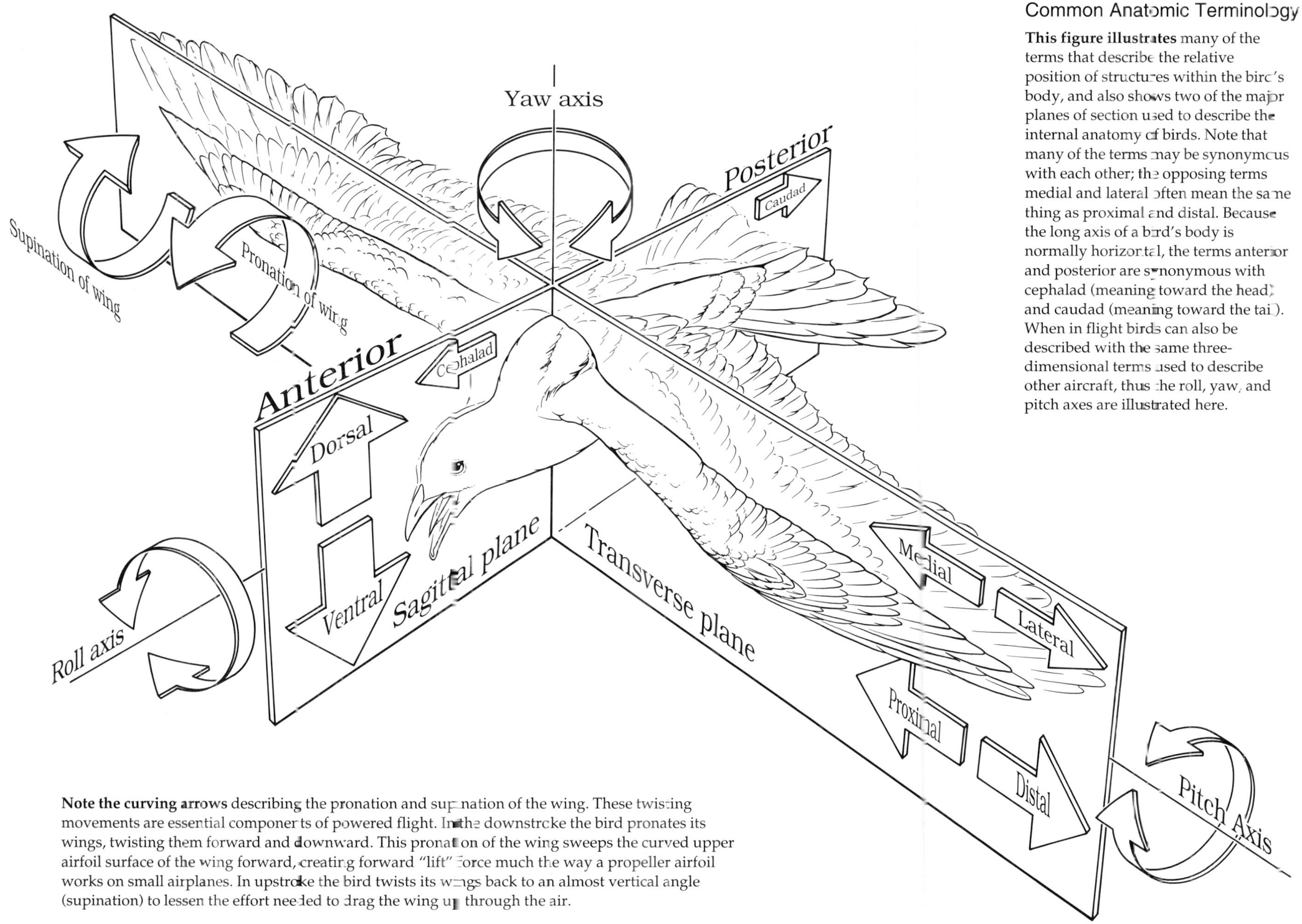

Note the curving arrows describing the pronation and supination of the wing. These twisting movements are essential components of powered flight. In the downstroke the bird pronates its wings, twisting them forward and downward. This pronation of the wing sweeps the curved upper airfoil surface of the wing forward, creating forward "lift" force much the way a propeller airfoil works on small airplanes. In upstroke the bird twists its wings back to an almost vertical angle (supination) to lessen the effort needed to drag the wing up through the air.

Common Anatomic Terms — Regions of the Body

Abdominal	Pertaining to the abdominal region, between the thorax and the pelvis.
Antebrachium	Referring to the forearm area, supported by the radius and ulna.
Axillary	Pertaining to the armpit area, called the **axilla**.
Brachial	Referring to the upper arm, that area supported by the humerus.
Buccal	Concerning the cheek area, usually the lateral walls of the oral cavity.
Carpal	Pertaining to the wrist or wrist area.
Celiac	Pertaining to the abdomen or stomach.
Cervical	Pertaining to the structures of the neck.
Costal	Referring to the ribs or rib cage.
Cranial	Also **cephalic,** pertaining to the head.
Crural	Pertaining to the leg, as in crural feathering.
Digital	Referring to the fingers; in birds, the remnants of four digits are present.
Dorsum	Pertaining to the back or entire dorsal surface of the body.
Occipital	Referring to the area where the spinal column meets the base of the skull at the nape.
Orbital	Pertaining to the eye sockets.
Pectoral	Concerning the ventral chest and breast regions between the sternum and the shoulder.
Sacral	Referring to the region between the crests of the pelvis, the fused bones of the synsacrum.
Sternal	Pertaining to the sternum.
Tarsal	Pertaining to the part of the lower leg that contains the tarsometatarsus.
Vertebral	Referring to individual vertebrae of the spinal column or, sometimes, to the entire spinal column.

References

American Ornithologists' Union. 1983. *Check-list of North American birds.* 6th ed. Lawrence, Kans.: American Ornithologists' Union.

Connor, J. 1988. *The complete birder.* Boston: Houghton Mifflin.

Corral, M. 1989. *The world of birds: A layman's guide to ornithology.* Chester, Conn.: Globe Pequot Press.

Fisher, J., and R. T. Peterson. 1964. *The world of birds.* Garden City, N.Y.: Doubleday.

Gill, F. B. 1990. *Ornithology.* New York: W. H. Freeman.

King, A. S., and J. McLelland, eds. 1979–89. *Form and function in birds.* 4 vols. New York: Academic Press.

Lucas, A. M., and P. R. Stettenheim. 1972. *Avian anatomy: integument.* Agric. Handb. 362:1–340. Washington, D.C.: U.S. Government Printing Office.

Marshall, A. J., ed. 1960–61. *Biology and comparative physiology of birds.* 2 vols. New York: Academic Press.

National Geographic Society. 1983. *Field guide to the birds of North America, ed.* S. L. Scott. Washington, D.C.: National Geographic Society.

Pasquier, R. F. 1977. *Watching birds: an introduction to ornithology.* Boston: Houghton Mifflin.

Peterson, R. T. 1963. *The birds.* New York: Time-Life Nature Library.

Peterson, R. T. 1980. *A field guide to the birds.* 4th ed. Boston: Houghton Mifflin.

Pettingill, O. S., Jr. 1985. *Ornithology in laboratory and field.* 5th ed. New York: Academic Press.

Sturkie, P. D., ed. 1986. *Avian physiology.* 4th ed. New York: Springer-Verlag.

Terres, J. K. 1980. *The Audubon Society encyclopedia of North American birds.* New York: Alfred A. Knopf.

Van Tyne, J., and A. J. Berger. 1976. *Fundamentals of ornithology.* 2d ed. New York: John Wiley and Sons.

Welty, J., and L. F. Baptista. 1988. *The life of birds.* 4th ed. New York: W. B. Saunders.

Systematics

2

Birds arose from reptilian ancestors some time in the Mesozoic era, about 150 to 200 million years ago. The reptilian origins of birds are evident in a number of ways, perhaps most obviously in the similarities between reptilian scales and the feathers and scales of birds. Birds retain the single occipital condyle also found in reptiles, and both birds and reptiles have a single middle ear bone (ossicle), whereas mammals have three. The lower jaw of birds and reptiles is made up of five fused bones; in mammals, the single dentary bone forms the lower jaw. The avian and reptilian ankle is located between the tarsal bones, whereas the mammalian ankle is jointed between the tibia and the tarsal bones. Both reptiles and birds have nucleated erythrocytes (red blood cells); in mammals, erythrocytes lack nuclei.

The avian urogenital and reproductive systems also display many reptilian characteristics, including remnants of a renal portal system and multi-lobed metanephric kidneys that excrete nitrogenous waste primarily as urates. Birds are the only major vertebrate group to have no members that produce live young (**vivipary**) (Blackburn and Evans 1986). All birds lay yolked eggs (**ovipary**) that are essentially similar to reptilian eggs, although the avian yolk contains much more fat than protein and the avian eggshell, unlike the leathery reptile egg, contains a brittle layer composed largely of calcium carbonate crystals.

In spite of such similarities with reptiles, modern birds have evolved a unique, remarkably homogeneous body plan, and as a group birds show much less variation in size, body form, and weight than do mammals or reptiles. All modern birds, members of the vertebrate Class Aves, share these fundamental characteristics:

Feathers

Feathers are the most diagnostic feature of birds. All birds have feathers, and no other animals, living or extinct, are known to have had feathers. Although there has been considerable speculation since the 1970s (see Bakker 1975, 1986) that some dinosaurs might have evolved feathers as insulation, to date no firm evidence exists that any dinosaur had feathers (Feduccia 1985; Ostrom 1987).

Lack of Teeth

Early in their evolutionary history birds evolved a horny bill to cover their jaws and lost their teeth. Avian bills are light in weight, much lighter than teeth and the heavy jaws needed to secure them. The bill is an extraordinarily plastic structure in bird evolution, having evolved into dozens of basic forms suitable for everything from cracking heavy nuts (grosbeak bills), probing mudflats (shorebird bills), and tearing flesh (hawk beaks) to

Three Subspecies of the Peregrine Falcon (*Falco peregrinus*)

Coloration patterns may vary widely within a species, as illustrated by these three geographic subspecies of the Peregrine Falcon (*Falco peregrinus*). The dark Peale's Peregrine (top pair, *F. p. pealei*, immature above, adult below) ranges through the Pacific Northwest coast and the Aleutian Islands. The continental subspecies (center pair, *F. p. anatum*) formerly nested throughout the continent, but currently nests only in western North America, western Canada, and interior Alaska. The palest subspecies is the tundra Peregrine, *F. p. tundrius* (bottom pair). As the name suggests, *tundrius* breeds in the Arctic tundra of North America. Pale *tundrius* birds are the subspecies most commonly seen in fall migration, but darker western *anatum* birds are also seen, even along the eastern coast of the United States. The subspecies of the Peregrine illustrate Gloger's Rule—that organisms living in more humid environments will tend to be darker than those in more arid regions. At one extreme, the dark Peale's Peregrine inhabits the most coastal environments of the Pacific Northwest, where the vegetation is dark and luxuriant in all seasons. The pale tundra Peregrine inhabits the cold, dry Arctic, where the tundra heaths and lichens form a pale palette of background colors (Clark and Wheeler 1987; Dunne et al. 1988).

straining algae from lake shallows (flamingo bills)

Bipedalism and Digitigrade Feet

Birds are **bipedal**, as presumably were their theropod dinosaur ancestors. Bipedalism has interesting implications for the development of avian flight. Most arboreal gliders are quadrupeds, such as bats, whose wings are anchored to both pectoral and pelvic limbs. As flying animals, birds are unique in remaining strong bipeds throughout their known evolution. The bipedalism of birds is often cited as evidence for the terrestrial theory of bird flight, which postulates that early birds may have evolved as runners, which later began to glide and fly weakly while chasing prey or escaping from predators (Gauthier and Padian 1985; A. Peterson 1985).

Birds are also **digitigrade**—that is, they walk only on the toes of their feet. In contrast, humans are **plantigrade**, with a foot composed of tarsals (ankle bones), metatarsals (the bones of the instep), and phalanges (the small bones of the toes). In birds, the ankle and foot bones have been extensively fused and reduced in number. An extended tarsometatarsus forms the lower-most section of the leg, and the phalanges of the three forward-pointing toes of the Blue Jay's foot articulate directly to the distal end of the tarsus. A single metatarsal at the base of the avian first digit (called the hallux, the toe that points backward) is all that remains of most of the other foot bones (Pough et al. 1989).

Fusion and Reduction of Bones

The avian skeleton has evolved into a highly rigid framework of bones that have been extensively fused. This fusion and reduction in the number of bones provides a strong, stable central platform for the flight muscles, the limbs, and the major flight feathers of the wing and tail. In particular, the pectoral and pelvic limbs show extensive fusion of the distal elements, particularly in the carpal, metacarpal, and digital bones of the avian manus, (equivalent to the hand). The axial skeleton is also extensively fused. In mature birds the skull is essentially one fused structure, showing few suture lines between individual elements of bone. The sacral vertebrae of the spine are thoroughly fused to each other and to the iliac bones of the pelvis, forming the synsacrum.

Pneumatic Bones

Birds are unique among living vertebrates in the extent and complexity of their pulmonary systems. In many birds the major bones of the body—particularly the humerus, vertebral column, femur, and vertebral ribs—are pneumatized, with direct connections to the air sacs and respiratory system (Gill 1990; Pough et al. 1989; Schmidt-Nielsen 1983).

Small Size

As a group birds are relatively small animals. Flightless birds such as the Ostrich (*Struthio camelus*, 150 kilograms) or the penguins (Family Spheniscidae) may be quite large and heavy, but they are very much the exceptions to the rule. The upper mass limit in flying birds seems to be about 16 kilograms (bustards, swans, and storks). The vast majority of the almost 9,700 modern bird species (Sibley and Monroe 1990) weigh well under a

Birds have lost the bony tail of reptiles, and have substituted a lighter tail composed almost entirely of feathers. The loss of a fleshy tail shifted the center of gravity forward between the wings, where it should be for a flying animal.

kilogram and are fewer than 50 centimeters in length.

Forelimbs Specialized for Flight

Except for the flightless ratites, the penguins, and secondarily flightless birds like the Galápagos Cormorant (*Nannopterum harrisi*), modern birds have pectoral limbs highly modified to support active flight. Even the penguins use their flipper like pectoral limbs to "fly" underwater. The highly developed avian neck is thought to be one result of this specialization in pelvic and pectoral limbs; without long, supple necks birds could not

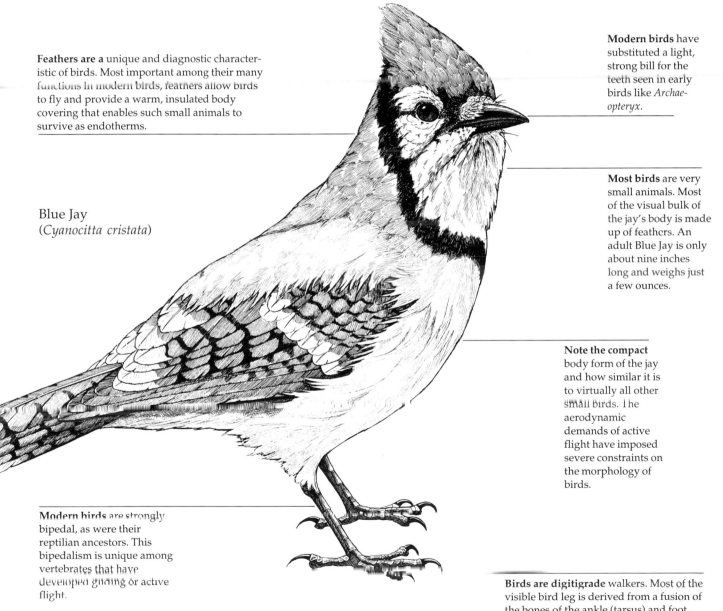

Feathers are a unique and diagnostic characteristic of birds. Most important among their many functions in modern birds, feathers allow birds to fly and provide a warm, insulated body covering that enables such small animals to survive as endotherms.

Blue Jay
(*Cyanocitta cristata*)

Modern birds are strongly bipedal, as were their reptilian ancestors. This bipedalism is unique among vertebrates that have developed gliding or active flight.

Modern birds have substituted a light, strong bill for the teeth seen in early birds like *Archaeopteryx*.

Most birds are very small animals. Most of the visual bulk of the jay's body is made up of feathers. An adult Blue Jay is only about nine inches long and weighs just a few ounces.

Note the compact body form of the jay and how similar it is to virtually all other small birds. The aerodynamic demands of active flight have imposed severe constraints on the morphology of birds.

Birds are digitigrade walkers. Most of the visible bird leg is derived from a fusion of the bones of the ankle (tarsus) and foot. Birds have a thigh, forward-pointing knee, and lower leg directly comparable to those of humans, but most of their upper leg is normally hidden under plumage.

adequately groom themselves and would have more difficulty in reaching and manipulating their food.

Centralized Body Mass

Birds are light in mass, but only because their bodies are small in size. When compared with mammals of the same overall size, avian bodies are not particularly light. Bird skeletons and other tissues weigh about the same as mammalian equivalents. But birds have evolved a compact, centralized body plan; by concentrating the mass of the body between the wings and around the center of gravity they achieve greater aerodynamic stability. Most of the muscles that control the wings and legs are located near the body centerline and control the limbs through long tendons that extend onto the limbs. In contrast, the limbs of most mammals are much more muscular than avian limbs. Active flapping flight would not be possible if the major flight muscles were located on the wings, because the heavy wings would have too high a moment of inertia to move effectively. By locating the main flight muscles at the body midline on the breastbone, birds can have both long wings and the powerful muscles to flap them.

Instead of teeth and heavy jaws, birds have a tough gizzard located near their center of gravity. The lack of a heavy skull also means that birds do not need a heavy tail as a counterbalance, and they have lost all but a short stub of the reptilian tail their bipedal ancestors possessed. The loss of the bony reptilian tail also shifted the body's center of gravity forward between the wings, where it should be for stable flight. The combination of a highly centralized body mass and a smooth outer coat of feathers produces the streamlined body form so characteristic of birds.

High Metabolism

Birds are endothermic, and they could not sustain their active lives without a high metabolism. As "warm-blooded" animals, birds consume as much as thirty times the amount of metabolic energy as reptiles of similar size (Gill 1990). The avian cardiovascular and respiratory systems are extremely efficient, enabling birds to withstand cardiopulmonary stresses far beyond what mammals can tolerate. For instance, migrating geese can fly fifty to sixty miles per hour at altitudes higher than 30,000 feet (Rüppell 1977). The world altitude record for birds is held (posthumously) by a Rüppell's Griffon (*Gyps rueppellii*), which was pulled into the jet engine of an airliner at 36,000 feet (Burton 1990). Although the vulture was undoubtedly soaring passively, no mammal of equivalent size could breath enough air even to remain conscious at that altitude.

Highly Developed Central Nervous System and Vision

As birds developed their physical adaptations to flight, they also developed the sophisticated central nervous system (CNS) necessary to control their bodies in the three-dimensional world they inhabit as flying animals. The balancing, coordinating, muscular control, and proprioceptive functions of the brain are all highly developed in birds, as is the keen sense of vision needed to inform the brain of a flying animal about the exact locations of objects in the surrounding environment.

Until the late 1970s virtually all of the major anatomic characteristics of birds were routinely described as adaptations to flight. These avian characters, including feathers, endothermy, air sacs, and pneumatic bones, were thought to have arisen well after birds split from their reptilian ancestors. Most vertebrate biologists believed the likely ancestors of early birds to be a group of small, unspecialized reptiles, the pseudosuchians. Because most of what is known about the pseudosuchians suggests that they were rather unspecialized early reptiles, all of the characteristically avian adaptations seen in modern birds (particularly feathers) would have arisen after the split from the pseudosuchians and would therefore be unique to birds (Feduccia 1980, 1985; Martin 1983, 1985; Martin et al. 1980).

In 1973 paleontologist John Ostrom published the first of a series of papers describing the close resemblance between *Archaeopteryx* and a group of small theropod dinosaurs, the coelurosaurs (see Gould 1982). Ostrom proposed that the remarkable similarities between *Archaeopteryx* and the small coelurosaurs could not be the result of convergent evolution alone, and that birds like *Archaeopteryx* must have arisen from small coelurosaurian dinosaurs similar to *Compsognathus* (Ostrom 1973, 1974, 1975, 1976). This dinosaurian theory of the origin of birds, along with earlier work Ostrom had done on the small, predatory coelurosaurs *Deinonychus* and *Compsognathus*, involved birds in the "hot-blooded dinosaur" debate that still rages in vertebrate biology today (Ostrom 1969, 1978). The majority of vertebrate paleontologists now accept the dinosaurian origin of birds as the most plausible theory of avian evolution (Carroll 1988; Gauthier and Padian 1985; Gauthier 1986; Gill 1990; Hecht et al. 1985; Padian 1986; Pough et al. 1989; Welty and Baptista 1988).

New fuel has recently been added to the debate by Texas paleontologist Sankar Chatterjee, who has found in Triassic sediments the remains of what he believes to be the earliest known bird, which he calls *Protoavis* (Chatterjee 1991). Chatterjee's controversial thesis rests on the similarities he sees between the skull and neck bones of *Protoavis* and the bones of modern birds. These structural similarities include a lightly built and pneumatized skull, evidence of cranial kinesis (moveable upper jaws, as found in modern birds), a temporal region similar to that of modern birds, saddle-shaped neck vertebrae, and a relatively large brain case, which suggests that *Protoavis* had begun to develop the highly specialized nervous system found in modern birds. Chatterjee's assertion that his *Protoavis* is a bird has drawn bitter criticism from those who advocate a theropod dinosaur origin for birds, for if *Protoavis* is a true Triassic bird, the origin of birds would be too early in evolutionary history for true dinosaurs to be the immediate ancestors of birds. Critics speculate that Chatterjee's poorly preserved specimen was completely disarticulated when it was found and that it is likely to be a collection of bones from several unrelated species of small dinosaurs rather than the bones of a single individual. Many of the crucial skull bones are mere fragments, providing ambiguous evidence for the supposedly birdlike features of the skull. Yet, some prominent avian paleontologists see *Protoavis* as important new evidence that birds arose as a distinct group before the theropod dinosaurs arose, and they strongly support Chatterjee's position that *Protoavis* has too many birdlike features to be written off as a misinterpretation of a poor specimen (see Anderson 1991; Zimmer 1992).

1 THE ORIGIN OF BIRDS

The past twenty years have been particularly interesting times in ornithology and vertebrate paleontology, as several competing theories have been elaborated to explain how birds evolved from their reptilian ancestors. Everyone concerned agrees that birds are clearly derived from some form of reptile, but the debate over *which* form of reptile and *when* the transition from reptiles to birds took place continues to generate controversy in the ornithological literature.

In 1927 Danish lawyer and paleontologist Gerhard Heilmann wrote *The Origin of Birds*, the first book-length exploration of avian evolution. Based on earlier work by Robert Broom (1913), Heilmann theorized that birds evolved in the early Triassic period from a group of small, generalized thecodont reptiles, the pseudosuchians. Heilmann and Broom based their theory on small pseudosuchians like *Euparkeria* and *Sphenosuchus*, primarily for two reasons: both were unspecialized enough to share many general skeletal features with birds, and the pseudosuchians were thought to be the only suitable ancestral reptiles that definitely possessed clavicles. Heilmann's work became the "mainstream" view of avian origins for the next fifty years (Feduccia 1980) and formed the basis for many current theories on the pseudosuchian origin of birds (Walker 1972, 1974, 1977). More recently, avian paleontologists Larry D. Martin, K. N. Whetstone, and others proposed a new variant on the pseudosuchian theory, stressing the similarities in inner ear and other skeletal structures among the pseudosuchians like *Cosesaurus*, modern crocodiles, and both modern and ancient birds (Whetstone and Martin 1979, 1981; Martin et al. 1980; Martin 1983). The theory supposes a common ancestor shared by both crocodiles and birds, an ancestor closely related to earlier pseudosuchian forms (Tarsitano and Hecht 1980). Chatterjee's Triassic *Protoavis* specimen (Chatterjee 1991) has recently strengthened interest in the pseudosuchian theory of bird origins. Whether *Protoavis* will be widely accepted as a true bird remains unclear.

The diagram at the far right sketches the theropod dinosaur theory of bird evolution, in which birds evolve some time in the early Jurassic period as an offshoot of the predatory coelurosaurian dinosaurs. The notion that birds might be the most highly derived form of the coelosaurian dinosaurs is not new; in the late 1800s British biologist Thomas Henry Huxley first noted the many similarities between *Archaeopteryx* and small coelurosaurian dinosaurs like *Compsognathus*. Gerhard Heilmann (1927) also appreciated the remarkable skeletal similarities, but he finally rejected the coelurosaurs as possible avian ancestors because of one crucial point: in Heilmann's time, coelurosaurs were thought to lack clavicles, and clavicles (the furcula, or "wishbone") are very prominent in all birds. Later discoveries showed that some coelurosaurs did indeed possess clavicles, which revived interest in the coelurosaurs as possible bird ancestors.

Such recent discoveries as Chatterjee's possible Triassic bird *Protoavis* (Chatterjee 1991) and new specimens of early Cretaceous birds from China (Rao and Sereno 1990) have further complicated the picture of avian evolution. Our understanding of the origins of birds is now based on remarkably few specimens and will be clarified only by additional discoveries of birds and bird ancestors from the Triassic and Jurassic periods.

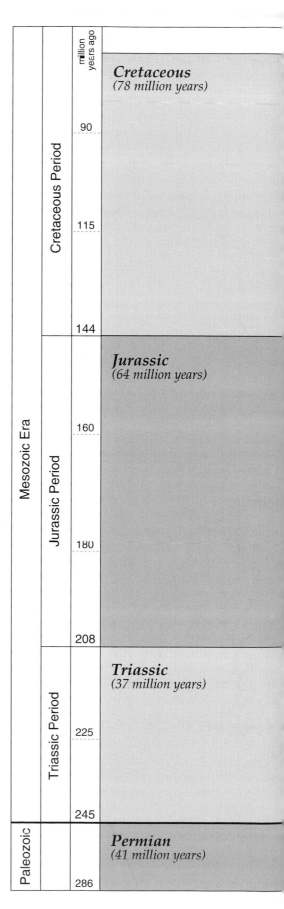

		million years ago	
Mesozoic Era	Cretaceous Period		**Cretaceous** *(78 million years)*
		90	
		115	
		144	
	Jurassic Period		**Jurassic** *(64 million years)*
		160	
		180	
		208	
	Triassic Period		**Triassic** *(37 million years)*
		225	
		245	
Paleozoic			**Permian** *(41 million years)*
		286	

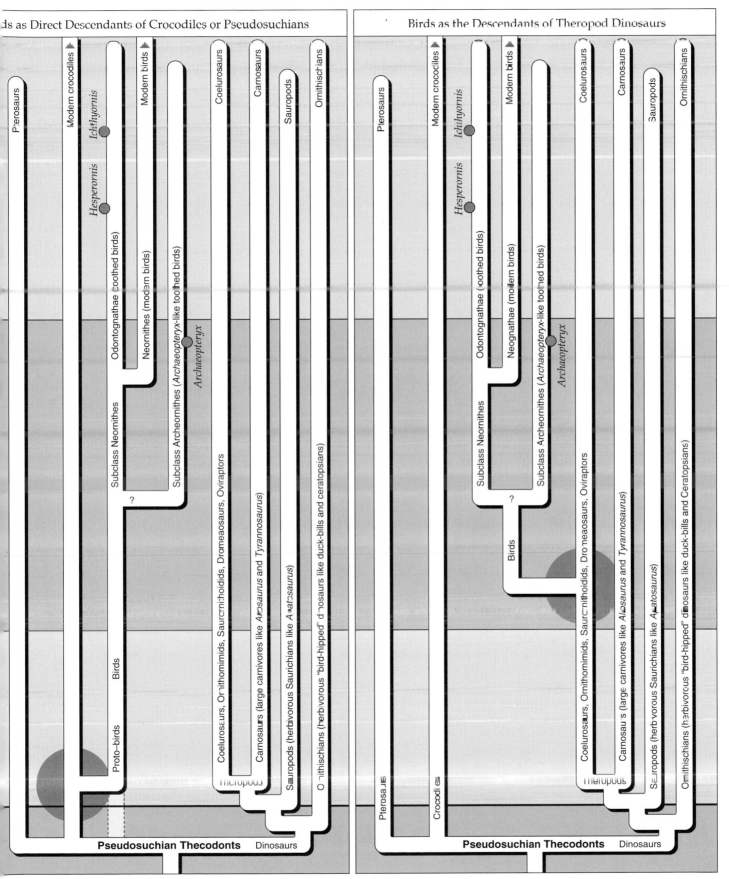

2 BIRDS AND THE GEOLOGIC RECORD

Some 150,000 bird species are estimated to have inhabited the earth since birds first evolved some time in the early Jurassic period (Brodkorb 1971). Although the avian fossil record is minute compared with that of mammals or reptiles, known fossils clearly indicate that bird evolution during the late Mesozoic and early Cenozoic eras was very rapid. All of the modern orders of birds are thought to have arisen from a common (but as yet uncertain) ancestral stock during the Cretaceous period, more than 65 million years ago. The major groups of toothed birds present during the Cretaceous (the gull-like Ichthyornithiformes and the loon-like Hesperornithiformes) did not survive the mass extinctions of the late Cretaceous, and these toothed aquatic birds were apparently not closely related to the ancestral groups from which modern birds arose (Carroll 1988).

The rise of the angiosperms (flowering plants) in the late Cretaceous and the development of the grasses during the Cenozoic era produced an abundant supply of foods for the rapidly radiating birds and mammals to exploit. The Paleocene and Eocene epochs (37 to 65 million years ago) were a time of great radiation in the birds, when as many sixteen new orders of birds developed worldwide. Most of the modern bird genera were present by the end of the Miocene epoch. During most of the Cenozoic the world's climate was much milder and more humid than it is today, but by the mid-Miocene (about 15 million years ago) the global climate began to cool and become much drier. Great grassland plains, savannas, and steppes developed in North America, Central Africa, and Central Asia. The rapid rise of the Himalayan mountain range dramatically altered the climate of Central Asia, and the Alps rose in southern Europe.

The Pleistocene epoch (starting about 1.5 million years ago) was a time of great climatic and biogeographic change across the globe. The Bering land bridge between Asia and North America and the rise of Central America linked the continents, but vast areas of the Northern Hemisphere were covered with glaciers during several Pleistocene ice ages. This climatic instability had an enormous impact on the diversity of bird species. At the beginning of the Pleistocene epoch there may have been as many as 21,000 species of birds, but by the end of the epoch fewer than 10,000 apparently remained (Brodkorb 1971).

Several major groups of flightless birds arose and disappeared during the Cenozoic. In North America the giant predatory bird *Diatryma* appeared early on but died out before the end of the Eocene. Much later Cenozoic giants include the elephant birds (Order Aepyornithiformes) of Madagascar and the giant moas of New Zealand (Order Dinornithiformes). Both moas and elephantbirds were apparent early casualties of *Homo sapiens*. They survived on isolated ocean islands into the Recent epoch (the past 11,000 years), until humans colonized the islands and wiped the birds out.

In the Recent epoch humans have become the dominant shapers of the environment. Through habitat destruction and over-hunting people have driven eighty-five species of birds to extinction since the seventeenth century (Fisher and Peterson 1964) and have threatened the viability of many hundreds of other endangered species.

Fossil Birds and the Geological Time Scale

Period		Epoch	Million years ago		Major Bird Groups Known from Fossils
Cenozoic Era	Quaternary	Recent Pleistocene	.01 2	A period of glaciation and massive climatic change, during which large numbers of bird species became extinct. Surviving modern species may be fewer than half those present at the beginning of the Pleistocene (as many as 21,000 as opposed to just 9,672 today; Bordkorb 1971).	All modern genera; virtually all modern species by Recent epoch (past 11,000 years). *Moas* become extinct.
Cenozoic Era	Tertiary	Pliocene	7	Virtually all modern genera are now established. Birds reach maximum diversity. Central American land bridge rises; Rocky Mountains form. Earliest known fossils of tinamous (Order Tinamiformes) and rheas (Family Rheidae).	Rheas, tinamous, larks, buntings, thrushes, fringillid sparrows, swallows, nuthatches
Cenozoic Era	Tertiary	Miocene	26	Most modern bird families and genera are now present in the fossil record. New groups include the Struthioniformes (Ostrich), Piciformes (woodpeckers), Podicipedidae (grebes), Coliiformes (mousebirds), Caprimulgidae (goatsuckers and nighthawks), Falconidae (falcons), Corvidae (crows and jays), Laniidae (the shrikes), and Parulidae (wood warblers). The Alps rise in Europe, and the Himalayas are formed by the collision of the Indian subcontinent with the south coast of Central Asia. World climate continues to dry and cool; grasslands dominate the central plains of Asia and North America.	Grebes, ostriches, falcons, woodpeckers, nighthawks, mousebirds, shrikes, North American wood warblers *Phororhacos* (giant flightless predatory birds) *Osteodontornis* (giant seabird)
Cenozoic Era	Tertiary	Oligocene	37	A relatively quiet period of avian evolution. Four major groups of modern birds make their first appearance: Cuculiformes (cuckoos), Passeriformes (songbirds), Anseriformes (ducks, geese, and swans), and Pelecanidae (pelicans). World climate dries and cools; great forests cover most land masses. Familiar modern genera present include *Anas* (dabbling ducks), *Sula* (boobies), *Puffinus* (puffins), and *Pelecanus* (pelicans).	Dabbling ducks, herons, storks, Old World warblers, sparrows, cuckoos, turkeys, gannets, boobies, petrels, modern owls, New World vultures
Cenozoic Era	Tertiary	Eocene	53	The major period of radiation in both birds and mammals. By the end of the epoch most modern orders of birds are present in the fossil record. The global climate is milder and more humid than today, and the rise of the flowering plants and grasses tremendously increases the food supply for Cenozoic birds and mammals. Important Eocene specimens include *Presbyornis*, an unusual duck with the long legs of a wader, and these modern genera: *Milvus* (kite), *Aquila* (golden eagle), *Haliaeetus* (sea eagles), *Phoenicopterus* (flamingo), *Charadrius* (plover), and many others.	Penguins, rheas, rails, herons, gulls, auks, avocets, bustards, cranes, kites, large eagles, hawks, shorebirds, ibises *Presbyornis* (duck-like bird) *Aepyornis* (elephant bird) *Haliaeetus* (sea eagles) *Aquila* (golden eagle)
Cenozoic Era	Tertiary	Paleocene	65	Two major modern groups appear in the Paleocene, the dawn of the Cenozoic: Coraciiformes (rollers, hornbills, kingfishers, and others), and Strigiformes (owls). A spectacular order of giant predatory birds, Diatrymiformes, appears, but dies out early in the Eocene. During the late Cretaceous and the Paleozoic many new forms of birds appear, but the group affinities of the relatively few known Paleocene avian fossils are not well understood (Martin 1983).	Rollers, hornbills, kingfishers, owls *Diatryma* *Gavida* (loon-like) *Gastornis* (ostrich-like) *Ogygoptynx* (owl)
Mesozoic	Cretaceous	Late Cretaceous		The two major groups of toothed birds, the loon-like diving Hesperornithiformes, and the gull-like Ichthyornithiformes, die out in the late Cretaceous extinctions. Only two modern groups, the Suborder Charadrii (shorebirds and gulls) and the Superfamily Procellarioidea (albatrosses and petrels) are known from the Cretaceous (Olson 1985).	Shorebirds, albatrosses, and petrels *Hesperornis* (loon-like) *Icthyornis* (gull-like) *Baptornis*

References: Brodkorb 1971; Carroll 1988; Feduccia 1980; Fisher and Peterson 1964; Howard 1955; Martin 1983; Olson 1985; Welty and Baptista 1988.

3 TAXONOMY AND CLASSIFICATION

Even the most casual observer of the natural world must wonder at the enormous variety of plants and animals on the planet. Humans seem to have an inborn need to associate and classify natural phenomena into systems that make a complex world more comprehensible. Recorded observations of birds are as ancient as our written records. In ancient Greece, Aristotle recognized and described more than 140 kinds of birds, no mean feat for a man without modern optics and any references but his own observations.

The most recent classification of the birds lists 9,672 species of living birds (Sibley and Monroe 1990). Faced with such a wealth of diversity, avian taxonomists use a hierarchical series of increasingly narrow and specific groupings to organize and classify birds. The taxonomic nomenclature system in use today is derived from the work of the eighteenth-century Swedish naturalist Carl von Linné, better known under his Latinized surname, Linnaeus. Linnaeus was the first to recognize that the local, common names given to organisms were highly variable and unreliable and that scientists needed a more precise method of identifying species. In his *Philosophia Botanica* (1752) and *Systema Naturae* (1757) Linnaeus developed the binomial system of Latin names for each species he listed and described in his surveys of plants and animals. Linnaeus gave each organism a unique combination of two names: a generic name and a specific name (or specific epithet). To describe the broader relations between groups of bird species Linnaeus devised and named a hierarchical structure of organization very similar to the modern scheme listed at the right. Although Linnaeus worked without knowledge of genetics or evolution, his system remains the backbone of modern taxonomy or systematics, the study and classification of the relations among organisms.

Morphological Systematics

The most thorough early attempt to establish the proper classification of birds was published by Hans Friedrich Gadow in 1892 and 1893, and until recently Gadow's outline formed the basis for most contemporary avian classification systems (Austin and Singer 1985; Storer 1971; Stresemann 1927–34; Wetmore 1960). Gadow and subsequent authors based their work on the careful study and comparison of shared anatomical characters within each bird group, characters that indicate a common ancestral lineage. These anatomic details, called "conservative" characters, form the basis for such comparative taxonomy because they are relatively constant across many species and over long periods of time. Bill size, leg length, or foot shape, for example, are not conservative characters. Bills, legs, and feet are often the first anatomic features to evolve in response to new feeding patterns or ecological conditions, and the superficial differences between such "plastic," changeable organs are thus less useful for studying relations among groups of birds. By contrast, the bony structure of the palate or the exact formulation of leg muscles are fundamental anatomic characters that are more stable and remain unchanged over long periods in the evolution of most bird groups. Therefore, differences or similarities in the details of the bony palate or leg musculature between species may be useful clues to how long it has been since two bird species last shared a common ancestor. Other conservative anatomic characters used in morphological taxonomy

Kingdom

|

Phylum

|

Class

|

Order

|

*Family**

|

Genus

|

Species

* Linnaeus did not use the "family" category in his original taxonomic scheme.

DNA-DNA
Hybridization Techniques

DNA samples from two species are
collected and purified.

Species 1
DNA Strands

Species 2
DNA Strands

The DNA is heated to disassociate
the molecules into individual strands.
The samples are then combined to
form a hybrid DNA.

The hybrid DNA is then
reheated and the melting
temperature is analyzed.

Mismatched areas
are weak and more
readily disassociate
when heated

Hybrid DNA
from both species

In addition to carrying immediately useful genetic instructions, the DNA (deoxyribo-nucleic acid) within the nucleus of each cell carries a record of the unique evolution-ary history of the species coded within its sequence of nucleotide bases. In 1962 Emile Zuckerkandl and Linus Pauling first proposed that protein molecules might evolve at a constant rate over time and could be used as "molecular clocks" to explore the evolution of complex molecules like DNA. Over the past thirty years this concept has been used to explore the relations among bird groups by analyzing differences and similarities between the DNA molecules of related species.

The double-helix of DNA is composed of two strands of linear sequences of nucleotides, each of which is formed from a sugar–phosphate group attached to one of four types of bases: adenine (A), cytosine (C), guanine (G), or thymine (T). The bases form matched pairs within the helix of DNA; adenine bonds only with thymine (A-T), while cytosine bonds only with guanine (C-G). In the DNA-DNA hybridiza-tion technique, pure DNA is obtained from the cell nuclei of a bird species and boiled at 100°C (212°F) to cause the two strands of material that form the DNA double-helix to disassociate (or "melt") into separate strands of nucleotides. If cooled to approxi-mately 60°C (140°F) and incubated for five days, strands of DNA from a single species will reassociate into stable double-helix molecules of DNA, because during incubation most single strands will reassociate with matching strands of complemen-tary base sequences.

To compare the evolutionary relations among two bird species, the DNA from each is disassociated into single strands, mixed, and incubated at 60°C to allow the strands to form double-helices again. Strands of hybrid double-helix DNA will form only where sequences of complementary base pairs exist. At 60°C about 80 percent of the bases must match to form a stable double-helical molecule of DNA. The occurrence and length of complementary nucleotide sequences common to the two species is directly related to how recently the two species shared a common ancestor. The bonding strength of hybrid DNA strands (a mixture of DNA from two species) is never as high as strands formed from the DNA of a single species, because hybrid DNA will always contain mismatched base pairs, reflecting genetic evolution and mutations since the two species last shared a common ancestor. In mismatched areas the links between nucleotide bases are weak, because inappropriate combinations of bases (for example, A-C or G-T pairs) do not form the hydrogen bonds that normally hold the double-helix of DNA together. Thus hybrid DNA helices are less stable and disassociate or "melt" at a lower temperature than DNA helices formed from a single species. This difference in the melting points between hybrid and single species DNA is the basis of the DNA-DNA hybridization technique, because it affords a quantitative measure of how closely related any two species are. The higher the melting point of a hybrid DNA, the more matches there are in the nucleotide sequences and the more closely two species are related.

Using the DNA-DNA hybridization technique to explore the similarity in the genomes of hundreds of pairs of bird species, Charles Sibley, Jon Ahlquist, and Burt Monroe constructed a classification system for birds based almost entirely upon genetic relations derived from DNA-DNA hybridization data (Sibley and Ahlquist 1983, 1986, 1990; Sibley and Monroe 1990). This classification system is currently under review by the American Ornithologists' Union. See particularly the *Scientific American* article by Sibley and Ahlquist (1986) for a lucid explanation of the techniques used in establishing the taxonomic relations among bird groups.

include the configuration of the internal and external nostrils, the exact configuration of the toe bones, the presence or absence of a fifth secondary feather, and the exact arrangements of scutes and scales on the legs. In the traditional morphological systems of avian classification these characters are carefully compared and used as the basis for grouping species into genera, related genera into families, and related families into orders within the Class Aves.

Biochemical Systematics

Avian morphologists have always been bedeviled by the many complex examples of convergence and divergence in the Class Aves and by the confusing effect these evolutionary trends can cause in taxonomic systems that rely solely on examining and comparing gross morphology. Over the past thirty years our increasing knowledge of genetic biochemistry and the properties of protein molecules has opened up exciting new possibilities for taxonomists. By systematically comparing the DNA of one bird species against other birds (see page 23 for an overview of the DNA-DNA hybridization technique), taxonomists have an additional alternative for clarifying the relations among bird groups. The biochemical classification of birds presented by Sibley and Ahlquist (1983, 1990) has yielded many surprising arrangements of bird families and has provided much new data on the evolution and biogeography of such major bird groups as the corvid (crow) complex (see also Sibley and Ahlquist 1986). Biochemical systematics remains controversial, and aspects of it have been critically reviewed by a number of authors (Houde 1987; Raikow 1985; Shields and Helm-Bychowski 1988).

Two major systems for classifying birds are outlined on the following four pages. The first spread shows the biochemical classification of Sibley, Ahlquist, and Monroe (1990), now under review by the A.O.U. as a method for classification of birds. A more detailed outline of the new Sibley and Monroe classification is given in appendix A. Because the biochemical system of Sibley and Ahlquist is so different from previous morphological systems, and because some of the biochemical taxonomy remains controversial, we have included for comparative purposes an outline of a strictly morphological classification of birds, based primarily on the work of Wetmore (1960) and Storer (1971). Some recent classifications of birds (for example, Gill 1990) are based primarily on the long-standing morphological model but incorporate new biochemical taxonomy for selected groups.

For a more comprehensive overview of the history of avian classifications systems than is appropriate here, see particularly Sibley and Ahlquist (1990:184–252), a review of the relative merits and controversies surrounding morphological, cladistic, and biochemical taxonomic systems.

How a Bird Species Is Classified

Kingdom
Animalia

Phylum
Chordata

Class
Aves

Infraclass
Neoaves

Parvclass
Passerae

Superorder
Passerimorphae

Order
Ciconiiformes

Suborder
Ciconii

Infraorder
Falconides

Parvorder
Falconida

Family
Falconidae

Genus
Falco

Species
columbarius

Subspecies
columbarius

Older morphological systems of avian classification were based on the classic hierarchical sequence of taxonomic categories: class, order, family, genus, and species. New biochemical classification methods provide a much more detailed picture of the relations and branching patterns within bird groups. To accommodate this newfound knowledge of how the major groups of birds are related to one another, biochemical taxonomies such as the Sibley and Monroe (1990) system need many more detailed levels of classification. Such new taxa as infraclass, parvclass, and parvorder are needed to show the levels of relations revealed through DNA-DNA hybridization. The levels of classification used in biochemical systematics (such as family, genus, and species) now are based on quantitative data on the relative disassociation points of DNA and not on the more subjective collections of structurally similar birds used in previous morphological classification systems. Each taxon level thus directly reflects a specific amount of structural difference in the nucleotide base sequences of the DNA of species at each taxonomic level.

The example species here is the most widespread or "nominate" subspecies of the Merlin, or "pigeon hawk" (*Falco columbarius columbarius*). Note that falcons and other birds of prey are no longer grouped into a discrete order (in older systems, Order Falconiformes). Biochemical analysis places the falcons into the large and extremely diverse Order Ciconiiformes (see the biochemical taxonomy chart on the following two pages).

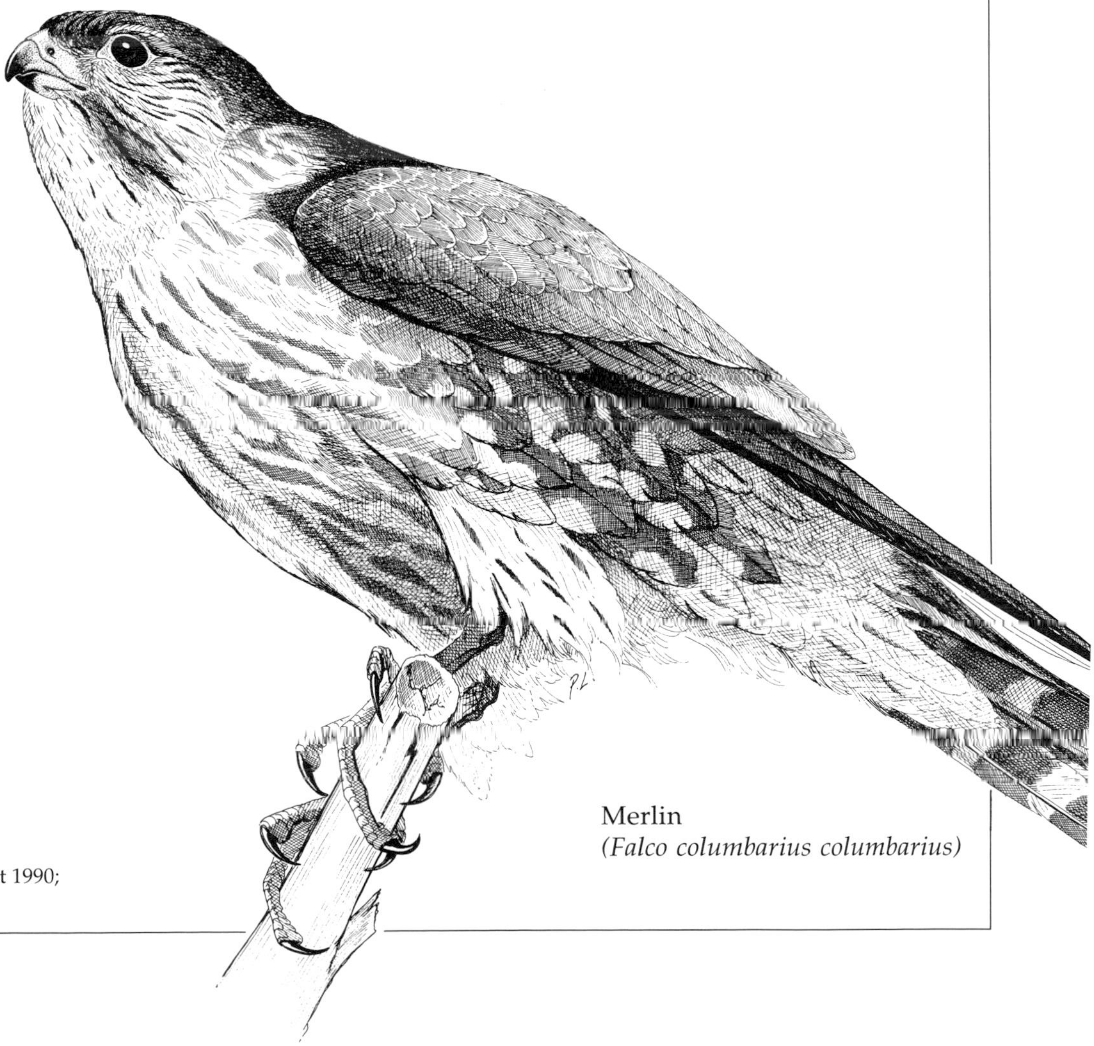

Merlin
(*Falco columbarius columbarius*)

References: Sibley and Ahlquist 1990;
Sibley and Monroe 1990.

4 BIOCHEMICAL CLASSIFICATION OF BIRDS

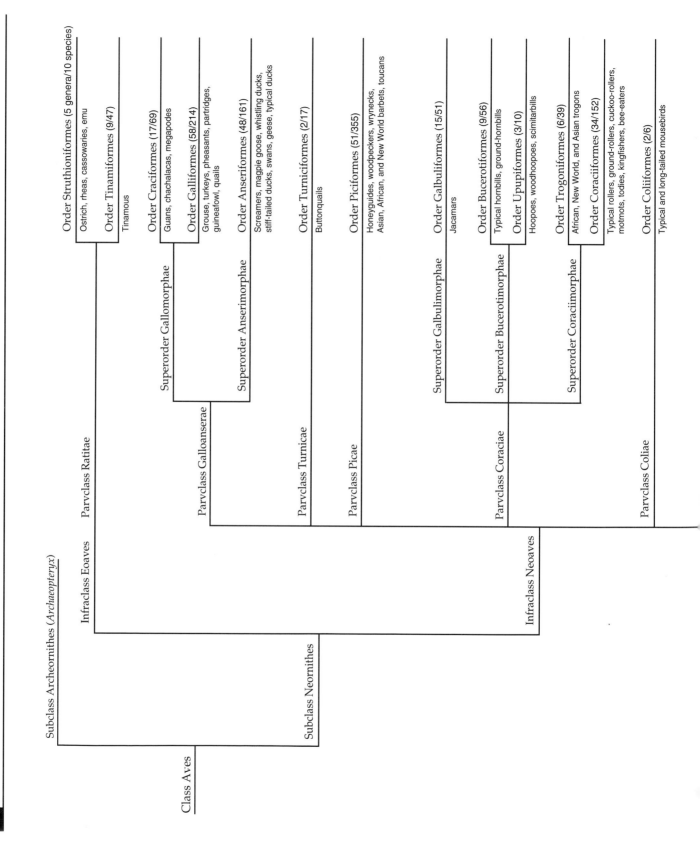

Parvclass Passerae

Superorder Cuculimorphae

Order Cuculiformes (30/143)
Old World cuckoos, coucals, American cuckoos, hoatzin, anis, guira cuckoo

Superorder Psittacimorphae

Order Psittaciformes (80/358)
Parrots and allies

Superorder Apodimorphae

Order Apodiformes (19/103)
Typical swifts, crested-swifts

Order Trochiliformes (109/319)
Hermits, typical hummingbirds

Order Musophagiformes (5/23)
Turacos, plantain-eaters

Superorder Strigimorphae

Order Strigiformes (45/291)
Barn and grass owls, typical owls, owlet-nightjars, frogmouths, oilbird, potoos, eared-nightjars, nighthawks, nightjars

Order Columbiformes (42/313)
Dodos, solitaires, pigeons, doves

Order Gruiformes (53/196)
Sunbittern, bustards, crowned-cranes, typical cranes, limpkin, sungrebes, trumpeters, kagu, seriemas, rails, gallinules, coots, mesites

Order Ciconiiformes (254/1027)
Suborder Charadrii
  Infraorder Pteroclides
    Sandgrouse
  Infraorder Charadriides
    Woodcocks, sandpipers, phalaropes, jacanas, avocets, stilts, plovers, skuas, skimmers, gulls, terns, auks, murres, puffins
Suborder Ciconii
  Infraorder Falconides
    Osprey, hawks, eagles, caracaras, falcons
  Infraorder Ciconiides
    Grebes, tropicbirds, gannets, anhingas, cormorants, herons, flamingos, pelicans, New World vultures, storks, frigatebirds, penguins, loons, petrels, albatrosses

Superorder Passerimorphae

Order Passeriformes (1161/5712)
Parvorder Corvida
  Superfamily Menuroidea
    Treecreepers, lyrebirds, bowerbirds
  Superfamily Meliphagoidea
    Fairywrens, grasswrens, honeyeaters, thornbills
  Superfamily Corvoidea
    Shrikes, crows, jays, birds-of-paradise, orioles
Parvorder Passerida
  Superfamily Muscicapoidea
    Waxwings, dippers, thrushes, starlings, nuthatches
  Superfamily Sylvioidea
    Nuthatches, creepers, wrens, swallows, kinglets
  Superfamily Passeroidea
    Larks, sparrows, tanagers, cardinals, blackbirds

One of the most surprising results of the new classification based on DNA-DNA hybridization is the changes in the Order Ciconiiformes, which had comprised herons, ibises, storks, flamingos, and other related wading birds. Under the new Sibley and Monroe classification system the Order Ciconiiformes now subsumes these former orders:

Order Sphenisciformes—Penguins

Order Gaviiformes—Loons

Order Podicipediformes—Grebes

Order Procellariiformes—Albatrosses, shearwaters, and petrels

Order Charadriiformes—Shorebirds, skuas, gulls, terns, skimmers, and auks

Order Falconiformes—New and Old World vultures, hawks, eagles, accipiters, kites, falcons, caracara, and secretarybird

Order Pelecaniformes—Pelicans, frigatebirds, shoebill, tropicbirds, gannets, boobies, anhingas, and cormorants

Order Ciconiiformes—Herons, hammerhead, flamingos, ibises and spoonbills

References: Sibley, Ahlquist, and Monroe 1988; Sibley and Ahlquist 1990; Sibley and Monroe 1990.

5 MORPHOLOGICAL CLASSIFICATION OF BIRDS

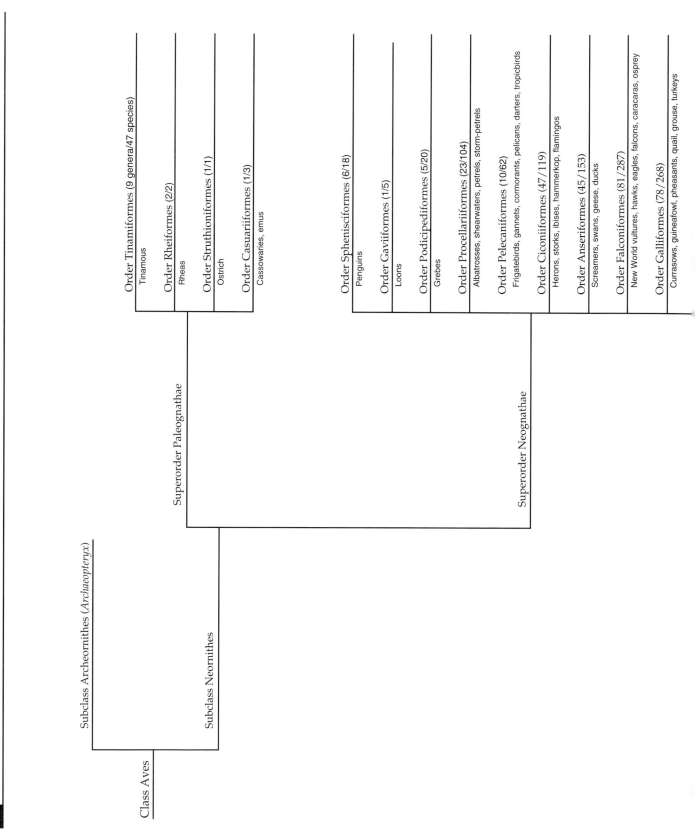

Order Gruiformes (81/215)
Rails, sunbittern, sungrebe, roatelos, kagu, buttonquail, cranes, limpkin, trumpeters, seriemas, bustards

Order Charadriiformes (75/332)
Jacanas, painted-snipes, woodcocks, oystercatchers, plovers, sandpipers, phalaropes, avocets, gulls, terns skuas, jaegers, skimmers, auks, murres, puffins

Order Columbiformes (42/307)
Dodos, solitaires (both now extinct), sandgrouse, pigeons, doves

Order Psittaciformes (81/339)
Parrots, macaws, lories, other parrot allies

Order Cuculiformes (43/140)
Turacos, cuckoos, coucals, roadrunners, anis

Order Strigiformes (30/145)
Barn owls, typical owls

Order Caprimulgiformes (24/95)
Oilbird, frogmouths, potoos, nighthawks, nightjars

Order Apodiformes (135/422)
Swifts

Order Coliiformes (1/6)
Colies (mouse-birds)

Order Trogoniformes (8/34)
Trogons, quetzal

Order Coraciiformes (44/194)
Kingfishers, todies, motmots, bee-eaters, rollers, hoopoe, hornbills, ground rollers

Order Piciformes (62/387)
Jacamars, puffbirds, barbets, honey-guides, toucans, woodpeckers, piculets, wrynecks

Order Passeriformes (1089/5076)
Broadbills, woodcreepers, ovenbirds, antbirds, tapaculos, cotingas, manakins, flycatchers, sharpbills, plantcutters, pittas, asities, New Zealand wrens, lyrebirds, larks, swallows, drongos, crow allies, bowerbirds, birds-of-paradise, titmice, nuthatches, creepers, babblers, bulbuls, dippers, wrens, thrashers, thrushes, Old World warblers, Old World flycatchers, accentors, pipits, waxwings, shrikes, starlings, sunbirds, white-eyes, vireos, Hawaiian honeycreepers, wood warblers, tanagers, finches, weaver finches, cardinals, grosbeaks, Old World sparrows

Morphological classification systems compare and contrast various anatomic features of birds to develop a scheme of relations among bird orders, families, genera, and species. As a group birds are remarkably homogenous, and this creates problems in separating convergent anatomic features from those that reflect true genetic relations among similar-looking species.

Note the relatively sparse structure of this morphological classification in comparison with that of the biochemical classification. Biochemical classifications, based on DNA-DNA hybridization data, yield a much more detailed picture of the ancestral relations of birds than do morphological systems and therefore typically include more taxonomic levels than morphological systems do. However, the morphological system has been the standard in ornithology for more than a century, and all but the most recent works follow this traditional system when classifying birds.

References: Austin and Singer 1985; Storer 1971; and Wetmore 1951.

6 CONVERGENCE AND DIVERGENCE

Early avian systematics classified birds on the basis of relatively superficial anatomic differences. Only in the last hundred years has it become possible to observe live birds routinely and to examine specimens from all over the planet, and thus to obtain a truly global perspective on the range of adaptations within the Class Aves. With this broader perspective on avian evolution it became obvious to taxonomists that unrelated bird species in widely separate geographic areas had often evolved similar ecological roles and even detailed resemblances in anatomic structure. This phenomenon is called **convergent evolution**, where because of behavioral and ecological similarities two completely unrelated groups of animals evolve a close physical resemblance. Convergence is one of the most interesting consequences of evolution, but it often led early taxonomists astray as they sought to define the true relations among groups of birds. In recent years taxonomists have added genetic analysis as a more reliable method for determining evolutionary relations. Though they remain controversial, DNA hybridization and sequencing techniques are revolutionizing avian taxonomy, providing a wealth of insights into the evolution and radiation of birds.

The Dovekie or Little Auk (*Alle alle*, top bird) of the Arctic and North Atlantic and the Magellan Diving-petrel (*Pelecanoides magellani*) of Cape Horn and Tierra del Fuego are remarkably similar in appearance, though they quite unrelated. Both species are small diving seabirds and share with most other pelagic divers the strong black-and-white counter-shaded plumage coloration (Fisher and Peterson 1964; Harrison 1983).

There are many examples of convergence in the Class Aves, principally because flying birds face such tight morphological constraints on their size and overall body form. Given these limitations, and the relatively homogeneous body plan of most birds, it is not so surprising that many unrelated types of birds have evolved similar physical adaptations to common ecological niches. The toucans (Family Ramphastidae) of the Neotropics (South and Central America) look similar to the African and Asian hornbills (F. Bucerotidae). Both hornbills and toucans have large curved bills for plucking fruit from the outermost branches of trees. Diving-petrels (F. Procellariidae) of the southern oceans look remarkably similar to the tiny Dovekies (F. Laridae) of the northern seas, yet the two groups of seabirds are not closely related. Both families are small black-and-white diving seabirds that lead similar lives in opposite hemispheres of the earth. Only the distinctive tubular nostrils of the diving-petrels provide an easy visual separation between the petrels and the dovekies.

One of the most interesting examples of convergence in the animal world is the detailed physical resemblance of the Eastern Meadowlark (*Sturnella magna*), a blackbird (F. Fringillidae) of North America, and the Yellow-throated Longclaw (*Macronyx croceus*) of southern Africa, a member of the pipit subfamily (F. Passeridae: Motacillinae). Both birds live in open grasslands, feed on the same types of foods, and have similar ecological niches and habits. When compared side-by-side the resemblance between the longclaw and meadowlark is astounding. Each has a yellow breast crossed by a wide black belt, mottled brown-and-black back patterning, and both birds even have similar white outer tail feathers. Only a slight difference in head patterning and the spur-like hallux claws of the longclaw distinguish these totally unrelated birds. When viewing such a remarkable display of the results of convergent evolution it is easy to see how early taxonomists could be led astray when using superficial physical resemblances to classify different bird groups (Fisher and Peterson 1964).

The power of convergent evolution to "mask" the true relations among bird groups was demonstrated by Sibley and Ahlquist (1983) when they sorted out the family affinities of the Australian "warblers" and Australian "flycatchers." Like the marsupial mammals, the crow family, Corvidae, arose in isolation on the island continent of Australia, and the corvids radiated into diverse forms to fill many ecological niches in the various habitats of the southern continent (Sibley and Ahlquist 1985). Through morphological systematics, various small Australian songbirds were first classified with Asian flycatchers (F. Muscicapidae) and Old World warblers (F. Sylviidae). DNA-DNA hybridization analysis demonstrated that the Australian "warblers" and "flycatchers" are in fact close relatives of the crows and only superficially resemble true flycatchers and warblers because they live and feed in the same manner. Even the marvelous New Guinea birds-of-paradise turn out to be just relatives of crows with exceptionally colorful plumages.

The other side of the coin is **divergent evolution**, where closely related species develop such distinct physical and behavioral differences that their true genetic affinities are not at all obvious. Again, DNA hybridization has turned up relations among bird groups that seem wildly unlikely, yet are proven by close genetic links. For example, it is hard to imagine an odder pairing than the carrion-scavenging New World vultures (formerly classed in Order Falconiformes) and the graceful waterbirds of the stork family (formerly grouped with herons in the Order Ciconiiformes), yet the two groups are quite closely related. Using DNA hybridization data, Sibley and Ahlquist (1990) classify both storks and New World vultures in the Family Ciconiidae, within a heavily revised Order Ciconiiformes that now includes such unlikely cousins as penguins, birds of prey, grebes, loons, shorebirds, and gulls, as well as herons and storks. With such a motley cast of characters, the new Order Ciconiiformes is surely the best example of divergent evolution ever devised.

Eastern Meadowlark
(*Sturnella magna*), of
eastern North America

Yellow-throated Longclaw
(*Macronyx croceus*), of
southern Africa

7 HYBRIDIZATION AND "SWAMPING"

During the early 1970s birdwatchers in the northeastern United States began to see American Black Ducks (*Anas rubripes*) that showed distinct white edges around the speculum (a highly colored area of secondary feathers near the base of the wing). This white edging is a characteristic of the closely related Mallard (*Anas platyrhynchos*). Hybridization between Black Ducks and Mallards in park ponds had been noticed for many years, and small-scale hybridization is a common occurrence in many closely related bird genera. However, Mallard characteristics spread rapidly in the Black Duck population, and by the early 1980s ornithologists felt that the Mallard was well on its way to overwhelming, or "swamping," the gene pool of the American Black Duck. Along much of the Eastern Seaboard it is now difficult to find Black Ducks that do not show at least some Mallard characteristics. Hunting restrictions on American Black Ducks were intensified in the 1970s, and the situation is being studied by the U.S. Fish and Wildlife Service.

The Blue-winged/Golden-winged warbler complex is another classic example of genetic swamping via hybridization. In the early 1800s the Blue-winged Warbler (*Vermivora pinus*) had a breeding range in southern North America, while the Golden-winged Warbler (*Vermivora chrysoptera*) was a northern-breeding species (Short 1963). When land began to be cleared in the mid-1800s the habitat conversion to second-growth vegetation boosted the populations of both species, and the Blue-winged Warbler began to move northward (Gill 1980). Where the ranges overlapped, hybridization occurred, creating the familiar Brewster's Warbler form (*V. pinus* x *chrysoptera*). These Brewster's Warbler hybrids remained fertile and further hybridized with either Golden-winged or Blue-winged warblers. This created yet more intermediate hybrid forms, including the rare Lawrence's Warbler. Gill (1980) found that within fifty years of contact in a locality, the Blue-winged Warbler will almost completely replace the Golden-winged Warbler. The swamping of local Golden-winged Warbler populations follows a predictable pattern. Blue-wings invade the Golden-wings' territory, and pure members of both species predominate at first. Then Brewster's hybrids begin to appear, and in succeeding years a mix of other intermediate forms appear along with many pure Blue-wings. As the swamping pattern continues, Lawrence's hybrids begin to occur, and the pure Golden-wings drop in number. Finally Blue-winged Warblers dominate, and only an occasional Lawrence's or Brewster's hybrid is seen.

Hybridization and swamping may be seen with other bird species where formerly separate populations are now part of the same gene pool. Baltimore Orioles (now *Icterus galbula galbula*) and Bullock's Orioles (now *I. g. bullockii*) are now lumped as subspecies under the Northern Oriole (*Icterus galbula*). Yellow-shafted Flickers (now *Colaptes auratus auratus*) and Red-shafted Flickers (now *C. a. cafer*) have become Northern Flickers (*Colaptes auratus*). In the Midwest, Rising (1983) reports that Indigo Buntings (*Passerina cyanea*) are replacing Lazuli Buntings (*P. amoena*). Coastal salt marshes in eastern North America have seen hybridization between Clapper Rails (*Rallus longirostris*) and King Rails (*R. elegans*) to the point where in some locations only hybrid forms are seen.

Hybridization between species must be analyzed over adequate periods of time and over the entire geographic range of the species before an accurate impression of the degree of interbreeding between two species can be gained. Often local instances of intense hybridization (called **hybrid swarms**) do not reflect the situation over the broader geographic range of the species. Hybrids like Lawrence's and Brewster's warblers may be numerous in an area for years, but more often than not they are soon reabsorbed into the original dominant gene pool (in this case, Blue-winged Warblers) and eventually vanish from the scene. The "lumping" of species like Bullock's and Baltimore orioles into the Northern Oriole or the Red-shafted and Yellow-shafted flickers into the Northern Flicker occupied years of debate and study, because for these and most other cases, years of observations, conducted over hundreds of miles of breeding range, must be conducted to assess hybridization or swamping events.

Blue-winged Warbler / Golden-winged Warbler
Hybridization Complex

Lawrence's Warbler
(*Vermivora pinus* x *chrysoptera*,
or *V. "lawrencei"*)

Brewster's Warbler
(*Vermivora chrysoptera* x *pinus*,
or *V. "leucobronchialis"*)

Blue-winged Warbler
(*Vermivora pinus*)

Golden-winged Warbler
(*Vermivora chrysoptera*)

References: Gill 1980; Short 1963.

8 SUBSPECIES AND GEOGRAPHIC VARIATION

Many species of birds inhabit wide geographic areas. The Song Sparrow (*Melospiza melodia*), for example, breeds throughout North America, from Mexico northward into Alaska. In such a wide-ranging species there are clearly geographic subsets within the population that show distinct local plumage patterns and song variants. When the distinct geographic forms of a species reach a point where 75 percent of the individuals are recognizably different from typical individuals of the "parent" species, the local group is formally designated a subspecies of the parent species. The subspecies is so named by attaching a third, subspecific name to the Latin name of the species. Thus the pale Song Sparrows of the southwestern deserts of North America are called *Melospiza melodia saltonis*, to distinguish that form from the parent species. With thirty-one recognized subspecies, the Song Sparrow ranks among the highest of North American birds in the number of its geographic varieties (Welty and Baptista 1988).

Often the subspecific group within a species is so distinct from the parent species that it can easily be recognized by birdwatchers, and obvious subspecies or races of birds are often depicted in field guides. Every major North American field guide, for example, illustrates distinct subspecies of the Horned Lark (*Eremophila alpestris*), particularly the dark northern form (*E. a. alpestris*) and the paler "prairie" form (*E. a. enthymia*). The Seaside Sparrow (*Ammodramus maritimus*) has a number of distinctive subspecies, including Florida's Cape Sable Sparrow (*A. m. mirabilis*), a Gulf coast form (*A. m. fisheri*), and the extinct Dusky Seaside Sparrow (*A. m. nigrescens*). Both the Cape Sable Sparrow and Dusky Seaside Sparrow were formerly recognized as separate species, and you may see them listed as such in older books and field guides.

Geographic variation within a species can have a considerable impact on the conservation and management of that species (O'Brien and Mayr 1991). In determining federal resource and funding priorities it matters a great deal whether the endangered bird is considered a unique species or merely an uncommon local variant of an abundant parent species. In the 1970s, Florida's Everglades Kite (*Rostrhamus sociabilis plumbeus*) was listed as an endangered species. The kite's numbers have fallen in recent decades as the populations of freshwater snails (*Pomacea*) have dwindled due to irrigation, drainage, and pollution in south Florida wetlands. Although conservation efforts continue, the kite is no longer listed as an endangered species, because taxonomists now consider it to be a subspecies of the abundant Snail Kite (*Rostrhamus sociabilis*) of Central and South America. In a similar reappraisal, the Dusky Seaside Sparrow was lumped as a subspecies of the abundant Seaside Sparrow in the late 1970s, removing it from the list of endangered species. O'Brien and Mayr warn, however, that federal guidelines for the protection of rare subspecies and endangered species may result in unfortunate management decisions or even the loss of genetically unique populations of endangered animals.

Many factors influence the evolution of bird subspecies. In most cases subspecies arise in areas where a breeding population is reproductively isolated at least in part from other populations of the same species. Over time random mutations and adaptations to local conditions accumulate

Pacific Coast Subspecies of the
Song Sparrow (*Melospiza melodia*)

Aleutian subspecies
Melospiza melodia maxima

Pacific Northwest subspecies
Melospiza melodia morphna

California coastal subspecies
Melospiza melodia heermanni

Desert Southwest subspecies
Melospiza melodia saltonis

AK

BC

WA

OR

CA

The Song Sparrow (*Melospiza melodia*) is found throughout North America, and is represented by thirty-one distinct subspecies. On the West coast, the Song Sparrow subspecies form a well-known cline in body size, plumage coloration, and song characteristics. There is a dramatic difference in appearance between the small, pale *M. m. saltonis* subspecies of the Southwestern desert region and the large, dark Aleutian subspecies aptly named *M. m. maxima*. Such a wide range of morphological variation in a single species makes it easy to see how geographic isolation can create opportunities for the formation of new species of birds. It seems doubtful that the desert *saltonis* sparrows would easily interbreed with the Alaskan *maxima* birds, even if the ranges of the two subspecies were to overlap someday. If a catastrophic occurrence were to eliminate the central West coast populations of the Song Sparrow, it seems likely that the north and south ends of the cline would quickly become two distinct species of sparrow.

changes in the gene pool of the isolated group, and this may be reflected in new song patterns, unique plumage coloration, and even distinctive behavioral patterns in the isolated group. Widely distributed species like the Song Sparrow are not a collection of discrete subspecies but a continuum of gradually varying genetic characteristics across the continent. Many intermediate forms of the Song Sparrow are present where the ranges of subspecies overlap. Because naming every possible local variation from the "typical" Song Sparrow would be impossible, the designation of exactly where a local population deserves subspecies recognition must be somewhat arbitrary. Taxonomists often must impose a simple set of discrete names over the enormously complex realities of nature.

Local environmental conditions directly affect the morphology of birds (James 1983) and play a long-term role in creating geographic subspecies. A number of general observations or "rules" illustrate some of the ways climatic factors influence body form:

Allen's Rule
Body extensions such as arms, tails, legs, and ears are proportionately longer in populations of a species living in warmer climates, and shorter in animals of the same species living in colder regions. Long appendages or limbs have a greater surface-to-volume ratio than shorter appendages, and long limbs therefore radiate more body heat. Shorter appendages have less surface area and help conserve body heat that might be lost through radiation from the skin surface, and therefore short appendages are presumably less vulnerable to frostbite.

Bergmann's Rule
Animals that live in cold regions tend to be larger than related animals in warmer climates. Large animals retain body heat better than smaller animals do because their larger bodies have proportionately less surface area to radiate heat. Alaskan subspecies of the Song Sparrow (*M. m. maxima*) and Fox Sparrow (*Passerella iliaca unalaschcensis*) are much larger than their southern counterparts.

Gloger's Rule
Animals that live in humid climates tend to be darker than animals from arid regions. This probably reflects the colors of vegetation, which tend to be dark and lush in humid climates like temperate and tropical forest and paler in arid climates like alpine tundra, grassland, or desert. Subspecies of the Song Sparrow illustrate this rule well: Song Sparrows in the humid forests of the Pacific Northwest are very dark (*M. m. morphna*), especially when compared to the pale subspecies of the desert Southwest (*M. m. saltonis*). Also see the note on the Peregrine Falcon subspecies on p. 13 (see James 1970).

Many subtler factors also influence geographic variation in bird species. Food size can affect overall bill size, as studies of subspecies of Fox Sparrows in western North America indicate (Zink and Remsen 1986). It has even been suggested that geographic variations in crest length in Stellar's Jays (*Cyanocitta stelleri*) may differ based on the openness of the habitats they frequent (Brown 1964).

Clines

When a species shows gradual genetic variation over a range of climatic factors (temperature, humidity, vegetation cover, altitude) or across geographical distance, the pattern of variation is called a **cline** (Huxley 1939). The spectrum of Song Sparrow plumages and body sizes along the Pacific coast of North America forms a well-known avian cline, where the plumage color and body size of the Song Sparrows vary continuously up the coastline from the desert Southwest to the humid forests of coastal Alaska. These variations in plumage and body size are thought to parallel gradients of decreasing temperature (gradually increasing body size of the sparrows) and increasing humidity (gradually darkening plumage in the sparrows) as one moves northward along the coast.

It is sometimes difficult to identify the environmental factor most responsible for clinal changes in species. In studying the relatively abundance of the red and gray color phases of the Screech Owl in eastern North America, Owen (1963) noted that the incidence of red birds changes as you move north, from 70 percent red-phase owls in Tennessee to less than 30 percent red-phase owls in southern Maine. Although temperature would seem to be the most obvious potential cause of the clinal variation in color phases, other factors may also be at work. In studies of winter roost sites chosen by Screech Owls in Connecticut, Proctor (in preparation) found that the availability of suitable tree roosts may affect the relative balance of the two color phases of the Screech Owl. Proctor found that 84 percent of red-phase owls chose to roost in oaks and beeches, trees that retain most of their dead leaves on their branches throughout the winter. Connecticut is near the northern limit of the distribution of the oak/beech tree family (Fagaceae), and north of Connecticut such trees become increasingly scarce. The northern distribution of the red-phase Screech Owl may thus be limited by its preference for oaks and beeches as winter roost sites, where the dead red-brown leaves offer suitable camouflage. A temperature gradient may control the Screech Owl cline indirectly (by limiting the distribution of oaks and beeches), but what governs it may be the availability of suitable winter roost sites for the red-phase birds. Gray-phase owls apparently are better camouflaged on bare, leafless trees or in evergreens.

Clinal Color Variation in
Two Fox Sparrow Subspecies
(*Passerella iliaca*)

Passerella iliaca fuliginosa
of the Pacific Northwest

Passerella iliaca unalaschcensis
of southwestern Alaska

9 COMPETITION AND SPECIATION

Hundreds of thousands of bird species have evolved to fill the wide variety of ecological niches found in almost every environment on earth. Bird species have evolved and disappeared, to be succeeded by new species branching from their ancestral stock. During the succeeding thousands of years, long periods of continent-wide glaciation and the general climatic cooling that followed the Pleistocene epoch have reduced the number of species to today's total of 9,672.

Competition and Genetic Variation

Evolution is a continuum. Bird species are constantly evolving, but even "rapid" vertebrate evolution happens within a time-frame so slow as to be almost undetectable in a human lifetime. In classic Darwinian evolutionary theory the most significant agent in changing the genetic makeup of a species is natural selection, a process often termed "the survival of the fittest." In forming his theory on evolution Charles Darwin (1859) noted two fundamental characteristics of wild species: organisms typically produce more offspring than their environment will support, and individual members of a species have unique attributes—subtle physical differences that distinguish them both from their parents and from other members of their species. While competing for limited food and for mating partners those individuals best adapted to their environment will survive and pass on their genes slightly more often than will less well-adapted individuals. Adaptive traits thus spread throughout the gene pool of a species, since successful individuals leave more and more offspring that carry these adaptive genes in their DNA. It is important to remember that genetic variation is an inherently random process occurring at the molecular level—neither individuals nor species ever "decide" what traits are adaptive or how best to "solve" the problems of competing with other animals in their environment. See Lewontin (1978) and Gould (1991a, 1991b) for lucid explanations of the complexities of adaptive evolution.

When Darwin wrote *On the Origin of Species* the formal concept and mechanisms of genetics were unknown, particularly the mechanisms of genetic variation inherent in sexual reproduction soon presented by Gregor Mendel in 1865. Unfortunately, Mendel's work did not fit into the then-accepted approach to inheritance, and his discoveries were largely unappreciated until early in this century (Sturtevant 1965).

Genetic Isolation and Speciation

Species are often defined as "groups of actually or potentially interbreeding natural populations, which are genetically isolated from other such groups" (Mayr 1942). A successful species must maintain a continuing population within its environment and cope well with its competitors within the local environment. To maintain its unique gene pool a species must evolve mechanisms that inhibit interbreeding with other similar species. This is called **genetic isolation**—all speciation ultimately springs from this phenomenon. Sexual reproduction guarantees a continuing supply of new variations in a gene pool, but the overall genetic characteristics of a species are normally stable because of the great numbers of individuals involved. In a large population it may take many thousands of generations for even a minor genetic change to spread to every member of the species. Speciation requires a mechanism to isolate a population from other members of the species. This isolation may be physical; mountainous terrain, periods of glaciation, and large bodies of water are the most common geographic barriers that isolate populations of animals. The isolating factor may also be behavioral or ecological: many bird species have evolved complex breeding behaviors and plumage patterns, creating social barriers that limit interbreeding with closely related species in their environments.

The term *gene pool* often refers to the total genetic material shared by all members of an interbreeding group. The size of this "interbreeding group" is relative; small, isolated populations of animals will have a small, isolated gene pool. The term may also be used more broadly to describe the total genetic makeup of all populations in a species.

Once a population is reproductively isolated, two mechanisms drive the process of speciation. The accumulation of random genetic variation, or **genetic drift**, gradually creates a unique gene pool in the isolated group, and selection processes favor variations that better adapt the group to the local environment. Populations that have been isolated for long periods evolve superficial but recognizable differences from the parent species. Given thousands of generations in isolation the accumulated differences between the isolated group and the parent species may be so great that the isolated group will no longer freely interbreed with members of its parent species. At this point the isolated group is recognized as a separate species, because it now maintains reproductive isolation from other closely related groups of animals.

Competition and Ecological Niches

Every species has relationships with other organisms in its environment, and the sum of those relationships among all species creates the **ecology** of the local environment. To compete successfully with other animals, a species must have a unique **ecological niche,** a singular strategy to obtain food and other resources from the environment. This concept helps explain some of the most perplexing phenomena of wildlife biology: how, for example, can three species of nearly identical boobies (Family Sulidae) reside on the same tropical island? Why do so many different species of alcids (F. Laridae) breed on the same Alaskan cliff face? How is it possible for many superficially similar species to coexist in the same area?

In these examples, differences in feeding strategies and reproductive behavior separate superficially similar birds into unique ecological niches, and so the birds do not compete directly for the same environmental resources. The Red-footed Booby (*Sula sula*) nests in trees; the other two boobies are ground-nesters. Unlike other boobies, the Red-footed Booby feeds only at dusk and at night. The Masked Booby (*S. dactylatra*) typically forages far out to sea for its primary food of flying fish, whereas the Blue-footed Booby (*S. nebouxii*), with its long tail, can brake rapidly when hitting the water and is therefore able to feed in shallow inshore waters. Though the three boobies look remarkably alike, each has a unique "job" or ecological niche that prevents head-on competition with the other booby species. In the case of the Alaskan alcids, the bills of the various species of auks, auklets, and puffins reveal that each bird feeds on different sizes and species of fish and therefore avoids direct competition with other species. Each species also has particular preferences in nesting areas. Some alcids prefer ledges on sheer rock faces. Other prefer burrows or wider vegetated areas on the cliff; some may nest only on or near the cliff rim. By avoiding direct competition for such scarce resources as food or nesting areas, five or six alcids may feed and nest in the same Bering Sea waters without undue conflict. Two directly competing species cannot live indefinitely within the same ecological niche in the same habitat. The better-adapted species, the one that more efficiently uses its habitat, will always out compete rival species and will eventually drive them out of the area (Gause 1934). The key to understanding how species coexist is to determine the unique ways each exploits its environment and avoids conflict with other species.

Mixed Foraging Flocks

Many migratory birds of North America illustrate a more complex pattern of niche separation and adaptation, because every year these birds must cope with two radically different sets of competitors: their summer neighbors in the forests of North America and the resident songbirds that live in the South American habitats where these birds spend the winter months. The Kentucky Warbler (*Oporornis formosus*) of the southern lowland forests of North America, for example, is a common ground-forager in this habitat and is often the dominant ground-feeder in its environment. But when the Kentucky Warbler migrates to the tropics of South America each winter, it assumes a peripheral role outside the normal foraging flocks of songbirds that follow the movements of army ants through the forests. No ground-feeding niche is available within the jungle flocks, because that "job" is held by permanently resident species, and the Kentucky Warblers must get by temporarily by exploiting what the resident birds overlook. In effect, the complex mix of resident tropical songbird foragers functions as a "family" unit, and the group excludes migrant species that may try to enter its ranks. If the Kentucky Warbler could not adapt to its peripheral winter niche in the tropical forests it would not survive as a species.

Mixed flocks of bird species can be an effective foraging and protection mechanism for the group. For uncommon species that lack the numbers to form protective flocks, joining groups of other species may greatly enhance their chances for survival. Foraging flocks also increase an individual bird's chances of finding food (Welty and Baptista 1988; Gill 1990). By observing such a foraging complex you get a feeling for niche availability and for the many ways bird species specialize to avoid direct competition. Russell (1983) describes the winter foraging of sparrow flocks in the Mojave Desert of the Southwest. As large aggregations of sparrows move through the bushes and grasses, the order of foraging species follows a distinct sequence. Brewer's Sparrows (*Spizella breweri*) and Chipping Sparrows (*S. passerina*) usually lead the flock, foraging in the high-to-midlevel portions of the bushes. Following close behind are the low-branch and ground-foraging Black-throated Sparrows (*Amphispiza bilineata*). These are followed by White-crowned Sparrows (*Zonotrichia leucophrys*) and Black-chinned Sparrows (*Spizella atrogularis*), usually feeding low in the bushes or at ground level on forb (herbs and "weed" species) seeds and grass seeds. House Finches (*Carpodacus mexicanus*) feed at all levels within the flock, and ground-feeding Dark-eyed Juncos (*Junco hyemalis*) bring up the rear of the flock. Green-tailed Towhees (*Pipilo chlorurus*) trail along the ground at the end of the foraging flock to clean up any items dropped or missed by the others. The flock has a distinct place for each species—if there is no place, there is no species.

Mixed Foraging Flocks

In tropical forests, flocks of birds often forage at the edges of army ant swarms. These birds feed on insects flushed by the activities of masses of ants but do not eat the ants themselves. Depicted here is a foraging flock of birds in a Costa Rican rain forest. Each species in the flock fills a specific ecological niche, a singular strategy for finding food that differs in some way from that of its neighbors in the feeding flock. This separation of "jobs" within the flock is what allows many superficially similar bird species to live in proximity without competing directly for the same food resources. Note that most of the species are permanent residents of the rain forest and do not migrate. When North American passerines like the Great-crested Flycatcher (1. *Myiarchus crinitus*) and the Kentucky Warbler (2. *Oporornis formusus*) migrate south to the rain forest of Central America, they must fit into peripheral roles within the feeding flocks of resident birds, catching those insects that the permanent residents miss.

Feeding flocks are often stratified by height. The large Black-faced Ant Thrush (3. *Formicarius analis*) feeds directly on the forest floor, while the Bicolored Antbird (4. *Gymnopithys bicolor*) and Spotted Antbird (5. *Hylophylax naeviodes*) are active in the area just off the ground. On a higher perch the Gray-headed Tanager (6. *Eucometis penicillata*) waits for flying insects escaping the ants below. As insects flee up the boles of the tree and vines, the heavy-billed Barred Woodpecker (7. *Dendrocolaptes certhia*) and small Plain Xenops (8. *Xenops minutus*) pick them off the tree bark. At the top of the flock, awaiting butterflies and moths that escape the other birds, is the Acadian Flycatcher (9. *Empidonax virescens*), the most peripheral species in this feeding flock.

References: Miller and Stebbins 1964; Russell 1983; Small 1975.

In the Field   COMPETITION AND MIXED FLOCKING

Competition, mixed flocking, and niche separation among birds may be observed virtually anywhere in the country at almost any time of year. In the fall, winter, and spring, mixed flocks of birds offer the best opportunity to observe competition and niche separation. You shouldn't need to go far afield: local parks or the grounds of your school will probably turn up suitable flocks to observe. If you or someone you know has an established bird feeder you can observe the interactions of different species there. During the winter many shorebird flocks exhibit interesting social behaviors, as the various species stake out feeding territories and defend them against rivals. The different species of gulls that congregate around open-air garbage dump sites also offer opportunities to observe competition and the hierarchical relationships among gulls. Some of the most interesting aspects of competition among birds may be seen in such seemingly unlikely places as fast-food parking lots, where sparrows, pigeons, and gulls compete for fallen scraps of food, handouts from passersby, and access to the best "look-out" sites above the parking lot. Your instructor can help you choose the best sites in your area and tell you which local bird species to observe.

Carefully note the behaviors of the species you observe and the way the birds interact with their environment:

Which species are present? Where does each species spend most of its time? What level of vegetation does each species use? Do any of the species seem to prefer particular foods or trees?

Are the birds aggressive toward each other? (These interactions are sometimes very subtle—watch carefully!) Is there a "pecking order" among species? Who is dominant? Does age or body size affect the intra-species hierarchy?

How do birds avoid each other or appease the more dominant birds around them? What is the most valuable local resource: food, perching sites, or something else?

Systematics

1. What ancestral group is now generally thought to have given rise to early birds? How were these ancestors similar to early birds like *Archaeopteryx*?

2. List five fundamental characteristics of birds:

a.

b.

c.

d.

e.

3. What evolutionary principle explains why the New World vultures (F. Ciooniidac) look so much like the Old World vultures (F. Accipitridae)? How are the body forms of both groups of vultures adapted to finding and scavenging dead animals?

4. Explain Bergmann's Rule. Can you think of any mammal or bird species that illustrates Bergmann's Rule? (Do not use any species listed in this chapter as your example.)

5. Define these terms:

  a. Nucleotide base

  b. Swamping

  c. Coelurosaur

References

Austin, O. L., Jr., and A. Singer. 1985. *Families of birds*. Rev. ed. New York: Golden Press.

Bakker, R. T. 1975. The dinosaur renaissance. *Sci. Amer.* 232(4):58–78.

Bakker, R. T., and P. M. Galton. 1974. Dinosaur monophyly and a new class of vertebrates. *Nature* 248:168–72.

Brodkorb, P. 1971. Origin and evolution of birds. In D. S. Farner, J. R. King, and K. C. Parkes, eds., *Avian biology*. 8 vols. New York: Academic Press, 1:19–55.

Carroll, R. 1988. *Vertebrate paleontology and evolution*. New York: W. H. Freeman.

Martin, L. D. 1983. The origin and early radiation of birds. In A. H. Brush and G. A. Clark, eds., *Perspectives in ornithology: essays presented for the centennial of the American Ornithologists' Union*. Cambridge: Cambridge University Press, 291–337.

Olson, S. L. 1985. The fossil record of birds. In D. S. Farner, J. R. King, and K. C. Parkes, eds., *Avian biology*. 8 vols. New York: Academic Press, 8:79–238.

Ostrom, J. H. 1976b. *Archaeopteryx* and the origin of birds. *Biol. J. Linn. Soc.* 8:91–182.

Ostrom, J. H. 1985. The meaning of *Archaeopteryx*. In M. K. Hecht, J. H. Ostrom, G. Viohl, and P. Wellnhofer, eds., *The beginnings of birds*. Eichstätt: Freunde des Jura-Museums, 161–76.

Sibley, C. G., and J. E. Ahlquist. 1983. Phylogeny and classification of birds based on the data of DNA-DNA hybridization. In R. F. Johnson, ed., *Current ornithology*. New York: Plenum Press, 1:245–92.

Sibley, C. G., and J. E. Ahlquist. 1986. Reconstructing bird phylogeny by comparing DNAs. *Sci. Amer.* 254(2):82–92.

Sibley, C. G., and J. E. Ahlquist. 1990. *Phylogeny and classification of birds: a study in molecular evolution*. New Haven: Yale University Press.

Sibley, C. G., and B. L. Monroe. 1990. *Distribution and taxonomy of birds of the world*. New Haven: Yale University Press.

Storer, R. W. 1971. Classification of birds. In D. S. Farner, J. R. King, and K. C. Parkes, eds., *Avian biology*. 8 vols. New York: Academic Press, 1:1–18.

Wetmore, A. 1960. A classification for the birds of the world. *Smithsonian Misc. Coll.* 139(11):1–37.

Topography

3

Every anatomic discipline requires a common language, a universal set of terms that describe both the general and the specific features of organisms in great detail. Even a casual birdwatcher must learn the basic topography (external anatomy) of birds to understand field guide descriptions of plumages. Unfortunately, to the public, and to new students of biology, the complexity of anatomic terminology often seems to hinder rather than help communication. But anatomic terminology is designed to be much more precise and descriptive than "plain English," and it necessarily covers a wide range of obscure anatomic structures with equally obscure names that simply have no equivalent in plain English. Try to imagine the confusion that might reign in ornithology books and journals if every author were obliged to write "those little bristlelike feathers around the corner of the mouth" when referring to "rictal bristles," the proper anatomic term for "those little bristlelike things" found at the corners of the mouth of many insectivorous birds.

Luckily for students of ornithology, avian external anatomy is remarkably uniform across the range of size, ecological specialization, and taxonomic grouping in the Class Aves (Dorst 1974; Gill 1990; Welty and Baptista 1988). As flying animals, birds are severely limited by the demands of aerodynamics, and most birds therefore share a common external anatomic arrangement of feathers that is readily recognizable from the smallest hummingbirds to the largest condors (Lucas and Stettenheim 1972; Rayner 1988). The only prominent exceptions to this topographic consistency are the nonflying groups, such as the penguins (Family Spheniscidae) and such ratites as the Ostrich. A detailed study of several representative bird species will thus provide a solid introduction to avian topography and a knowledge of anatomic terminology that may be readily extended to most birds (Berger 1961; Van Tyne and Berger 1976; Pettingill 1985).

Although the basic avian anatomic plan is fairly monotonous, birds have evolved an amazing range of variations on it, adaptations that have allowed them to inhabit successfully virtually every habitat on the planet (King and King 1979; Pouch et al. 1989). These variations on the fundamental avian motif, both obvious and subtle, are often the best clues to how a particular species makes its way in the world and offer the ornithologist a wealth of information on the present ecology and previous evolutionary history of the species. A thorough knowledge of the basics, therefore, is essential to recognize and interpret unusual and complex anatomic nuances found throughout the avian world, and this knowledge will continue to reward the careful student with insights into the natural history, ecology, and evolution of birds.

In this plate of twelve thrushes of the Genus *Turdus* it is easy to see how crucial a knowledge of surface topography is in ornithology and taxonomy. Mexico alone has eight species of *Turdus* thrushes—all nearly identical in body form. Only the colors and patterns of the plumage vary significantly from one species to another. Without a solid understanding of topographic terms it would be difficult to interpret field descriptions of the species, and impossible to describe your observations accurately to another ornithologist or birdwatcher. From the top, left to right, the thrushes from different parts of the world illustrated here are: American Mountain Thrush, *Turdus plebejus;* San Lucas Robin, *T. migratorius confinis;* Fieldfare, *T. pilaris;* Red-legged Thrush, *T. plumbeus;* White-eyed Thrush, *T. jamaicensis;* Eyebrowed Thrush, *T. obscurus;* Rufous-collared Robin, *T. rufitorques;* Black Thrush, *T. infuscatus;* White-throated Thrush, *T. assimilis;* American Robin, *T. migratorius;* Clay-colored Thrush, *T. grayi;* and Rufous-backed Thrush, *T. rufopalliatus.*

1 GENERAL TOPOGRAPHY

In any introduction to a subject with so much unique vocabulary it is easy to get lost in the details and lose sight of the general features of the organism. Before plunging into your study of the figures at the right, consider the major features of the bird portrayed, the Blue Jay (*Cyanocitta cristata*). Note, for example, the overall shape of the Blue Jay, a common bird of forests and wooded edge habitats in most of North America (Bent 1946; Hardy 1961). The most notable features of the jay are its short, broad wings and its long tail. This general form is common to most woodland birds; wide, oval-shaped wings allow rapid acceleration and deceleration in short bursts of flight. The jay's long tail feathers act as a third airfoil surface and help it to steer around foliage and branches, giving the jay great maneuverability. When fanned, the large tail acts as an airbrake when the Blue Jay is landing.

Note the jay's sturdy but relatively unremarkable beak. The Blue Jay is a generalist of the forest, an omnivorous bird ready to exploit a wide variety of animal and vegetable food sources within its habitat (Laskey 1958). Its beak is both slender enough to be an effective probe and chisel in hunting insects and other small animals and sturdy enough to crack and tear many types of seeds and other plant foods. The jay's legs and feet are also those of a woodland generalist, adapted primarily to absorbing the shock of landing, providing a strong jumping take-off, and gripping tree branches. The jay's feet allow it to hop along the ground, but the jay does not usually walk as, for example, shorebirds or ducks routinely do.

Another obvious feature of the jay is not really related to habitat but is a distinctive departure from the structural monotony of songbirds: the crest on its head. The crest gives the jay a distinctive profile that helps in identification and plays an important role in communication. By raising and lowering its crest the Blue Jay can communicate its intentions in courtship displays and show dominance or subordination when interacting with other Blue Jays (Brown 1964).

Observe the following structures

Study the diagrams here and on the following pages to acquire the basic topographic terminology used in describing avian surface anatomy. Pay particular attention to the nomenclature of the chest, abdomen, and back, and study the other species diagrammed on the following pages to see how this basic naming scheme is applied in birds with differing body shapes.

The topographic details of the wings, head, legs, and feet are covered later in the chapter.

General Topography
Blue Jay (*Cyanocitta cristata*)

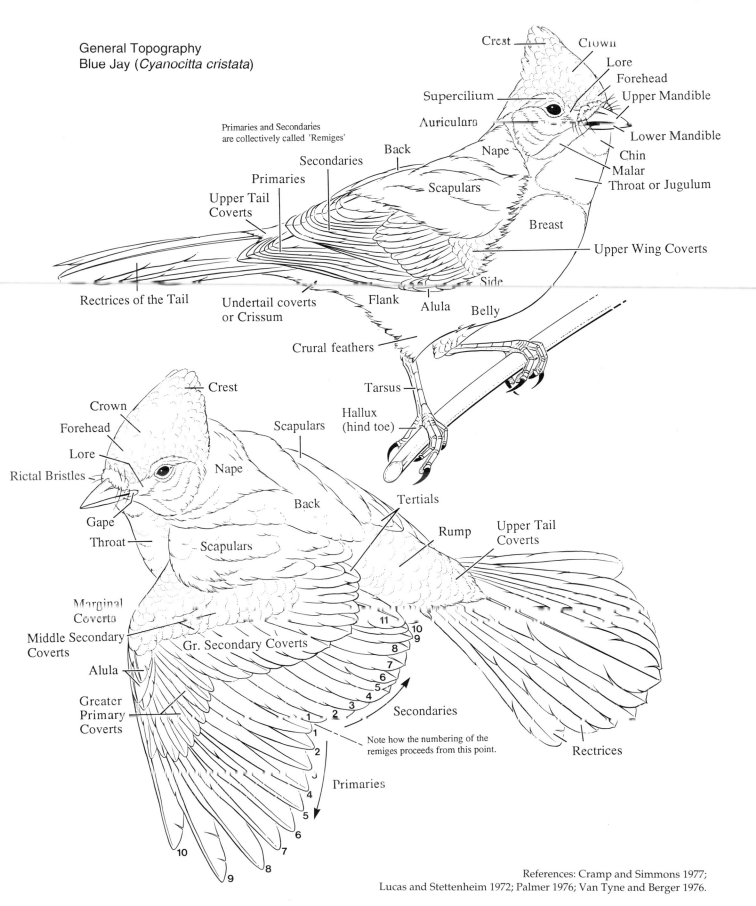

Crest ——— Crown
Lore
Forehead
Supercilium ——— Upper Mandible
Auriculars ——— Lower Mandible
Chin
Nape ——— Malar
Scapulars ——— Throat or Jugulum

Primaries and Secondaries
are collectively called 'Remiges'

Back
Secondaries
Primaries
Upper Tail
Coverts
Breast

Upper Wing Coverts

Rectrices of the Tail
Undertail coverts
or Crissum
Flank
Alula
Belly
Side

Crural feathers

Tarsus

Hallux
(hind toe)

Crest
Crown
Forehead
Lore
Rictal Bristles
Gape
Throat

Scapulars
Nape
Back
Scapulars

Tertials
Rump
Upper Tail
Coverts

Marginal
Coverts
Middle Secondary
Coverts
Alula
Greater
Primary
Coverts

Gr. Secondary Coverts

11
10
9
8
7
6
5
4
3
2
1
1
2
3
4
5
6
7
8
9
10

Secondaries

Note how the numbering of the
remiges proceeds from this point.

Primaries

Rectrices

References: Cramp and Simmons 1977;
Lucas and Stettenheim 1972; Palmer 1976; Van Tyne and Berger 1976.

Topographic Terminology for the Thorax, Abdomen, and Back

Name	Definition	Notes
Ventral Regions		
Throat, also called the **upper foreneck** or **gular region**	Ventral region of the neck extending from just under and below the lower mandible to the anterior margins of the sternum and breast musculature.	The throat area of many species is brightly colored. The lateral malar areas on either side of the throat are also sometimes distinctly colored.
Jugulum	The most ventral, midline region of the throat.	The jugulum is the middle, ventral area of the throat, flanked by the *sides* of the throat.
Breast, also called the **chest** or **pectoral region**	Area of feathering that extends over the musculature of the breast, covering about $3/4$ of the surface of the breast muscles.	The breast usually refers to the ventral area of the bird's thorax most visible from the front as the bird stands while at rest.
Abdomen, or **belly**	The most ventral area along the midline of the body, extending from the posterior $1/4$ of the sternum to the vent, or cloacal area.	In flying birds the sternum is so long that the "abdominal" topographic region actually overlies part of the posterior portion of the sternum and breast muscles.
Crural feathers	The feathers that cover the tibial portion of the leg; visually contiguous with the feathers of the abdomen.	In arctic birds like the Snowy Owl (*Nyctea scandiaca*), the crural feathers extend distally to the tips of the bird's toes.
Side, or **flank**	The lateral portions of the bird's trunk, extending from the abdominal region up to the base of the wings.	The sides are contiguous with the *axillary* area at the base of the wing. The side region lies entirely under the wing when the wing is extended in flight.
Axillary region	The base of the ventral wing, extending from the side out onto the ventral wing lining.	The "armpit" area of the bird. It is distinctly marked in some species, such as the Black-bellied Plover (*Pluvialis squatarola*).
Flank	The lateral area posterior to the side region, lying below the pelvis and extending back to the base of the tail.	The flank area is the lateral area posterior to the wing when the wing is extended in flight.
Undertail coverts, or **crissum,** also called the **circumcloacal region**	The loose feathers that surround the cloaca, including the undertail coverts that cover the ventral base of the tail.	Can be distinctly colored in some species, such as the Grey Catbird (*Dumetella carolinensis*).
Anal pteryla, or **cloacal circlet**	In most birds, two rows of feathers arranged in concentric circles around the cloaca.	Mostly covered by the feathers of the crissum.

References: Cramp and Simmons 1977; Lucas and Stettenheim 1972; Palmer 1976; Pettingill 1985; Van Tyne and Berger 1976.

General Topography
Northern Pintail (*Anas acuta*)

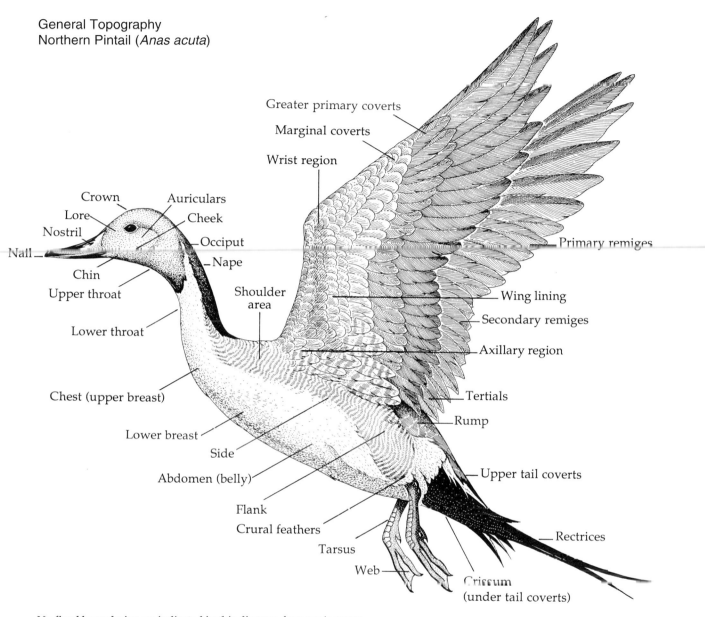

Greater primary coverts

Marginal coverts

Wrist region

Crown

Auriculars

Lore

Cheek

Nostril

Occiput

Nail

Nape

Chin

Upper throat

Shoulder area

Lower throat

Primary remiges

Wing lining

Secondary remiges

Axillary region

Chest (upper breast)

Tertials

Rump

Lower breast

Side

Upper tail coverts

Abdomen (belly)

Flank

Crural feathers

Tarsus

Web

Rectrices

Crissum (under tail coverts)

No fixed boundaries are indicated in this diagram because in many species the boundary lines between anatomic regions in the trunk and abdomen are not absolute. In such cases it is important to recognize simply the general area the name refers to rather than an arbitrary set of boundary lines that are often difficult to apply consistently from species to species.

References: Cramp and Simmons 1977; Lucas and Stettenheim 1972; Palmer 1976; Van Tyne and Berger 1976.

More Topographic Terminology for the Thorax, Abdomen, and Back

Name	Definition	Notes
Dorsal Regions		
Nape, also called the **upper hindneck** or **nuchal region**	The dorsal surface of the neck, from the occipital region of the skull posterior to the base of the neck and beginning of the thoracic vertebrae.	The nape of the neck is often a dark color that contrasts with the lighter colors of the ventral neck.
Back, also called the **interscapular region** or **dorsum**	The dorsal region of the thorax; roughly, the area of the back between the wings.	This region is laterally bounded by the **scapular feathers,** which cover the dorsal bases of the wings.
Scapulars, also called the **humeral region**	The feathers that overlie the scapula bone at the base of the dorsal wing.	When the bird is perched or standing, the scapulars often cover much of the folded wing (note the dorsal view of the jay, below).
Rump, also called the **uropygium, uropygial region,** or **lower back**	The region that overlies the pelvic bones (synsacrum and ilia); lateral boundaries are the **flanks** along the side of the body just anterior to the tail.	The rump is sometimes colored in contrast to the back, as in the Northern Flicker (*Colaptes auratus*) and Northern Harrier (*Circus cyaneus*).
Upper tail coverts	A single row of feathers covering the bases of the tail feathers (rectrices).	

References: Cramp 1977; Lucas and Stettenheim 1972; Palmer 1976; Pettingill 1985; Van Tyne and Berger 1976

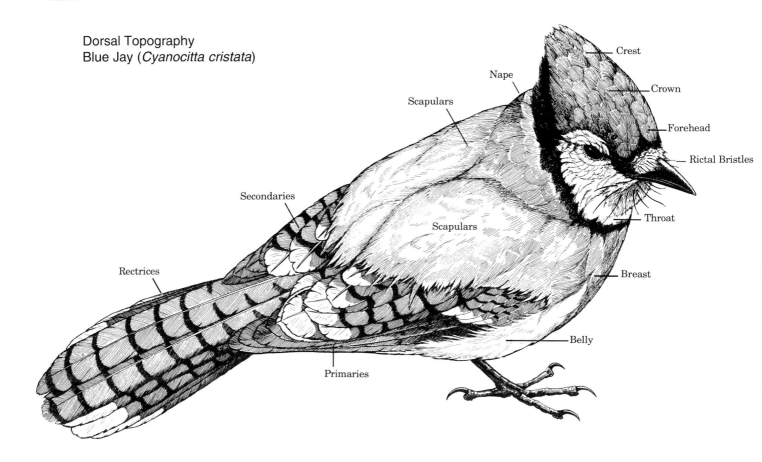

Dorsal Topography
Blue Jay (*Cyanocitta cristata*)

Topographic regions are often more distinctive on larger birds, such as the Great Egret (*Casmerodius albus*). In songbirds, such as the House Finch (*Carpodacus mexicanus*), the body is often so small that many of the line distinctions between body regions are impractical. The structure of the plumage, for example, is simpler in small passerines (songbirds), with fewer rows of feathers covering the wings and back. If your lab specimens are primarily songbirds you may be unable to make fine distinctions between median and lesser covert rows on the wings or between such small lateral regions as the flank, side, and axilla. Concentrate on the general locations of regional names and on the specialized vocabulary of avian topography.

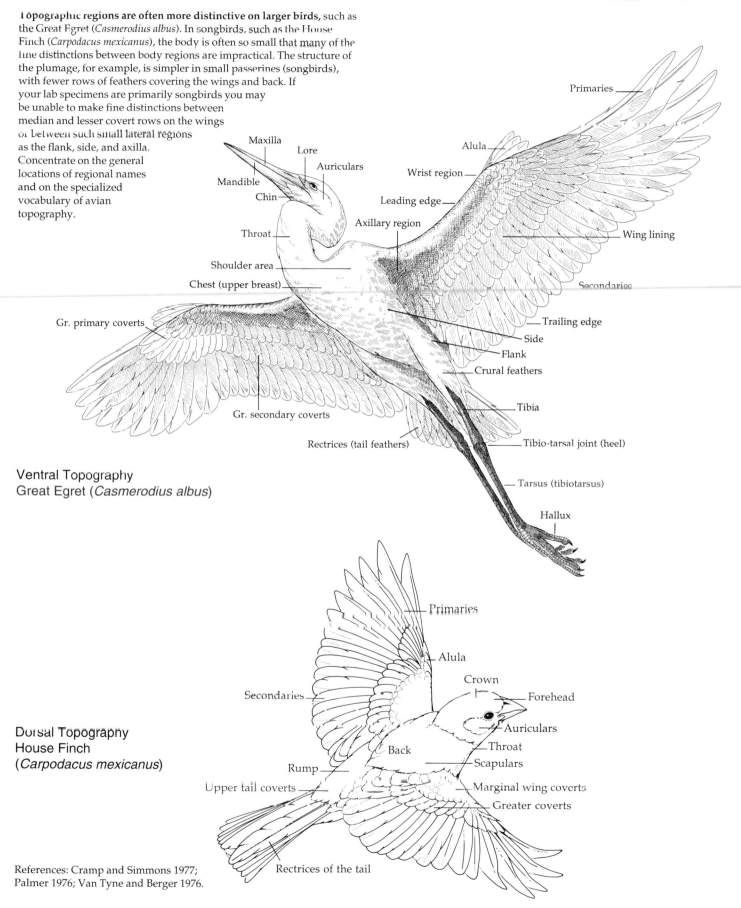

Ventral Topography
Great Egret (*Casmerodius albus*)

Dorsal Topography
House Finch
(*Carpodacus mexicanus*)

References: Cramp and Simmons 1977; Palmer 1976; Van Tyne and Berger 1976.

1 GENERAL TOPOGRAPHY, *continued*

The Rock Dove (*Columba livia*) is the common pigeon of urban areas and rural farms in North America and most of Europe. Like the Blue Jay, the pigeon is an avian generalist, with a sturdy but anatomically unremarkable body suited to exploit a wide variety of habitats and food sources. As a bird adapted to relatively open country, however, the pigeon is swift and agile in flight, with gracefully pointed wings, and it can out-fly most birds of prey in a chase. The pigeon's tail is moderate in length and quite broad, because the pigeon has less need of the tight turning ability so crucial to a woodland bird such as the Blue Jay. Most perching birds, like the jay, can only hop or hobble awkwardly when on the ground, but the pigeon's feet are adapted to permit true walking, with toes long enough to grasp branches when the bird perches in trees.

General Topography
Common Pigeon, or Rock Dove (*Columba livia*)

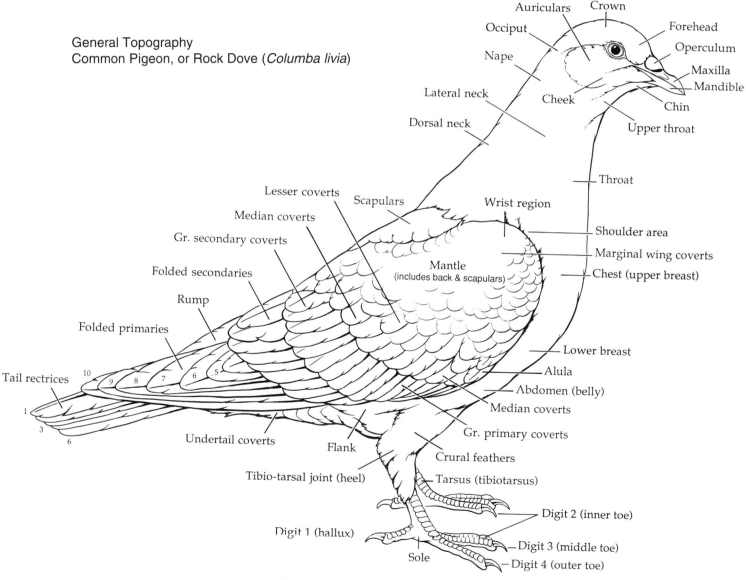

References: Lucas and Stettenheim 1972; Pettingill 1985.

General Topography
House Wren (*Troglodytes aedon*)

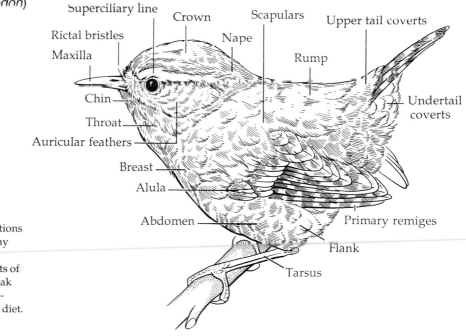

Superciliary line
Crown
Scapulars
Upper tail coverts
Rictal bristles
Nape
Rump
Maxilla
Undertail coverts
Chin
Throat
Auricular feathers
Breast
Alula
Abdomen
Primary remiges
Flank
Tarsus

Notice how different are the body proportions of the House Wren and Rock Dove. The tiny wren seems to be all head and body, with stubby round wings adapted to short bursts of flight in heavy foliage. Its long, delicate beak can probe for the small insects and invertebrates that make up the bulk of the wren's diet.

Form and Function

Where did the wings go?

Where are the wings on that duck? Novice birders are sometimes puzzled when trying to apply field guide descriptions to the way birds appear in the wild. Most birds, like the American Black Ducks pictured here or the House Wren depicted above, can tuck their folded wings into lateral "pockets" formed by the dense covert feathers of the breast and scapular regions. The resting duck shown here has tucked its wings almost entirely under the feathers of the breast and side, with only a few secondaries and primaries to show where the wing emerges from under the covert feathers. This gives the resting or standing bird a sleek, aerodynamic profile that keeps cold winds away from the body.

American Black Duck
(*Anas rubripes*)

2 | TOPOGRAPHY OF THE WING

As in most anatomic structures, the surface topography of the wing is determined largely by the supporting structures of the skeleton and underlying integument. The skeletal structure of the wing also provides the basic definitions of the two major flight feather groups of the wing, the primary feathers that are attached to the bones of the manus (hand and finger bones) and the secondary feathers that are attached to the trailing edge of the ulna. Several sheets of tough, tendinous tissue form significant parts of the wing structure: the **patagium** and the **patagialis longus** muscle and tendon actually form much of the leading edge of the wing, and the **postpatagium** envelops the proximal quill shafts of the **remiges** (the main flight feathers of the wing). The postpatagium provides much of the elastic strength of the wing and keeps the flight feathers properly aligned and firmly attached to the wing skeleton. The structure of the wing skeleton is covered in greater detail in chapter 5, The Skeleton.

Observe the following structures

Humerus — The short, thick bone at the base of the wing. As in reptiles and mammals, the avian humerus supports the other bones of the pectoral limb.

Radius and **ulna** — The supporting bones of the forearm portion of the wing. Note the larger size of the ulna and the attachment of the secondary quills to the posterior surface of the ulna.

Carpometacarpus — In birds the distal wing skeleton is a much reduced and simplified version of the standard vertebrate forelimb plan. The carpometacarpus is a fusion of carpal ("wrist") and metacarpal ("hand") bones. Note how the carpometacarpus supports most of the primary remiges. Bony remnants of the **second** and **third digits** form the most distal portion of the wing skeleton.

First digit, or **alula** — Notice the tiny first digit at the bend of the wrist area of the wing. The first digit **phalanges** (finger bones) support three small flight feathers, collectively referred to as the alula.

Patagium, and **patagialis longus muscle** and **tendon** — The tough, fibrous sheet of tissue that connects the shoulder area of the wing to the carpal bones at the wrist. Note that the patagialis longus actually forms much of the leading edge of the wing; the only bones at the leading edge of the wing are the **carpal bones** and **carpometacarpus** at the bend of the wrist.

Postpatagium — A tough band of tendinous tissue that envelops and supports the quills of all the wing remiges, from elbow to wingtip. The fleshier **humeral patagium** connects the elbow to the thorax.

Quills of the **primary** and **secondary remiges** — Note how the two groups of flight feathers relate to the bones and tendons of the wing. The **primaries** attach only to the bones of the manus (hand). The **secondaries** attach only to the posterior surface of the ulna.

Internal Structures of the Wing
Rock Dove (*Columba livia*)

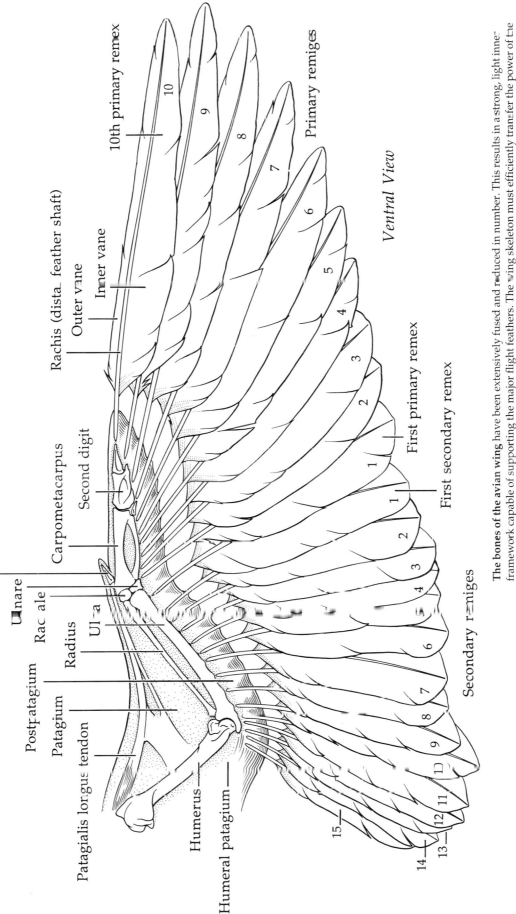

10th primary remex

Rachis (distal feather shaft)

Outer vane

Inner vane

Pollex, or "thumb" (first digit)

Carpometacarpus

Second digit

Ulnare

Radiale

Radius

Ulna

Postpatagium

Patagium

Patagialis longus tendon

Humerus

Humeral patagium

Primary remiges

Ventral View

First primary remex

First secondary remex

Secondary remiges

10
9
8
7
6
5
4
3
2
1

1
2
3
4
5
6
7
8
9
10
11
12
13
14
15

The bones of the avian wing have been extensively fused and reduced in number. This results in a strong, light immobile framework capable of supporting the major flight feathers. The wing skeleton must efficiently transfer the power of the flight muscles to drive active flight. Note how short and thick the humerus is in the modern pigeon wing. The humerus must bear the main force of the pectoralis major, the massive breast muscle that powers the downstroke phase of active flight. If the humerus were less sturdy it could not bear up against the huge lever forces generated by flight.

References: Cramp and Simmons 1977; Palmer 1976; Van Tyne and Berger 1976.

Topographic Terminology — The Dorsal Surface of the Wing

Name	Definition
Primaries, also called **primary remiges** (singular, **remex**)	The major flight feathers of the wing (remiges) that are attached to the manus of the forelimb. The primaries form the main propulsive area of the outer wing, providing most of the forward thrust in active flight. Note the strong asymmetry of the primary remiges—each remex acts as an individual airfoil. Most birds have ten primary remiges, though many songbirds have just nine.
Secondaries, also called **secondary remiges**	The remiges that are attached to the ulna of the forearm. Each secondary remex is connected to the trailing edge of the ulna. The secondaries form the trailing edge of the wing's airfoil. In larger soaring birds the secondaries make up most of the surface area of the wing. The number of secondary remiges in a species varies with wing length, ranging from a low of nine in most songbirds to twenty-five in the larger vultures.
Tertiaries, also called **tertial** or, sometimes, **humeral feathers**	A group of (usually) three to four feathers proximal to the innermost secondaries. True tertiaries are not arranged in the same row as the secondary remiges, but the innermost three or four secondaries of some species are sometimes mistakenly called tertiaries.
Scapulars, also called the **humeral region**	The feathers that overlie the scapula at the base of the dorsal wing.
Alular quills, collectively called the **alula**	Three (usually) small, stiff quills arising from the first digit (**pollex**) of the manus. The alula acts as an aerodynamic slot and spoiler, aiding and disrupting air flow over the wing in flight.
Alular quill coverts	In larger birds, a separate area of coverts protecting the base of the alular quills may be distinguished from other marginal coverts.
Marginal coverts, or **wing coverts**	In larger birds, the marginal coverts cover a significant portion of the anterior dorsal surface of the wing. In smaller birds, the marginal coverts may be reduced to two to four rows of tiny feathers overlying the leading edge of the patagium. The marginal coverts are often more hairlike, with more flexible vanes than those of the greater, median, and lesser coverts.
Greater coverts of the primaries and secondaries	A single row of relatively large covert feathers adjacent to the primary and secondary flight feathers. These coverts shield the bases of the remiges. In most species each remex is matched by one greater covert feather, although the greater covert of the most distal primary is very small and may appear to be missing in many groups.
Median coverts of the primaries and secondaries	A single row of coverts just anterior or proximal to the greater coverts. In smaller birds, the secondary coverts may be difficult to distinguish from the marginal coverts.
Lesser coverts of the primaries and secondaries	In larger birds, the last two to three distinct rows of covert feathers between the larger coverts and the small, soft feathers of the marginal coverts.

References: Cramp and Simmons 1977; Lucas and Stettenheim 1972; Palmer 1976; Pettingill 1985; Van Tyne and Berger 1976.

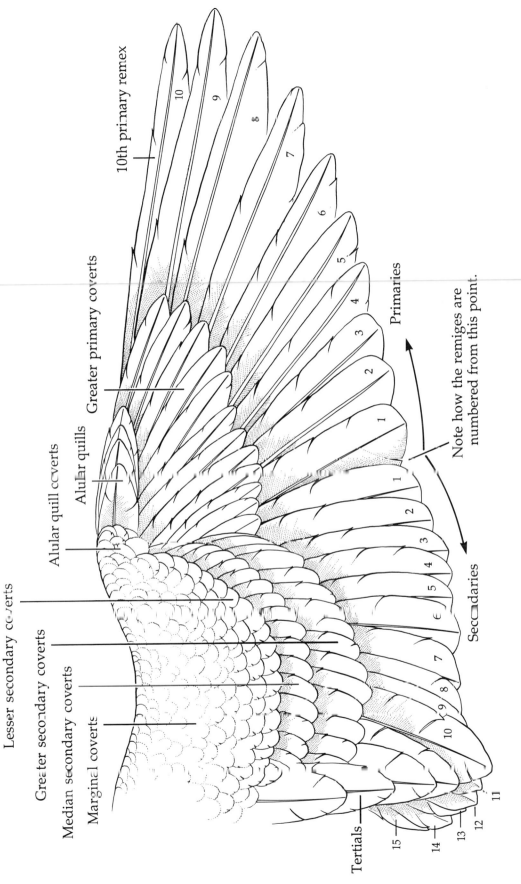

Topography of the Dorsal Wing
Rock Dove (*Columba livia*)

10th primary remex

Greater primary coverts

Alular quill coverts

Alular quills

Lesser secondary coverts

Greater secondary coverts

Median secondary coverts

Marginal coverts

Tertials

Primaries

Note how the remiges are numbered from this point.

Secondaries

Note how the remiges of the wing are numbered: the primary remiges are numbered from the most proximal remex attached to the manus outward to the outermost primary; the secondary remiges are numbered in the opposite direction, with the numbers increasing in the more proximal secondary remiges.

Topographic Terminology — The Ventral Surface of the Wing

Name	Definition
Primaries, also called the **primary remiges** (singular, **remex**)	The major flight feathers of the wing (remiges) that are attached to the manus. The primaries form the main propulsive area of the outer wing, providing most of the forward thrust in active flight. Note the strong asymmetry of the primary remiges—each remex acts as an individual airfoil. Most species have ten primary remiges, though many songbirds have just nine.
Secondaries, also called the **secondary remiges**	The remiges that are attached to the ulna of the forearm. Each secondary remex is attached to the trailing edge of the ulna. The secondaries form the trailing edge of the wing's airfoil. In larger soaring birds the secondaries make up most of the surface area of the wing. The number of secondary remiges in a species varies with wing length, ranging from a low of nine in many songbirds to twenty-five in larger vultures.
Tertiaries, also called **tertial** or, sometimes, **humeral feathers**	A group of (usually) three or four feathers just proximal to the innermost secondaries. True tertiaries are not arranged in the same row as the secondary remiges, but the innermost three or four secondaries of some species are sometimes mistakenly called tertiaries.
Greater coverts of the primaries and secondaries	A single row of relatively large covert feathers adjacent to the primary and secondary flight feathers. These coverts shield the bases of the remiges. In most species each remex is matched by one greater covert feather, although the greater covert of the most distal primary is very small and may appear to be missing in some groups.
Wing lining, or **marginal coverts**	Very soft feathers that form a smooth, featureless surface on the anterior edge of the ventral wing. Although the wing lining is made up of rows of coverts, in most species the individual rows of coverts cannot be readily distinguished.
Axillaries, or **axillar region**	The relatively long and stiff covert feathers covering the ventral base of the wing, in the "armpit" area. In most species these feathers are white, but in a few species, such as the Black-bellied Plover (*Pluvialis squatarola*) and the Prairie Falcon (*Falco mexicanus*), the axillaries are distinctly dark and are a good field mark.

References: Cramp and Simmons 1977; Lucas and Stettenheim 1972; Palmer 1976; Pettingill 1985; Van Tyne and Berger 1976.

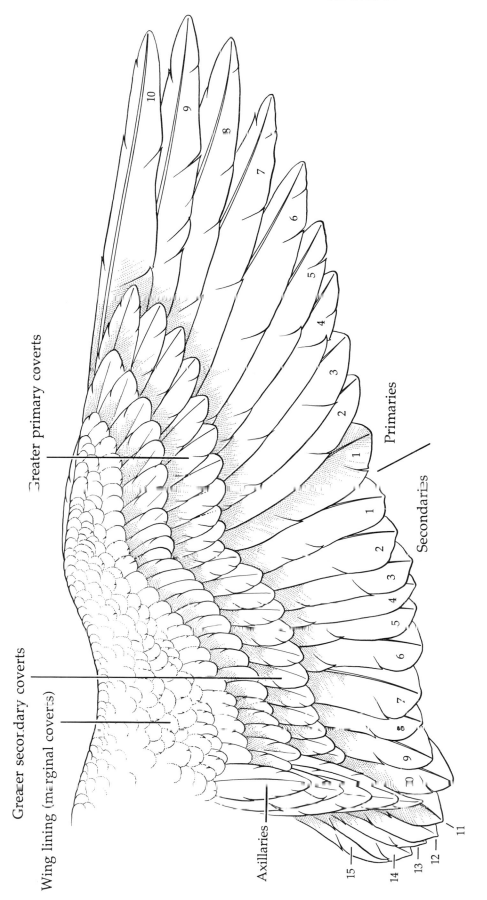

Topography of the Ventral Wing
Rock Dove (*Columba livia*)

Greater secondary coverts

Wing lining (marginal coverts)

Greater primary coverts

Primaries

Secondaries

Axillaries

Nothing about the construction of the bird's wing is arbitrary. Notice how the primaries and secondaries overlap. The most distal (outermost) remiges overlap the inner remiges. This arrangement of the primaries and secondaries is fundamental to active flight—the overlapping remiges form a solid surface in the downstroke phase, to push as much air as possible. In the upstroke the outer feathers twist and open like a venetian blind, allowing air to slip between the vanes of the primaries and reducing the work required to raise the wing.

3 TOPOGRAPHY OF THE HEAD

In spite of the remarkable consistency of avian anatomy from one taxonomic group to another, each species of bird has evolved unique physical adaptations to its environment and ecological niche. As both the focus of the nervous system and the entry point of the digestive system, the head is usually the most individual and idiosyncratic region of an animal's body. The shape and details of the bird's head and bill are often the best anatomic indicators of its normal diet and natural history. The head is also the focus for communication between members of a species, and most birds have adapted plumage and soft tissue coloration, contrasting patterns, crests, bill ornamentation, and other devices for communication that increase the complexity of the topography of the head. The head is divided into many topographic regions to handle the complexity of these plumage color and contrast patterns, which are often important field marks.

The avian head exhibits a number of adaptations to flight and the increased demands on the nervous and sensory systems that flight demands. An obvious adaptation is the loss of teeth and the corresponding reduction in the weight of the jaws. Aside from feathers, the bill is perhaps the most evident characteristic of birds. This light but effective horny covering of the jaws compensates for the loss of teeth. Birds tear, rip, and crush food in their bills, but the difficult work of chewing is performed by the gizzard in the stomach. Birds have thus moved much of the heavy business of chewing food toward their center of gravity, where the weight of the gizzard is less important to aerodynamics and balance in flight.

The eyes dominate the avian head more than in any other vertebrate group. In proportion to their body size, birds have evolved the largest eyes of all the vertebrates, and the avian skull has been extensively modified to handle the huge globes of the eyes, which occupy most of the volume of the head. The braincase has been pushed back and rotated upward to accommodate the eye sockets. This rotation causes the ear openings of most birds to lie well under and posterior to the eye, under the area covered by the auricular feathers.

Observe the following structures

Use the diagrams opposite as well as the diagrams and topographic tables on following pages to become familiar with the basic anatomic terminology of the head, eye, and bill. Note the wide range of adaptations and some of the unique anatomic structures of the avian head.

Note the major differences in the form of the head and bill in these two species. The heavy conical bill of the Grasshopper Sparrow is adapted for cracking small seeds from open grasslands. As a ground feeder and nester, the sparrow needs good camouflage to avoid predators. The wide, contrasting superciliary and crown stripes are part of a disruptive camouflage color scheme that breaks up the outline of the sparrow's head. The combination of grass-colored plumage and these bold color contrasts makes it hard to pick out the profile of the sparrow against the background of grass, and from a few yards away these small birds are virtually invisible.

The Sharp-shinned Hawk is a woodland predator of small songbirds. Its beak is adapted to tearing flesh from prey, and the sharp, hooked upper beak (**maxilla**) is used to pick the carcass clean. The distinct notch in the **tomium** (cutting edge) of the maxilla is thought to aid the bird in delivering a coup-de-grace bite that severs the spine of its prey (Brown 1976). The tomium behind the notch extends out in a structure called a tomial "tooth."

Topography of the Head
Grasshopper Sparrow (*Ammodramus savannarum*)

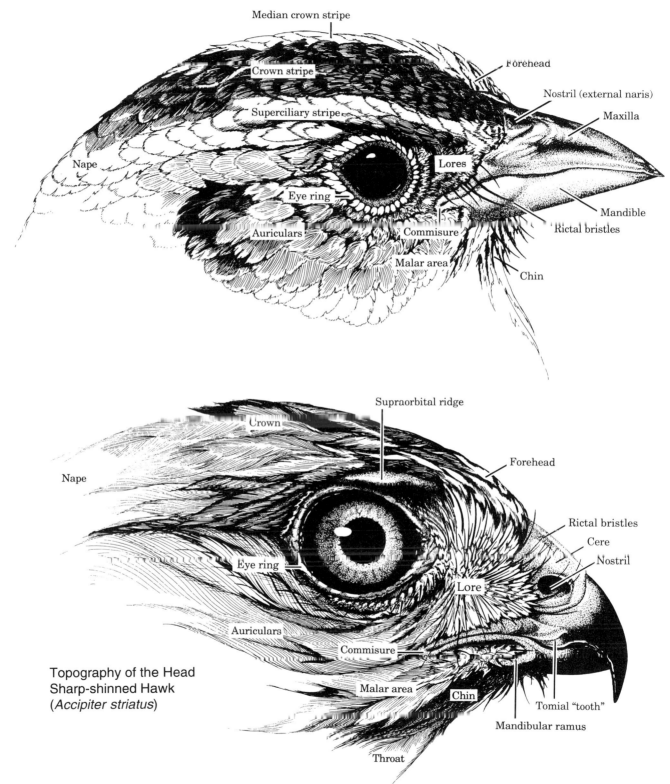

Median crown stripe

Forehead

Crown stripe

Nostril (external naris)

Superciliary stripe

Maxilla

Nape

Lores

Eye ring

Mandible

Auriculars

Commisure

Rictal bristles

Malar area

Chin

Supraorbital ridge

Crown

Forehead

Nape

Rictal bristles

Cere

Eye ring

Nostril

Lore

Auriculars

Commisure

Topography of the Head
Sharp-shinned Hawk
(*Accipiter striatus*)

Malar area

Chin

Tomial "tooth"

Mandibular ramus

Throat

References: Blake 1956; Lucas and Stettenheim 1972; Palmer 1976.

Topographic Terminology — Plumage of the Head

Name	Definition
Forehead, or **frontal region**	The region from the base of the beak posterior to a line drawn between the anterior angle (**nasal canthus**) of each eye. The ventral margin is a line drawn between the nostril and the nasal canthus of the eye.
Crown; in crested birds, the **crest**	The region just posterior to the forehead. The crown extends from the forehead back to the beginning of the cervical vertebrae of the neck. In crested birds the crown feathers are lengthened.
Pileum of the head	The *entire* top of the head, including the forehead, crown, and occipital regions.
Lore (plural, **lores**), or **loral region**	The lateral area of the head just posterior to the bill and anterior to the nasal canthus of the eye. The commissure of the mouth forms its lower border. Fleshy and naked of feathers in some birds, often brightly colored, especially during breeding season, as in the Snowy Egret (*Egretta thula*).
Superciliary line, or **supercilium**	A contrasting stripe of feathers above the eye along the side of the head found in many birds (such as most sparrows). The **supercilium** is the region immediately above the eye area.
Eye ring, or **rimal feathering**	Two to four tiny concentric bands of feathers surrounding the eyes at the edge of the eyelids. Often colored in contrast to the plumage that surrounds the eye, forming a distinctive eye ring, as in Swainson's Thrush (*Catharus ustulatus*). Absent in the pigeon family (Columbidae).
Eyeline	A line extending from the posterior angle of the eye (**temporal canthus**) and running posteriorly toward the nape of the neck. Often an important field mark.
Narial feathers	Long feathers at the base of the **maxilla** (upper bill) and extending anteriorly to partially cover the nostrils; present in some groups, such as the crows (Family Corvidae).
Rictal bristles	Bristlelike feathers around the corner of the mouth. Present and well developed in many passerines (songbirds) and other groups that feed on flying insects, although this has not been confirmed. May serve a tactile, sensory function, particularly in nocturnal species, where they act like a cat's whiskers. Absent in the pigeon family (Columbidae).
Auricular feathers, also called **auriculars** or **ear coverts**	A wide lateral patch just ventral and posterior to the eye and covering the ear opening. Auriculars are arranged in concentric bands extending down and back from the eye, covering the lateral "cheek" area of the head below the eye.
Malar region, also called **malars** or **mustache feathers**	A patch that extends posteriorly and ventrally from the mandibular ramus, ventral to the commissure of the mouth, and between the auricular feathers of the cheek and the throat feathers. Can be boldly marked, as in the male Northern Flicker (*Colaptes auratus*) or the Prairie Falcon (*Falco mexicanus*).
Commissure, also called the **commissural point** or, sometimes (incorrectly), the **gape**	The angle at which the maxilla and mandible meet. **Gape** more properly refers to the gap between the open maxilla and mandible when the bill is opened wide.

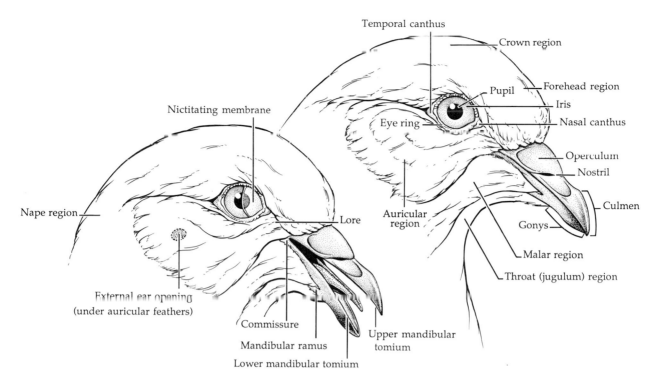

Name	Definition
Nape region, also called **upper hindneck** or **nuchal region**	The dorsal surface of the neck from the occipital ridge of the skull or first cervical vertebrae posterior to the base of the neck.
Upper mandible, or **maxilla**	The combined upper jaw and bill.
Lower mandible	Technically the lower jaw or the lower jaw and lower bill. In ornithological literature, however, the mandible may be *either* the upper *or* the lower bill.
Mandibular tomium (plural, **mandibular tomia**)	The cutting edge of either the upper or the lower mandible.
Culmen	The central midline ridge running from the tip of the upper bill back to the base of the bill.
Gonys	The central midline ridge running from the tip of the lower bill back to the anterior end of the head.
Mandibular rami (singular, **ramus**)	The two lateral halves of the lower jaw between the anterior synthesis and the quadrate articulation.
Rhamphotheca	The horny sheath that covers the bill.
Nasal canthus, temporal canthus	The angles of the eye where the upper and lower eyelids meet. The nasal canthus is the anterior angle of the eye; the temporal canthus is the posterior corner of the eye.
Operculum	A soft, fleshy structure at the base of the bill in pigeons and such other birds as starlings that covers the **external naris** (nostril).

References: Blake 1956; Cramp and Simmons 1977; Lucas and Stettenheim 1972; Palmer 1976; Pettingill 1985; Van Tyne and Berger 1976.

4 | THE BILL

Perhaps no other structure in the vertebrate world has been as variously adapted and modified as the avian bill.* Birds depend on their bills not only to obtain food but also to preen their feathers, build their nests, perform courtship displays, and defend themselves from predators or rivals. As an adaptation bills are not unique to the birds. Such dinosaurs as the duck-billed hadrosaurs (Family Hadrosauridae) and the Ostrich-like ornithomimids (F. Ornithomimidae) had horny, birdlike bills. But no other group of vertebrates has adapted and exploited the bill to the extent that birds have, and aside from feathers the bill is surely the most quintessentially birdlike feature of the avian body.

The bill is composed of a bony framework covered by a tough jacket of keratin that forms the visible shape of the bill. The upper bill is supported by the maxilla and other bones of the skull. Unlike the jaw of mammals and most reptiles, the upper jaw of all birds is at least slightly mobile. The pterygoid, quadrate, and zygomatic arch bones (see chapter 5, The Skeleton) that support the maxilla can slide forward or backward, allowing the upper jaw to extend upward. At the forehead the maxilla bones join the skull at a thin, flexible sheet of nasal bones called the **nasofrontal hinge**. In most bird species the upper bill is thus much more mobile and flexible than it may appear (Bock 1964).

Observe the following structures

Use the diagrams on the opposite page and the tables on the previous pages to locate the features of the bill in the species illustrated. Review the bills of other birds your instructor has provided, noting these characteristics of each species:

What is the **typical diet** of the species?

Does the shape of the bill seem consistent with the **natural history** and **ecological niche** of the species?

What **special** or **unusual features** does the bill exhibit? Can you guess why the feature might have developed? Might the special feature play a role in the bird's life history beyond the gathering of food?

Does the bill seem to be adapted to **courtship display** or **defense against predators**?

The various bills of eleven species of birds: the White-throated Sparrow (*Zonitrichia albicollis*), with its light bill adapted to small seeds and plant material; the heavy seed-cracking bill of the Evening Grosbeak (*Hesperiphona vespertina*); the tall but thin beak of the Atlantic Puffin (*Fratercula arctica*), adapted to catching and holding small fish; the probing bill of the Golden-winged Warbler (*Vermivora chrysoptera*), adapted to feeding on small insects; the long bill of the Whimbrel (*Numenius phaeopus*), adapted to probing mudflats for small invertebrates; the Pileated Woodpecker (*Dryocopus pileatus*), with its long, heavy bill for chopping away wood to expose insects; the Keel-billed Toucan (*Ramphastos sulfuratus*), which feeds on the fruits of rain forest trees; the Great Blue Heron (*Ardea herodias*), with its long bill for stabbing and seizing small animals and fish; the unique bill of the Greater Flamingo (*Phoenicopterus ruber*), which houses a complex filtration system to strain small invertebrates from shallow mud flats and lakes; the Sword-billed Hummingbird (*Ensifera ensifera*), with its extremely long bill for probing deep into the bells of tubular tropical flowers; and the sharp, decurved beak of the Red-tailed Hawk (*Buteo jamaicensis*), adapted to tearing flesh from the small mammals on which it feeds.

*Do birds have bills or beaks? In modern ornithology the terms **bill** and **beak** are often used synonymously. Originally, **beak** referred to the sharp, decurved bills of birds of prey like hawks and falcons.

Specialized Bills and Bill Shapes

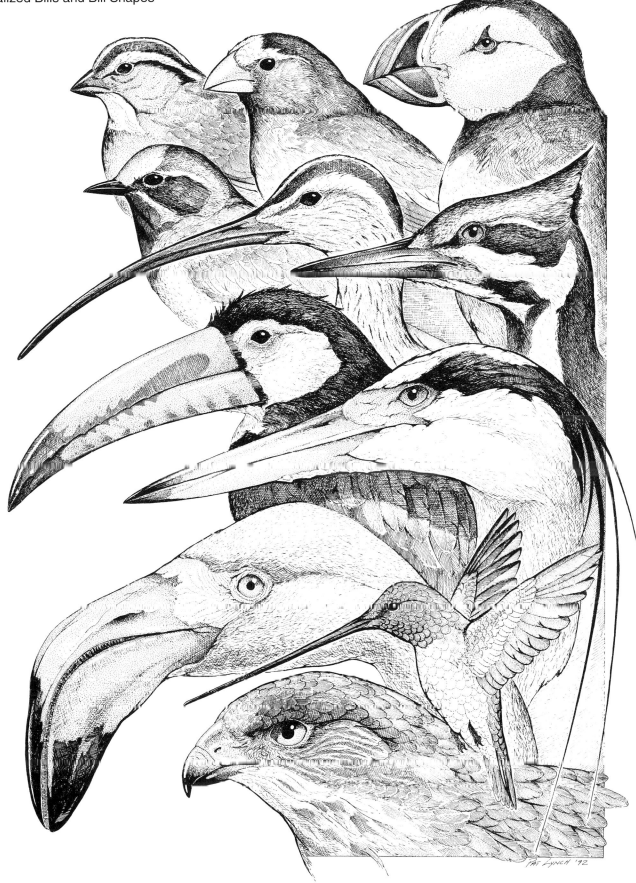

The mouth is the entry point of the bird's digestive system, and it must perform a variety of functions related to the typical diet of the species and the needs of the digestive system. The overall size and shape of the bill closely reflect the physical demands of the bird's diet. Songbirds that feed primarily on small insects typically have the most delicate beaks. More omnivorous birds, such as the thrushes and jays, have sturdy, generalized bills, well suited to a variety of plant and animal foods. Grosbeaks and parrots have huge conical beaks well suited to crushing the hard outer shells of nuts. Most raptors and vultures have short, decurved bills with sharp cutting edges for dismembering their prey. The long, thin bills of the fish-eating mergansers are distinctly serrated along their cutting edges to help hold onto slippery fish. In each species the bill is used to capture and sometimes dismember or pull apart the food items. Once the pieces of food are small enough to fit down the esophagus, the bird swallows them whole without chewing. All further "chewing" takes place within the muscular gizzard stomach.

The bill is also the entry point of the respiratory system, and the surface of the upper bill (**maxilla**) is perforated by two external **nares,** or nostrils. Air is inhaled through the nostrils, and then travels through the nasal passages at the base of the bill, enters the mouth cavity, and flows posteriorly through the larynx and into the trachea. A few birds, such as the gannets and boobies (Family Sulidae), have no external nares and breathe through a narrow gap between the rear portions of their upper and lower bills. Because gannets and boobies dive from great heights while fishing, this lack of external nostrils prevents water pressure from blowing out the nasal passages as the birds hit the water at high speed.

A wonderful bird is the pelican
His bill can hold more than his belican
    He can take in his beak
    Food enough for a week
But I'm damned if I see how the helican.

Dixon Lanier Merritt, 1910

Brown Pelican
(*Pelecanus occidentalis*)

Merritt's famous limerick on the pelican's bill is correct in at least one respect—the massive fleshy pouch suspended from the pelican's mandibles *can* hold more volume of fish and water than the pelican's stomach can (Austin 1961). Contrary to popular belief, however, the pelican never stores or transports fish in its pouch. The pouch is used only to scoop up fish after the bird dives down upon them through the ocean surface. After grabbing a mouthful of fish and water, the pelican immediately strains out the water in the pouch through the sides of its bill and swallows the fish trapped in its pouch and jaws. The pelican may store the fish for a time in its crop before digesting the meal but apparently never travels any distance with fish stored in the fleshy pouch.

Form and Function
Hawaiian honeycreepers

Because of their famous role in the formation of his evolutionary theories, Darwin's finches (Subfamily Geospizinae) of the Galápagos Islands are probably the best-known examples of divergent evolution and adaptation. But the Hawaiian honeycreepers (Family Fringillidae, Tribe Drepanidini) are actually a much more complete example of how isolation in virgin habitat can accelerate the process of speciation, where a single ancestral species living in isolation from competitors rapidly gives rise to a multitude of variations. The Hawaiian honeycreepers exhibit a remarkable divergence in bill shape and size, which allows them to exploit a variety of ecological niches.

The ancestral Hawaiian honeycreeper is thought to have arrived in the Hawaiian Islands when the islands were fully vegetated but uninhabited by perching birds. Extensive examination of the anatomy of the Hawaiian honeycreepers shows that the ancestral species was probably a relatively primitive finch (Raikow 1974, 1976) rather than a member of the honeycreepers (F. Fringillidae, Tribe Thraupini) of Central and South America. The many mountain peaks and secluded valleys of the Hawaiian Islands may also have contributed to **allopatric speciation,** in which small, isolated pockets of honeycreepers quickly diverged in form and behavior from their ancestral stock to create new species.

Over many millennia (no one knows exactly when the colonization took place) the ancestral honeycreeper began to form divergent populations of birds that evolved bills and behavior patterns mimicking many of the families of birds seen in woodland habitats on the continents. Some Hawaiian honeycreepers even evolved roles similar to true honeycreepers. Through convergent evolution the Akohekohe, or Crested Honeycreeper (*Palmeria dolei*), and the Iiwi (*Vestiaria coccinea*) have evolved a grooved tongue for removing nectar from tubular flowers that is remarkably similar to the tongues of the true honeycreepers of Central and South America. The Akiapolaau (*Hemignathus wilsoni*) and the even more remarkable Akialoa (*Hemignathus obscurus*) of Kauai have evolved probing bills to work deep into the crevices of bark and rotting wood in search of insects. The Akepa (*Loxops coccineus*) functions like a typical woodland warbler and has evolved a warbler's sharp and delicate bill to probe vegetation for insects, buds, and soft seeds. Other species, such as the Nihoa Finch (*Telespiza ultima*), have bills that may be close to the bills of the finches that gave rise to the Hawaiian honeycreepers. Note that in the evolution of each honeycreeper the primary physical adaptation has been a modification of the bill and that the bill is the primary anatomic clue to how each honeycreeper has evolved into its unique ecological niche on the Hawaiian Islands.

6 TOPOGRAPHY OF THE FOOT

Birds are the most aerial of all vertebrates, but even species like the swifts that spend most of their lives in the air must come down to earth occasionally to rest. The feet are usually the bird's most immediate physical contact with the environment, and apart from the bill, the feet are often the most obvious physical manifestation of how species lives in its world. In birds of prey the feet are the primary means of catching and killing prey animals, and in many other groups the feet are used along with the bill to grip and pull apart food items. The feet of aquatic birds must do triple duty, in walking, swimming, and absorbing the shock of landing. The environment and ecological niche of each species provide a different substrate and impose varying requirements on the feet. From stony alpine hillsides to mid-ocean waves, each substrate has influenced the evolution of avian feet and legs.

In adaptation to the diverse environments and needs of birds, the standard vertebrate leg arrangement has undergone extensive modification. In fact, the modifications of the leg and foot bones are excellent clues to the evolutionary history of a species, since many groups of birds have independently evolved similar solutions to the problems of perching, swimming, and walking. Most perching birds, for example, have the classic **anisodactyl** ("chicken foot") toe arrangement, with digits 2, 3, and 4 pointing forward and digit 1 (the **hallux**) pointing backward. However, several major groups of tree-dwelling birds, such as owls, parrots, and woodpeckers, have X-shaped **zygodactyl** feet, where digits 2 and 3 point forward and digits 1 and 4 point to the rear. Both the zygodactyl foot pattern and the more common anisodactyl foot are well adapted to grasping branches and aiding the bird in moving around on branches. But closer examination reveals many subtle differences between the bony anatomy of the zygodactyl feet of owls and the superficially similar feet of parrots. This suggests that owls, parrots, and probably the other zygodactyl groups all evolved the zygodactyl foot pattern independently. Evolutionary biologists call this kind of coincidence an instance of **convergent evolution,** where two unrelated groups independently develop a similar solution to similar problems (Bock and Miller 1959). The zygodactyl foot grips branches well, so it is perhaps not surprising that such an obviously adaptive design should arise independently in many groups of birds.

Observe the following structures

Use the diagram on the opposite page and the tables on the following pages to locate and identify the basic features of the avian foot. Pay particular attention to the various **types of feet, toe configurations,** and **leg scutellation patterns**. The appearance of the legs and feet are important in the taxonomic classification of bird species, and the color and appearance of the legs and feet can be important field marks in identifying many species in the wild.

Birds are called **digitigrade** because they actually walk on their toes and not on all of the foot bones (as humans do). The bones and flesh of the foot are covered with a tough plating of scales that further strengthens the foot and resists the wear and tear of walking and perching. In the pigeon the anterior tarsus is covered with a row of large scales called **scutes**. This scale pattern is found in many species of birds and is called a **scutellate tarsus**.

Topography of the Foot
Rock Dove (*Columba livia*)

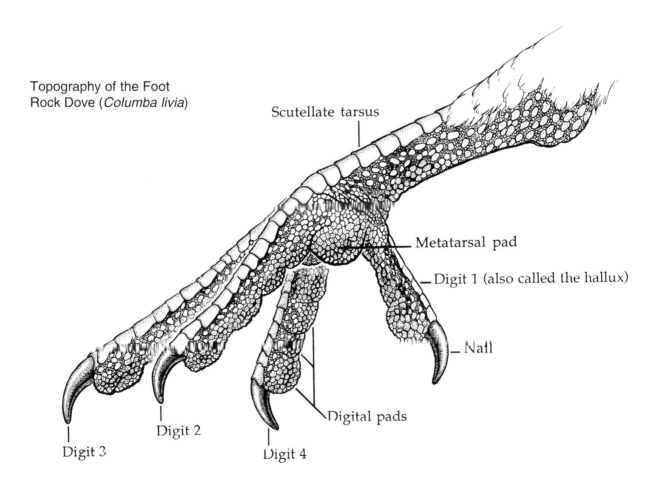

Scutellate tarsus

Metatarsal pad

Digit 1 (also called the hallux)

Nail

Digital pads

Digit 3

Digit 2

Digit 4

References: Lucas and Stettenheim 1972; Palmer 1976; Pettingill 1985.

Topography of the Foot — Toe Arrangements

Type of Foot	Toe Configuration	Notes

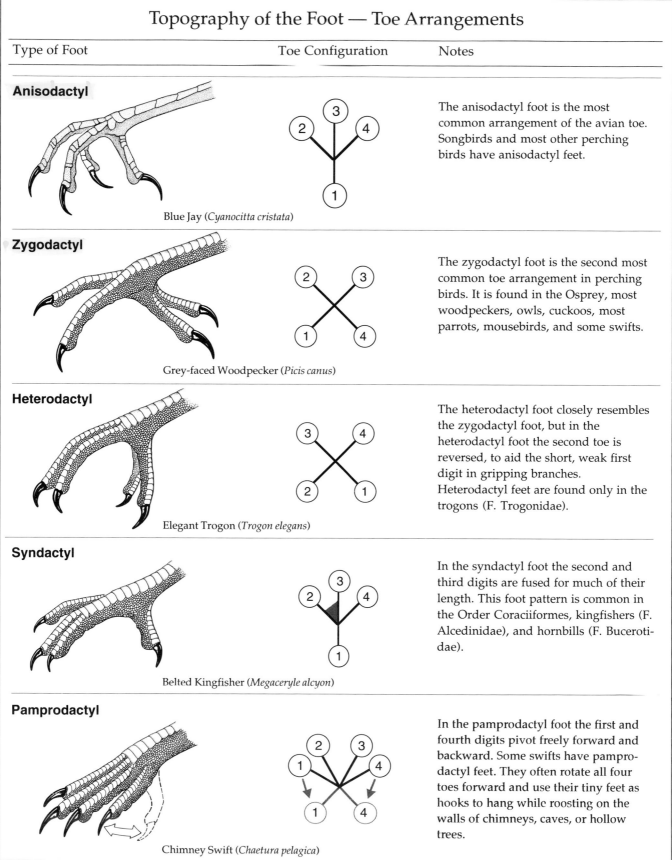

Anisodactyl

Blue Jay (*Cyanocitta cristata*)

The anisodactyl foot is the most common arrangement of the avian toe. Songbirds and most other perching birds have anisodactyl feet.

Zygodactyl

Grey-faced Woodpecker (*Picis canus*)

The zygodactyl foot is the second most common toe arrangement in perching birds. It is found in the Osprey, most woodpeckers, owls, cuckoos, most parrots, mousebirds, and some swifts.

Heterodactyl

Elegant Trogon (*Trogon elegans*)

The heterodactyl foot closely resembles the zygodactyl foot, but in the heterodactyl foot the second toe is reversed, to aid the short, weak first digit in gripping branches. Heterodactyl feet are found only in the trogons (F. Trogonidae).

Syndactyl

Belted Kingfisher (*Megaceryle alcyon*)

In the syndactyl foot the second and third digits are fused for much of their length. This foot pattern is common in the Order Coraciiformes, kingfishers (F. Alcedinidae), and hornbills (F. Bucerotidae).

Pamprodactyl

Chimney Swift (*Chaetura pelagica*)

In the pamprodactyl foot the first and fourth digits pivot freely forward and backward. Some swifts have pamprodactyl feet. They often rotate all four toes forward and use their tiny feet as hooks to hang while roosting on the walls of chimneys, caves, or hollow trees.

Topography of the Foot — Webbing and Other Adaptations

Type of Foot	Toe Configuration	Notes

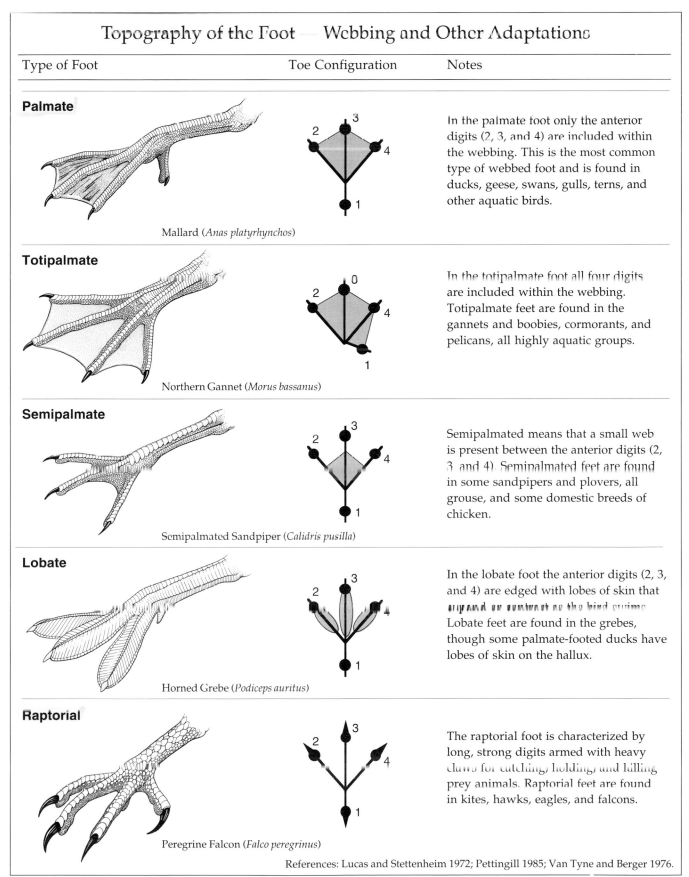

Palmate

Mallard (*Anas platyrhynchos*)

In the palmate foot only the anterior digits (2, 3, and 4) are included within the webbing. This is the most common type of webbed foot and is found in ducks, geese, swans, gulls, terns, and other aquatic birds.

Totipalmate

Northern Gannet (*Morus bassanus*)

In the totipalmate foot all four digits are included within the webbing. Totipalmate feet are found in the gannets and boobies, cormorants, and pelicans, all highly aquatic groups.

Semipalmate

Semipalmated Sandpiper (*Calidris pusilla*)

Semipalmated means that a small web is present between the anterior digits (2, 3, and 4). Semipalmated feet are found in some sandpipers and plovers, all grouse, and some domestic breeds of chicken.

Lobate

Horned Grebe (*Podiceps auritus*)

In the lobate foot the anterior digits (2, 3, and 4) are edged with lobes of skin that expand or contract as the bird swims. Lobate feet are found in the grebes, though some palmate-footed ducks have lobes of skin on the hallux.

Raptorial

Peregrine Falcon (*Falco peregrinus*)

The raptorial foot is characterized by long, strong digits armed with heavy claws for catching, holding, and killing prey animals. Raptorial feet are found in kites, hawks, eagles, and falcons.

References: Lucas and Stettenheim 1972; Pettingill 1985; Van Tyne and Berger 1976.

Topography of the Foot — Investments of the Tarsus and Foot

Type of Foot	Notes

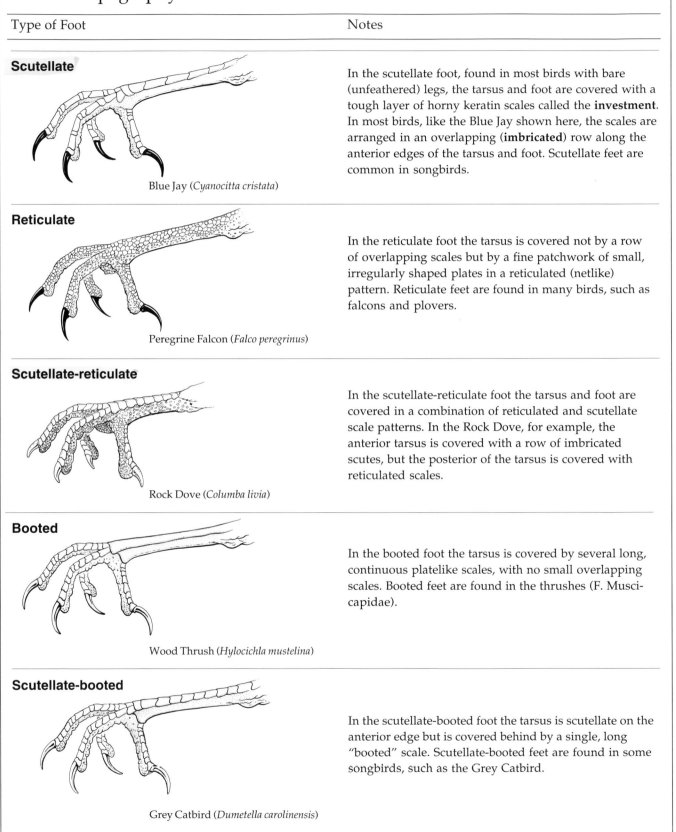

Scutellate

Blue Jay (*Cyanocitta cristata*)

In the scutellate foot, found in most birds with bare (unfeathered) legs, the tarsus and foot are covered with a tough layer of horny keratin scales called the **investment**. In most birds, like the Blue Jay shown here, the scales are arranged in an overlapping (**imbricated**) row along the anterior edges of the tarsus and foot. Scutellate feet are common in songbirds.

Reticulate

Peregrine Falcon (*Falco peregrinus*)

In the reticulate foot the tarsus is covered not by a row of overlapping scales but by a fine patchwork of small, irregularly shaped plates in a reticulated (netlike) pattern. Reticulate feet are found in many birds, such as falcons and plovers.

Scutellate-reticulate

Rock Dove (*Columba livia*)

In the scutellate-reticulate foot the tarsus and foot are covered in a combination of reticulated and scutellate scale patterns. In the Rock Dove, for example, the anterior tarsus is covered with a row of imbricated scutes, but the posterior of the tarsus is covered with reticulated scales.

Booted

Wood Thrush (*Hylocichla mustelina*)

In the booted foot the tarsus is covered by several long, continuous platelike scales, with no small overlapping scales. Booted feet are found in the thrushes (F. Muscicapidae).

Scutellate-booted

Grey Catbird (*Dumetella carolinensis*)

In the scutellate-booted foot the tarsus is scutellate on the anterior edge but is covered behind by a single, long "booted" scale. Scutellate-booted feet are found in some songbirds, such as the Grey Catbird.

Form and Function
The Horned Grebe

Horned Grebe
(*Podiceps auritus*)

As the leg is pulled forward, the lobes of the toes collapse down to a narrow blade just an inch wide. This minimizes the effort the bird must make to pull its leg forward against the water.

In the power stroke the lobes of the toes flare out to make a wide paddle as the bird pushes back against the water to propel itself forward.

Grebes (Family Podicipedidae) are an ancient group of birds well adapted to life in aquatic environments. Note that a grebe's feet are placed far back on the body, much farther than in most other groups of birds. This allows the bird to gain the maximum benefit from its powerful legs while underwater, but there is a drawback—grebes are virtually unable to walk or move quickly on dry land. The legs are so far back on the body that they can't support the bird or allow it to stand on land. This is why a grebe's nest is always constructed as a floating platform in shallow water; the grebe can simply hop into the nest from the water surface.

In the Field   TOPOGRAPHY AND BIRDWATCHING

Topography plays a crucial role in the accurate identification and description of birds in the field. Many excellent field guides, covering almost every region of the earth, are available today for the birdwatcher. Unfortunately, with this wealth of reference material many inexperienced birdwatchers spend more time "birding" their field guides than looking at the birds. Many experienced birders find that their heightened powers of observation are one of the most useful side benefits of birdwatching. These exercises may make you a more careful observer not just of the birds around you but of everything else in your world as well.

1. Take your binoculars and a notebook into the field, but do not bring a field guide or any other reference along. Your goal is to find and describe at least two species entirely in your own words and sketches. Choose any convenient birds you can find, preferably those you can easily watch for a few minutes while you sketch and make notes. Carefully note every topographic detail you see, including overall size and shape, colors and plumage patterns, length of wings and tail, length of wings relative to length of tail, and color of visible soft tissue, such as legs, feet, bill, lores, and eye ring. Record any characteristic or interesting behaviors that might aid in identifying the species. Summarize your notes and sketches into final detailed notes on the two species.

2. Working with your instructor, pick a good birdwatching site in your area and spend at least two hours noting every bird species you observe. Count each bird of every species you see, or make careful estimates of the numbers of flocking birds if there are too many to count individually. Prepare a detailed summary of your observations that includes the location and time of your outing, the weather and temperature, the species and numbers of individual birds seen, and notes on their behavior. The goal is to see and record as much detail as possible on the birds within the area at that time, just as you might if you were a field biologist assessing and cataloging the diversity of animal life of an exotic location where no references are available.

There is a catch: you cannot use a field guide, and all of your identifications of each bird species must be based on your own notes and observations. Record just what you see, make up names for the different species if that helps keep them straight in your mind, but don't worry about identifying each species until you are home. Then consult your field guides to identify any species that were unfamiliar. Here your field notes and visual memory will be important in accurately recording which species you saw.

Topography

1. Each of the four birds below has a distinctly shaped bill. For each bird, briefly describe the type of bill portrayed and what that head and bill anatomy suggest about the habitat and ecological niche of the species.

Chapter Worksheet, continued

2. Label all of the topographic landmarks and feather groups visible on the falcon
   drawn below.

3. Explain why forest birds tend to have short wings. What other topographic features
   are common in woodland birds?

4. To which bones or area of the skeleton are the following feather groups anchored?

   a. Primaries

   b. Secondaries

   c. Alula

   d. Rectrices

References

Amadon, D. 1950. The Hawaiian honeycreepers. *Bull. Amer. Mus. Nat. Hist.* 95:151–262.

Austin, O. L., Jr. 1961. *Birds of the world.* New York: Golden Press.

Blake, C. H. 1956. Topography of a bird. *Bird-Banding* 27:22–31.

Bock, W. J., and W. DeW. Miller. 1959. The scansorial foot of woodpeckers, with comments on the evolution of perching and climbing feet in birds. Amer. Mus. Nat. Hist. *Novitates* 193.

Cramp, S., and K. E. L. Simmonds, eds. 1977. *Handbook of the birds of Europe, the Middle East, and North Africa: the birds of the western Palearctic.* Vol. 1: *Ostrich to ducks.* Oxford: Oxford University Press.

Ehrlich, P. R., D. S. Dobkin, and D. Wheye. 1988. *The birder's handbook.* New York: Simon and Schuster.

Grant, P. R. 1986. *Ecology and evolution of Darwin's finches.* Princeton, N.J.: Princeton University Press.

Guravich, D., and J. E. Brown. 1983. *The return of the Brown Pelican.* Baton Rouge: Louisiana State University Press.

Lucas, A. M., and P. R. Stettenheim. 1972. *Avian anatomy: integument.* Agric. Handb. 362:1–340. Washington, D.C.: U.S. Government Printing Office.

Palmer, R. S., ed. 1962. *Handbook of North American birds.* Vol. 1: *Loons through flamingos.* New Haven: Yale University Press.

Perrins, C. 1987. *New generation field guide to the birds of Britain and Europe.* Austin: University of Texas Press.

Peterson, R. T. 1980. *A field guide to the birds.* 4th ed. Boston: Houghton Mifflin.

Pettingill, O. S., Jr. 1985. *Ornithology in laboratory and field.* 5th ed. New York: Academic Press.

Raikow, R. J. 1974. Species-specific foraging behavior in some Hawaiian honeycreepers (*Loxops*). *Wilson Bull.* 86:471–74.

Raikow, R. J. 1976. The origin and evolution of the Hawaiian honeycreepers (Drepanididae). *Living Bird.* 15th annual. Ithaca, N.Y.: Cornell University.

Terres, J. K. 1980. *The Audubon Society encyclopedia of North American birds.* New York: Alfred A. Knopf.

Wilson, B., ed. 1980. *Birds: readings from* Scientific American. San Francisco: W. H. Freeman.

It is difficult to think of an animal with more innate grace and beauty than the Barn Swallow (*Hirundo rustica*), its flight feathers shaped by millions of years of air flowing over wing and tail. Ironically, birds are so well adapted to their aerial way of life that few evolutionary traces are left to show us *how* they evolved their many adaptations to flight.

Feathers

4

Feathers are the most characteristic feature of birds. All birds have feathers, and no animals but birds have yet been proven to have had feathers (Feduccia 1985; Ostrom 1987). Feathers form a strong, light, warm, and flexible integument and are a marvel of evolutionary form and function.

The Origin of Feathers

The evolutionary origins of feathers are obscure. The known facts often appear contradictory, and the subject is surrounded by controversy in the scientific literature (Regal 1975; Martin 1983; Ostrom 1985). All the fossils of *Archaeopteryx lithographica*, the earliest unambiguous fossil bird, have been quarried from limestone beds near Solnhofen, Germany. In 1860, shortly before fully feathered *Archaeopteryx* specimens were discovered, quarry workers unearthed a single feather from the limestone. This earliest known fossil feather is so modern-looking as to be indistinguishable from the feathers of birds flying today (Feduccia 1980). Unfortunately, the very modernity of the Solnhofen feather specimen, as well as the fully developed feathers of the *Archaeopteryx* fossils, sheds no light on just how the earliest birds came to develop feathers (Regal 1975, 1985).

It is clear even to the most casual observer that birds share many characteristics with reptiles, foremost among them a scaly integument. Scales and feathers are composed of **keratin,** and both form an integument composed of overlapping elements that shield the thin skin beneath from mechanical damage and extremes of temperature (Lucas and Stettenheim 1972). Although no one disputes that feathers are in some way derived from reptilian scales, the adaptive value of a structure intermediate between a scale and a feather is not obvious. Yet feathers could not have evolved without some plausible adaptive value in all of the intermediate steps necessary between a reptilian scale and a fully formed feather such as *Archaeopteryx* had (Peters and Gutmann 1985; Regal 1975, 1985).

Two theories dominate the debate on the origin of feathers. Birds are the only animals with feathers, and birds are also the most aerial of the vertebrates, so it is reasonable to connect the two characteristics and theorize that the origin of feathers was intimately linked to the development of passive, gliding flight in early birdlike reptiles (Heilmann 1927; Parkes 1966). In this flight theory of feather evolution, each tiny elongation in the scales of the arboreal ancestors of birds made them progressively better gliders, much like the flying squirrels that glide through modern forests (Peters and Gutmann 1985). But feathers are also the most efficient natural insulators known, and birds as a group are the smallest homoiothermic (warm-blooded) animals. Small homoiotherms are always in danger of excess heat loss or gain, because small animals have more surface area in relation to their body volume than do larger animals. Many authors now believe that feathers arose along with homoiothermy in the ancestors of modern birds, primarily as an insulation against excess heat or cold (Ostrom 1974; Bakker 1975, 1986; Regal 1975). In this physiological view of feather evolution, feathers evolved from scales primarily to protect early birds from heat or cold, and only later did feathers become elongated into flight-related structures.

1 STRUCTURE OF THE FEATHER

A typical contour feather is composed of a long central **shaft** and a broad flexible **vane** on either side of the shaft. The bare, proximal part of the central shaft is called the **calamus** or **quill**. In the mature feather the calamus is a hollow tube with a small opening at the proximal tip of the feather called the **inferior umbilicus**. In most mature feathers the inferior umbilicus is sealed with a plate of keratin. Within the lumen of the calamus are several **internal pulp caps**, remnants of the pulp tissue that nourished the growing feather early in its development. The calamus area of the central shaft runs distally from the inferior umbilicus to another small opening, the **superior umbilicus**. The superior umbilicus is a remnant of the distal end of the epithelial tube that forms early in feather development (see Feather Development, below). Distal to the superior umbilicus is the central feather shaft, the **rachis**. In the major flight feathers the rachis is distinctly grooved on the ventral surface. Roughly square in cross-section, this grooving of the rachis gives the outer feather shaft added strength in bending. In the Rock Dove the **afterfeather** is absent from most feathers or present only as a few strands of down near the superior umbilicus. Your instructor can provide you with examples of other feathers with more substantial aftershafts present.

Two vanes emerge from either side of the rachis. If your specimen is a large primary wing flight feather (remex), note the **asymmetry of the vanes:** the forward or anterior vane of the primary remiges is always narrower than the trailing or posterior vane. This gives the feather an **airfoil cross-section**, and each primary or secondary feather can act as an individual airfoil in flight. The smaller contour feathers that cover the body need not act as airfoils, and so the vanes of body feathering are symmetrical. Note the differences in the texture of the proximal and distal ends of the vanes. The proximal end of the vane is **plumaceous**, or soft and downy. The distal vane is firm and bladelike and is said to be **pennaceous**.

Structure of a Typical Contour Feather

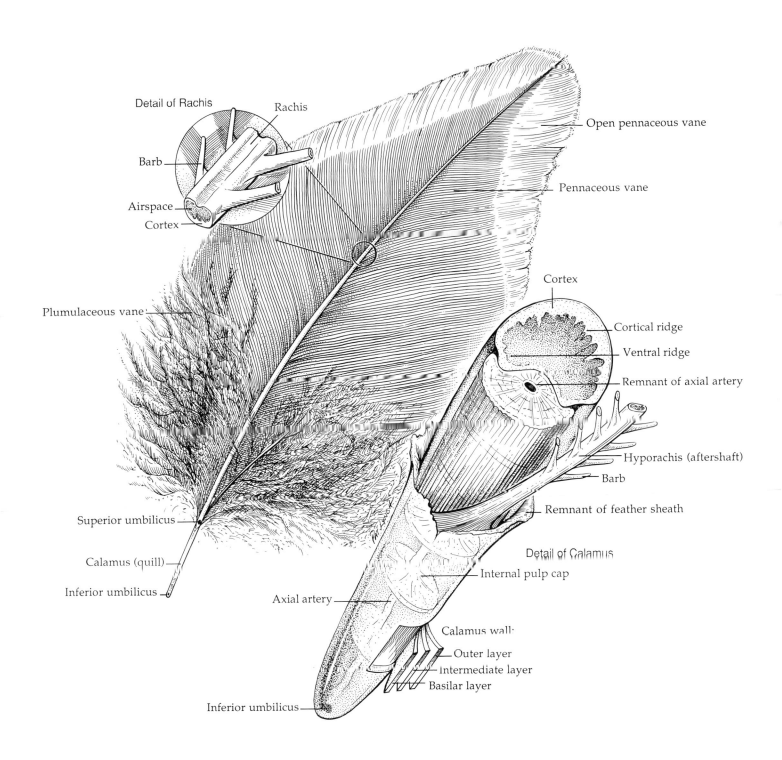

Detail of Rachis

Rachis

Barb

Airspace

Cortex

Open pennaceous vane

Pennaceous vane

Cortex

Cortical ridge

Ventral ridge

Remnant of axial artery

Plumulaceous vane

Hyporachis (aftershaft)

Barb

Remnant of feather sheath

Detail of Calamus

Internal pulp cap

Superior umbilicus

Calamus (quill)

Inferior umbilicus

Axial artery

Calamus wall

Outer layer

Intermediate layer

Basilar layer

Inferior umbilicus

Redrawn after Lucas and Stettenheim 1972.

2 STRUCTURE OF THE FEATHER VANE

The feather vane is made up of hundreds of tiny parallel **barbs** that branch off from either side of the central rachis of the feather. The barbs have a comma-shaped cross-section, with the rounded portion of the "comma" facing the dorsal surface of the feather. This asymmetry in the cross-section of the barbs is what gives the outward-facing dorsal vane surface a smooth, glossy face that resists wear. The ventral or inner-facing vane surface has a more linear, textured look. In the flight feathers of the wings and tail the linear texture of the ventral vane and special barbules that project from it produce friction that helps keep the major flight feathers of the wing and tail from separating too much during flight or other activities.

Each barb is further divided into tiny branches called **barbules**. The barbules are barely visible to the unaided eye, but if you separate the barbs of a large flight feather you should be able to see the hairlike barbules projecting from the lateral faces of each barb. The central shaft of each barbule is called the **ramus**. Each barbule has dozens of tiny projections collectively called **barbicels,** that fasten together the surface of the feather vane. The **proximal barbules** (the barbules on the inner side of each barb) have a cross-section like a curled-over comma, with a distinct **flange** along the edge. The **distal barbules** (the barbules on the side of the barb that faces the feather tip) each have four to five tiny **hooklets** or **hamuli** along their length. The hooklets of the distal-facing barbules catch on the flanges of the proximal-facing barbules, similar to the way the hooked and looped parts of Velcro fasteners work together. When you "zip up" a feather with your fingers to mend a split in the vane it is the tiny hamuli hooks on each barbule that fasten the barbs of the feather together again.

Ducks, geese, and swans (Order Anseriformes) and a few other groups of birds have specially modified feather vanes. In ducks the cross-section of each barb on the flight feathers is not comma-shaped but more L-shaped, with a wide shelf called a **tegmen** that projects from the ventral side of each barb. This overlapping of the special tegmen barbs gives the underside of a duck primary feather a distinctive glazed sheen. The tegmen area of the vane seems to strengthen the feather and apparently helps keep air from slipping through the vane while in flight.

A primary feather of an American Crow (*Corvus brachyrhynchos*). Note the strong asymmetry in the size of the anterior and posterior vanes in this primary remex on the page opposite. The asymmetry of the vanes forces the feather to twist open like venetian blinds when the crow raises its wings in flight. As the wing is brought down, the feathers twist back into a flat wing surface again. Like many large birds the crow has **emarginate primaries,** where the anterior vane sharply narrows near the outer tip of the feather. Birds with emarginate primaries, like the *Buteo* hawks, vultures, and crows, have a ragged "fingered" flight silhouette, because each of the outer primaries is isolated from its neighbors. This narrowing of the anterior vane gives each primary a classic airfoil cross-section, and each primary acts both as an individual airfoil and as part of the larger airfoil of the wing. The notch also acts as an aerodynamic slot, forcing air to rush into the gap between each primary feather tip with greater speed. This increases the lifting power of each primary and causes the feather tips to bend upward with the increased lift (Rüppell 1977).

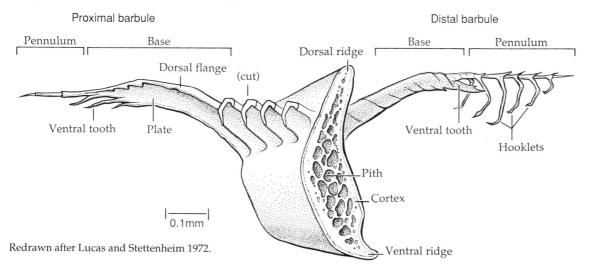

Proximal barbule

Pennulum   Base   Dorsal flange   (cut)   Dorsal ridge   Base   Pennulum

Ventral tooth   Plate   Ventral tooth   Hooklets

Pith

Cortex

0.1mm

Ventral ridge

Distal barbule

Redrawn after Lucas and Stettenheim 1972.

Structure of the Feather Vane
Primary Remex
American Crow (*Corvus brachyrhynchos*)

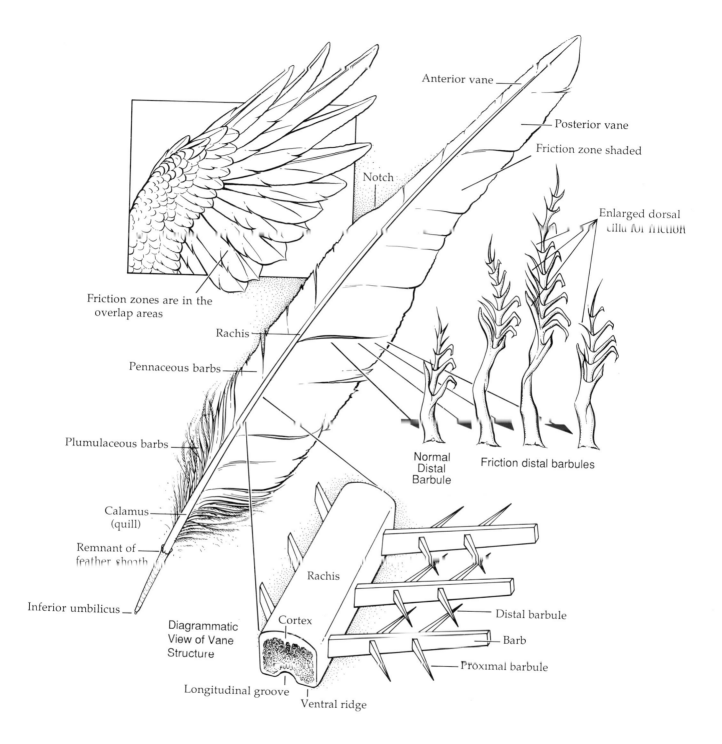

Anterior vane

Posterior vane

Friction zone shaded

Enlarged dorsal cilia for friction

Notch

Friction zones are in the overlap areas

Rachis

Pennaceous barbs

Normal Distal Barbule

Friction distal barbules

Plumulaceous barbs

Calamus (quill)

Remnant of feather sheath

Inferior umbilicus

Diagrammatic View of Vane Structure

Rachis

Cortex

Distal barbule

Barb

Proximal barbule

Longitudinal groove

Ventral ridge

References: Espinasse 1964; Lucas and Stettenheim 1972.

3 DEVELOPMENT OF THE FEATHER

Like hair, nails, and horns, feathers are inert structures when mature. Feathers are an outgrowth of special **feather follicles,** which are derived from the bird's **epithelium** and **dermis** layers of the skin. Early in the growth of the embryo, small **feather papillae** begin to form on the skin of the feather tracts. These papillae quickly increase in size, and the epidermis surrounding each papilla begins to invaginate to form a deep ring fold around the feather papilla. The papillae then begin to point toward the posterior of the bird.

Epidermal cells in the outside layers of the papilla then begin to differentiate into structures that become the protective sheath surrounding the growing feather and into the feather itself. This layer of epidermis within the papilla is called the **Malpighian layer** and is composed of all four layers of the normal avian epidermis: the **stratum corneum,** the **stratum transitivum,** the **stratum intermedium,** and the **stratum basale.** Within the cone of the Malpighian layer is a central core of dermis tissue. The dermis tissue becomes the pulp at the root of the growing feather and supplies the feather with nutrients during development.

The Malpighian layer subdivides into three distinct structures. The outer Malpighian layer forms the **feather sheath,** a capsule of cornified tissue around the developing feather. The middle Malpighian layer (composed of the transitivum, intermedium, and basale layers of the epidermis) forms the rachis and vane structure of the feather. The dermal core of the feather papilla is then gradually encased in a growing tube of cornified feather and feather sheath produced by the Malpighian layer. As this feather tube develops further, the pulp fills the entire cavity of the shaft, creating a series of **pulp caps** that may be seen intact within the mature calamus of the feather.

As the middle Malpighian layer begins to form the rachis and vane of the feather, a complex series of events gives the feather its character as either a soft, downy feather or a flight feather with a strong vane. The barbs of the future feather begin to form at the proximal end of the Malpighian layer, at a ring of stratum basale cells called the **epidermal collar** or **germinal collar.** The dorsal side of the germinal collar produces the central rachis of the main feather shaft, and the ventral side produces the rachis of the afterfeather. The distal ends of the barbs emerge from the germinal collar and gradually form a complex cone of barbs that will form the feather vane (see the figure inset at right). In pennaceous feathers the bases of the barbs migrate toward the central rachis of the feather. If this does not happen, the barbs remain attached only to the calamus at the base of the feather, and a plumulaceous or downy feather results. As the feather develops the vane rolls within the feather sheath, and the whole feather sheath pushes up through the skin. In the chick embryo the feather sheath simply splits to release the rachis and the rolled-up feather vane. As older birds molt, the new feather papilla pushes up into the base of the older feather, eventually forcing the older feather out of the follicle altogether (Simmons 1964; Lucas and Stettenheim 1972).

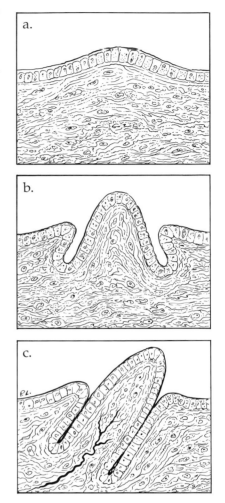

Development of the feather papilla and follicle: (a.) Early formation of the feather papilla. (b.) The papilla invaginates, forming a shallow collar around the papilla called the **feather follicle.** (c.) Later stage in the development of the feather papilla and follicle, just before the stage illustrated opposite. After Lucas and Stettenheim 1972.

Diagrammatic Reconstruction
of a Developing Contour Feather

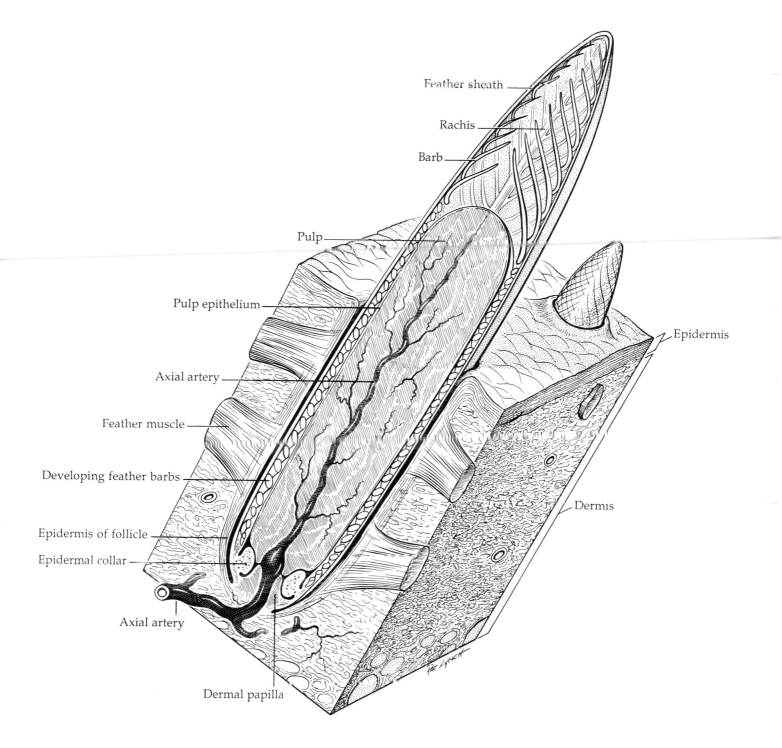

Feather sheath

Rachis

Barb

Pulp

Pulp epithelium

Axial artery

Feather muscle

Developing feather barbs

Epidermis of follicle

Epidermal collar

Axial artery

Dermal papilla

Epidermis

Dermis

References: Espinasse 1964; Lucas and Stettenheim 1972.

Form and Function
Feathers and the origin of flight

Feathers are a little too perfect—that's the problem. Birds are so thoroughly adapted to their aerial way of life, and their feathers are so exquisitely designed for both flight and thermal protection, that feathers no longer bear any of the small but telling flaws that allow evolutionary biologists to see how feathers developed from reptilian scales. There are no obvious traces of how feathers came to be, so until evidence is found we can only look for the simplest explanation that accounts for all the known facts. Scientists call that simplicity of explanation **parsimony**, the theory that a lean, economical explanation is much more likely to be correct than a complicated story with many unlikely elements. Unfortunately for evolutionary theory, feathers are very complicated--and very unlikely. They are by far the most complex derivative of the vertebrate integument (skin), much more elaborate than mammalian hair, for example. Fur is almost as good an insulator as feathers, and bats fly well without a complex, birdlike wing. So why didn't birds simply develop a coat of fur and membranous wings like a bat or a pterosaur?

Any explanation of the origin of feathers takes us deep into the larger debate on how feathers, homoiothermy, and flight developed in birds. Until about fifteen years ago the most popular theories of bird evolution placed early birds like *Archaeopteryx* up in the trees as weak fliers or gliders. This **arboreal theory**, championed by many paleontologists and ornithologists (Bock 1985; Feduccia 1980, 1985; Heilmann 1927; Martin 1983; Parkes 1966), explains feathers primarily in aerodynamic terms, though there is not general agreement about how or why homoiothermy might have evolved in arboreal bird ancestors. According to the classic arboreal theory, the ancestors of birds were tree-dwelling bipedal animals that gradually developed long scales to allow them first to "parachute" (a weak glide) from limb to limb and then to glide and fly as the long scales became true feathers. Heilmann's "Proavis," illustrated here, remains the basic model for current arboreal theories, and the argument does have a powerful intuitive appeal: feathers, particularly the stiff, vaned feathers of flying birds, are and seemingly always were primarily about flight and not about homoiothermy or anything else.

One of the central problems with the arboreal theory is that modern arboreal gliders look nothing like birds (and, come to think of it, neither does Heilmann's gliding Proavis). Arboreal gliders, such as tree squirrels or the gliding agamid lizard *Draco*, are quadrupeds with narrow, elongated bodies and membranous gliding planes between the pectoral and pelvic appendages (Peters and Gutmann 1985). If all you simply want to do is glide from one tree to another, the easiest "solution" is a fold of skin extending from the sides of the body and supported by all four limbs (or extended ribs in *Draco*). Minimal steering is required, and indeed modern gliding animals show little specialization for maneuvering in the air (Peterson 1985). In many respects the arboreal theory better explains how a bat or pterosaur might have evolved, with relatively simple membranous wings connected to both pectoral and pelvic limbs. Some arboreal theorists see homoiothermy as a separate issue, not bound to the development of feathers and flight (Feduccia 1980). Others say that feathers and homoiothermy were intimately linked and that arboreal, short-feathered proto-birds only later developed the elongated feathers necessary for gliding and flight (Bakker 1986; Paul 1988). Neither arboreal-homoiothermy theory really addresses the complexity of feathers, since both bats and pterosaurs adapted well to flight with membranous wings and hairy or hairlike thermal protection, rather than anything as complex as feathers.

In a series of papers published through the 1970s, John Ostrom (1969, 1974, 1975, 1976a, 1976b) laid out a very different **cursorial theory** of avian evolution, one that stresses the skeletal similarity of *Archaeopteryx* to a group of bipedal, predatory theropod dinosaurs, the coelurosaurians (see chapter 2, Systematics). After describing the swift, agile coelurosaurian *Deinonychus*, Ostrom noted many similarities between its anatomy and that of *Archaeopteryx*. Without its feathers no one would have recognized *Archaeopteryx* as a bird, and indeed in 1973, F. X. Mayr "discovered" a new specimen of *Archaeopteryx* that had been lying since 1950 in the Eichstatt Museum in Germany, misidentified as a

Heilmann's 1927 reconstruction of a hypothetical "Proavis," proposed as an arboreal gliding ancestor to *Archaeopteryx* and modern flying birds, from his classic book *The Origin of Birds*.

specimen of *Compsognathus* (Wellnhofer 1974; Ostrom 1975). Ostrom's cursorial theory places *Archaeopteryx* on the ground as a active, homoiothermic predator, with a natural history and anatomy quite like its near relatives, the coelurosaurs. Ostrom views *Archaeopteryx* and its ancestors as running and gliding predators that were poor fliers, with feathered wings adapted primarily for sweeping small prey animals or insects into the range of their snapping jaws. In this cursorial view, feathers are present first as thermal protection and later become elongated fly swatters, but the theory does not explain why feathers are structurally so complex. Most ground-dwelling birds have hairlike feathers, not the strongly vaned feathers seen in flying birds (Feduccia 1985). Later elaborations on the cursorial theory (Caple et al. 1983; Gauthier and Padian 1985; Peterson 1985) suggest that feathers and the modern avian wings and tail arose as a means to help these running predators turn quickly in pursuit of prey or to aid them in making short escape flights from predators (Harrison 1976). Significantly, the cursorial theory also fails to explain fully why the primary feather vanes of *Archaeopteryx* are so asymmetrical, a condition seen only in modern flying birds (Feduccia and Tordoff 1979; Feduccia 1985), and several authors have questioned whether the feet of *Archaeopteryx* really show strong cursorial adaptations (Martin 1983; Feduccia 1980).

In discussing these theories, biologist Philip Regal (1985) sounds a warning for all those who might predict any animal's natural history on the basis of anatomy alone. Regal notes that the Gray Foxes (*Urocyon cinereoargenteus*) of California's Channel Islands are entirely insectivorous, a fact you would never guess from their anatomy or the natural history of other foxes. North America's mightiest predator, the Grizzly Bear (*Ursus arctos*), subsists almost entirely on berries and ground squirrels through much of its range, again a fact its anatomy would never suggest. Until we have more fossil evidence we will never know how birds developed such complex adaptations to flight.

Unlike birds, which are uniformly bipedal, most arboreal gliders developed as quadrupeds, with membranous patagia stretched between all four limbs and the tail. After Peters and Gutmann 1984.

A reconstruction of a hypothetical cursorial "Proavis" that might have had elongated scales adapted primarily for predation, running, and maneuverability, not for gliding. After Ostrom 1976.

4 FEATHER COLORS

Birds are among the most colorful of all vertebrates. Only the coral reef fish of tropical oceans show as wide a range of bold colors and patterns. Most terrestrial animals wear dull, inconspicuous colors; camouflage to avoid predators seems to be the most important selective force acting on animal coloration. As the most mobile of the vertebrates (only fish rival flying birds in three-dimensional mobility) birds are under less selective pressure to avoid attracting predators (Welty and Baptista 1988).

The colors seen in bird plumage are produced by a variety of pigments and structural adaptations of the feather. The principal pigments found in bird feathers are of three types. **Melanins,** the most common pigment, produce black, grays, and browns. **Carotenoids** produce intense reds and yellows, as in the vivid red of the male Northern Cardinal (*Cardinalis cardinalis*) and the bright yellow of the male American Goldfinch (*Carduelis tristis*). A third class of pigments, **porphyrins,** produce a range of reds, browns, and greens, notably the intense green and red hues of the African turacos (Family Musophagidae) and the brown pigments of many owls (F. Strigidae). Feather type also influences coloration: melanins can occur in all types of feathers, but carotenoids are usually seen only in the contour and semiplume feathers of the body.

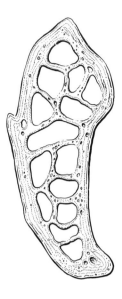

Cross-section of a feather barb showing the distribution of melanin granules within the keratin of the barb walls. After Lucas and Stettenheim 1972.

Most blue and green birds have no blue or green pigment in their plumage at all; they produce their (apparent) colors by creating complex patterns of reflection and refraction in the cell walls at the surface of the barbs and barbules of each feather. The Blue Jay (*Cyanocitta cristata*) isn't really blue. Its feathers reflect only the blue light wavelengths, and these give the appearance of blue coloration. Iridescent colors, such as those seen in the Indian Peafowl (*Pavo cristatus*), the Wood Duck (*Aix sponsa*), or the Common Grackle (*Quiscalus quiscula*), result from a complex interaction of the microstructure of the feather and melanin granules embedded in the barbules of each feather. In some species, iridescence results from many laminations of layers of keratin, each of which reflects different wavelengths of light, just as a soap bubble or oil slick shows many colors in reflection. In hummingbirds, an elaborate layering of reflective melanin granules shines with a rainbow of colors that are determined by the angle of these melanin layers relative to the viewer's eye.

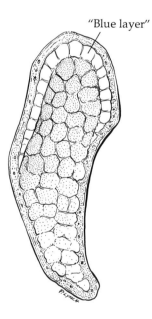

Cross-section of a Blue Jay feather barb showing the thin layer of blue-producing cells on the dorsal surface of each barb. After Gower 1936.

The color of feathers is determined by a range of environmental conditions and evolutionary pressures. Plumage colors and patterns aid birds in many ways besides mating displays, sexual dimorphism, or the need for concealment. Like all animals that spend time in the sun, birds must protect their skin from the ultraviolet components of sunlight. Most birds have dark back plumage that absorbs the ultraviolet light before it reaches the skin. White seabirds often have skin that contains dark pigments or a grayish underplumage of down and semiplume feathers that absorb ultraviolet light before it reaches the skin. Most birds, even predominantly white birds like gulls and pelicans, have dark primary feathers or feather tips. The melanin granules that make feathers black or dark brown are always associated with increased amounts of keratin, which strengthens feathers. Dark areas of feathers thus are stronger and resist abrasion more than lighter or white areas (Lucas and Stettenheim 1972).

Pigments and Structural Colors of Feathers

	Pigment	Notes
MELANINS		
Black Gray Dark brown	Eumelanin	Melanins are the most common pigments in feathers. They are synthesized by the bird through the oxidation of the amino acid tyrosine. Melanin granules produce color in direct proportion to their presence in the feather; the more melanin, the darker the color. Melanin occurs in all types of feathers, notably the major flight feathers.
Light brown Brick red Dull yellow, tan	Phaeomelanin	
CAROTENOIDS		
Bright yellow	Lutein (a xanthophyll) zeaxanthin, beta-carotene	Carotenoids are derived exclusively from the bird's diet, principally from yellow carotene pigments in grains, seeds, and other vegetable matter. Once eaten, the color and chemical structure of the carotenoid pigments may be modified in the bird's body. Most of the bright red, orange, and yellow colors seen in birds are carotenoid pigments. Carotenoids are rarely seen in flight feathers, but occur conspicuously in back and breast plumage.
Bright red	Astaxanthin, rhodoxanthin, canthaxanthin	
PORPHYRINS		
Green	Turacoverdin	Porphyrins are feather pigments related to hemoglobin and other bile pigments formed by the breakdown of hemoglobin by the liver. Porphyrins occur in many groups of birds, including pigeons, owls, and gallinaceous birds. The most common porphyrins produce brown pigments, but porphyrins can also produce the bright reds and greens seen in turacos and a few other species.
Red	Turacin (uroporphyrin)	
Brown, red-brown	Coproporphyrin III	
STRUCTURAL COLORS		
White	Structural only	Structural colors result from the modification or separation of the components of white light by the structure of the feather. In white feathers the feather structure simply reflects back the whole spectrum. Blues and greens result when light is scattered through fine layers of cell walls in the barbs. Combinations of structural and pigment colors are common, particularly in yellow-green, green, and blue-green feathers. Iridescence is caused by the complex layering of cell walls or melanin granules in the barbules of feathers. These structures selectively absorb or reflect varying wavelengths of light. The exact colors seen depend on the viewpoint of the observer.
Blue	Structural, rarely caused by pigment	
Green	Usually structural, but sometimes caused by combinations of yellow carotenoids and black melanins	
Iridescence	Primarily structural, though melanin granules are almost always abundant in iridescent feathers	

Reference: Lucas and Stettenheim 1972.

5 TYPES OF FEATHERS

There are six major types of feathers: contour or vaned feathers (including the major flight feathers of the wing and tail), down feathers, semiplumes, filoplumes, powder feathers, and bristles. Together these feathers provide a remarkably wide range of services: furnishing insulation from heat and cold; protection from abrasions to the skin; protection for the eyes, ears, and nostrils; sensory "whiskers" to judge flight speed; a fine powder to groom other feathers; screening for the skin from ultraviolet light; waterproofing in aquatic species; protective coloration or display plumages; and smooth body contouring for flight or swimming underwater. And, of course, feathers make up the extremely flexible aerodynamic lift and control surfaces necessary for avian flight (Simmons 1964; Lucas and Stettenheim 1972; Welty and Baptista 1988).

Observe the following structures

Use the drawings and tables on the following pages and view examples of each of the following types of feathers:

Contour feathers — Note both the smaller vaned feathers of the body and the larger flight feathers of the wings and tail. If available, view contour feathers from a gallinaceous bird, which will show a very prominent **hypoptile,** or afterfeather. In the pheasant contour feather illustrated opposite, the afterfeather is almost as long as the main shaft, giving each contour feather a double shaft.

Semiplumes and **down feathers** — Note the soft **plumaceous** barbs, which lack the hooked cilia that keep vaned feathers together. If possible, view both a vaned feather and a semiplume or down feather under a microscope (at about 100x), and note the great differences in the structure of the barbs in the two types of feathers. If specimens are available, note the heavy down insulation under the contour feathers of a diving (*Athya* sp.) or dabbling (*Anas* sp.) duck specimen, and contrast this with the lighter down layers under the feathers of a songbird or other nonaquatic specimen.

Bristles — These hairlike feathers are easier to view and understand when seen in situ on preserved bird skins or mounted specimens. If possible, view the prominent **rictal bristles** around the mouth of a Whip-poor-will (*Caprimulgus vociferus*), Chuck-will's-widow (*Caprimulgus carolinensis*), or other nightjar specimen (F. Caprimulgidae); these may be used as tactile structures to help the birds capture flying insects and may also help protect the eyes. Note the smaller bristles around the eyes and lores of other specimens, or see your lab instructor for other specimens of bristles.

Filoplumes — These whiskery feathers are scattered throughout the plumage. They help the bird determine the location of its main flight feathers, provide it with a sense of touch within the plumage, and may even help the bird judge its airspeed when flying. The heavy coat of contour, semiplume, and down feathers shields the bird from heat, cold, and mechanical damage but also prevents it

References: Clark and de Cruz 1989; Espinasse 1964; Lucas and Stettenheim 1972; Necker 1985; Van Tyne and Berger 1976.

The Major Types of Feathers

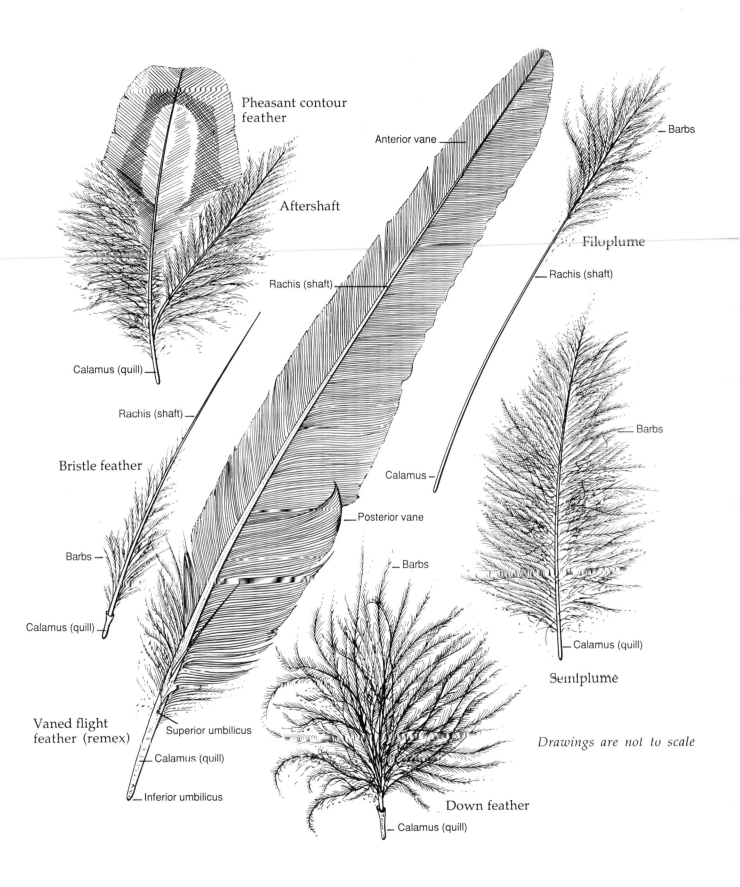

Pheasant contour feather

Aftershaft

Anterior vane

Barbs

Filoplume

Rachis (shaft)

Rachis (shaft)

Calamus (quill)

Rachis (shaft)

Bristle feather

Barbs

Barbs

Calamus

Calamus (quill)

Posterior vane

Barbs

Semiplume

Vaned flight feather (remex)

Superior umbilicus

Calamus (quill)

Inferior umbilicus

Calamus (quill)

Drawings are not to scale

Down feather

Calamus (quill)

Types of Feathers

Name	Notes
Contour feathers	Contour feathers are the basic vaned feathers of the body and wings, including the large flight feathers of the wing and tail (see below). The smaller contour feathers that cover the body have symmetrical vanes divided between a firm, pennaceous distal vane area and a soft, plumulaceous inner vane area. In some birds the contour feathers of the body tend to have more prominent afterfeathers than do the flight feathers.
Remiges	Remiges are the flight feathers of the wing, including the primaries, secondaries, and tertiaries. Remiges (singular, remex) are pennaceous contour feathers with prominent, often asymmetrical vanes. In ducks, gallinaceous birds, and owls the ventral vane surface is partially modified into a shiny, firm structure formed by specialized **tegmen** feather barbs, which are believed to strengthen the vane and resist the flow of air upward through the vane surface.
Rectrices	The rectrices (singular, rectrix) are the large, vaned flight feathers of the tail. The rectrices are similar in structure to the remiges of the wing, and also have asymmetrical vanes. In some groups, like woodpeckers, the rectrices have been adapted and strengthened to act as props, helping birds remain vertical as they forage on tree trunks. Swifts use similar stiff rectrices as an aid in perching on vertical surfaces.
Semiplumes	Semiplumes are intermediate in form between the more pennaceous contour feathers and the strictly plumulaceous down feathers, which lack a central rachis. Semiplumes always have a distinct rachis that is longer than any of the barbs. They are seldom exposed but lie under the surface contour feathers, insulating the body and forming smooth, aerodynamic body contours.
Adult down, or **definitive down**	Down feathers of adult birds are extremely plumulaceous feathers that provide a layer of insulation underneath the contour feathers. Down feathers either lack a central rachis or sometimes have a very short rachis that is shorter than the longest barbs. The barbs sometimes attach directly to the basal calamus of the feather. Down is not evenly distributed, and some groups (sea ducks, for example) have much heavier down coats than other groups (such as songbirds).

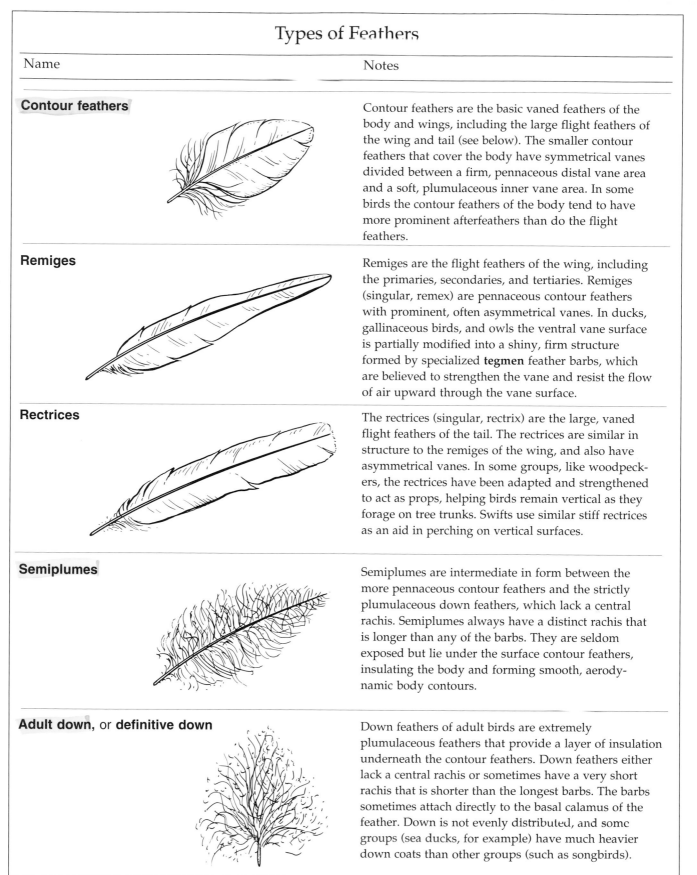

More Types of Feathers

Name	Notes

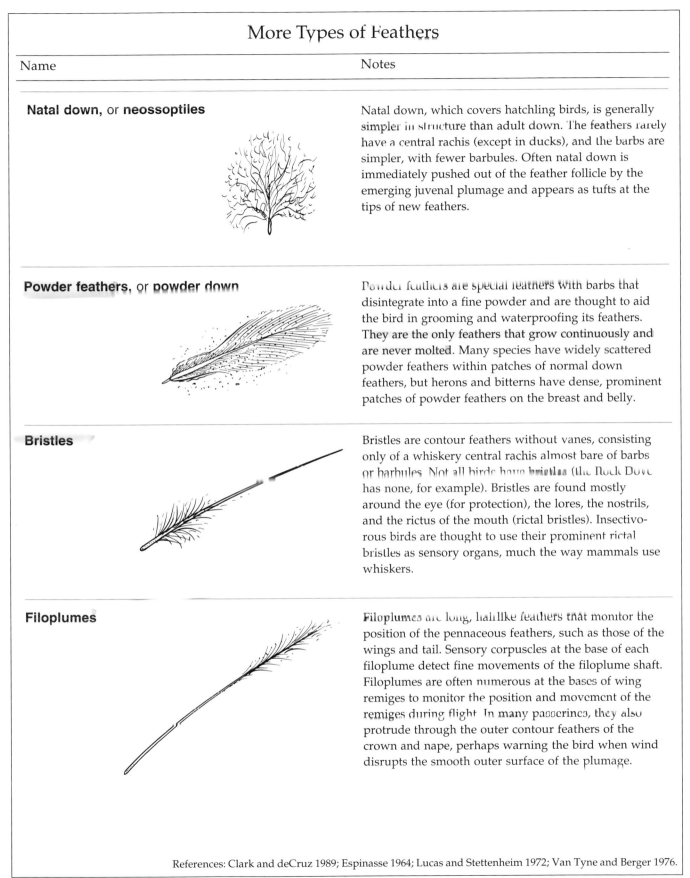

Natal down, or **neossoptiles**

Natal down, which covers hatchling birds, is generally simpler in structure than adult down. The feathers rarely have a central rachis (except in ducks), and the barbs are simpler, with fewer barbules. Often natal down is immediately pushed out of the feather follicle by the emerging juvenal plumage and appears as tufts at the tips of new feathers.

Powder feathers, or **powder down**

Powder feathers are special feathers with barbs that disintegrate into a fine powder and are thought to aid the bird in grooming and waterproofing its feathers. They are the only feathers that grow continuously and are never molted. Many species have widely scattered powder feathers within patches of normal down feathers, but herons and bitterns have dense, prominent patches of powder feathers on the breast and belly.

Bristles

Bristles are contour feathers without vanes, consisting only of a whiskery central rachis almost bare of barbs or barbules. Not all birds have bristles (the Rock Dove has none, for example). Bristles are found mostly around the eye (for protection), the lores, the nostrils, and the rictus of the mouth (rictal bristles). Insectivorous birds are thought to use their prominent rictal bristles as sensory organs, much the way mammals use whiskers.

Filoplumes

Filoplumes are long, hairlike feathers that monitor the position of the pennaceous feathers, such as those of the wings and tail. Sensory corpuscles at the base of each filoplume detect fine movements of the filoplume shaft. Filoplumes are often numerous at the bases of wing remiges to monitor the position and movement of the remiges during flight. In many passerines, they also protrude through the outer contour feathers of the crown and nape, perhaps warning the bird when wind disrupts the smooth outer surface of the plumage.

References: Clark and deCruz 1989; Espinasse 1964; Lucas and Stettenheim 1972; Van Tyne and Berger 1976.

from sensing anything that touches the plumage. Filoplumes are thought to help restore a sense of touch within the plumage, acting much as the small hairs on human skin do to enhance the sense of touch. Filoplumes have **sensory corpuscles** around the feather follicle at the base of the shaft (the calamus). These sensory structures register fine movements of the filoplume feather shaft just as whiskers and fine hairs on the skin act as sensory organs in mammals. Filoplumes are especially numerous around the bases of the major flight feathers. As the remiges are moved in flight, the filoplumes help the bird to judge what position its flight feathers are in and give the bird tactile feedback as it shifts the position of the flight feathers. In many species of songbirds the tactile sense provided by filoplumes protruding through the contour feathers of the nape may also help resting birds sense disruptions of the plumage surface caused by wind (Clark and de Cruz 1989; Lucas and Stettenheim 1972).

Powder feathers — A specialized feather, sometimes called **powder down**. The barbs of this feather disintegrate into a fine powder as the feather matures. Powder feathers are thought to aid the bird in grooming its plumage (acting like a talcum powder), although the exact role of the powder feathers is not well understood. Patches of powder feathers shed water, and this has led to the suggestion that powder downs help waterproof the plumage. Highly developed yellow **powder feather patches** are best viewed under the breast plumage of heron or bittern specimens (F. Ardeidae). Powder feathers are present in most bird species but are scattered sparsely throughout the plumage and are difficult to isolate in most birds (Lucas and Stettenheim 1972).

Woodcock primaries

During mating display flights the American Woodcock (*Scolopax minor*) produces a soft, twittering "call" that isn't entirely a vocalization. The three outer primary feathers on the woodcock's wings are specially stiffened and narrowed to produce sound when the wing is flapped rapidly. During its spectacular courtship flights the woodcock male flies about three hundred feet upward into the evening sky above fields and marshes before spiraling downward in an erratic, zigzagging pattern, flapping its wings to produce a soft, almost insectlike twittering "call." This whistling or twittering may also help the woodcock to startle predators that flush the bird from its usual habitat on the banks of streams, swamps, and other shallow wetlands. The woodcock prefers to stand still and will not flush until a predator (or birder) is almost on top of it; then it bursts forth in a mad dash, its wings twittering as it goes. This combination of sudden flight and sound probably startles predators (it definitely startles *birdwatchers*) and helps the woodcock avoid capture (Sheldon 1967).

American Woodcock
(*Scolopax minor*)

Form and Function
Feathers and insulation in sea ducks

Common Goldeneye
(*Bucephala clangula*)

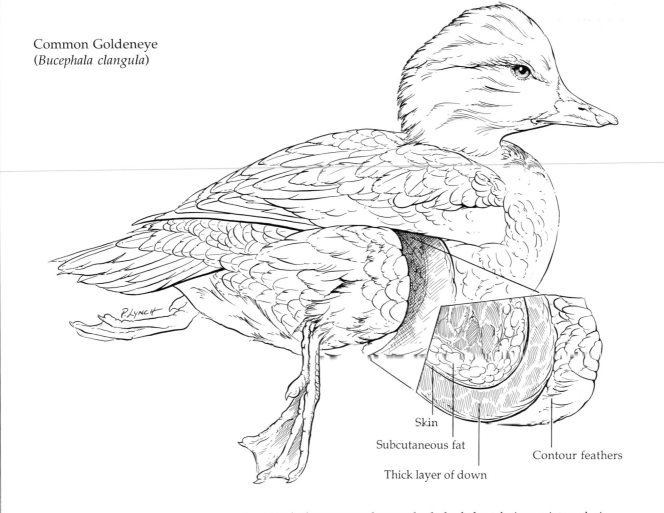

Skin

Subcutaneous fat

Contour feathers

Thick layer of down

Anyone who has ever stood out on the deck of a boat during a winter pelagic birding trip will remember the bone-cracking cold of the winter wind at sea. Yet the sea ducks (goldeneyes, eiders, scoters, and mergansers) spend more than half of their lives exposed to such conditions, and apparently thrive in weather that defeats the thickest (human-made) down coats and gloves. Like other homoiothermic ("warm-blooded") aquatic animals, such as seals and whales, many sea ducks have a thick layer of subcutaneous fat to help insulate them from the cold sea water. But in sea ducks the primary barrier to cold is an almost impervious layer of outer contour feathers that overlies a $^1/_3$-to-$^2/_3$-inch thick layer of extremely dense down feathers. The down layer covers the entire thorax and abdomen, and extends in thinner layers over the rest of the duck's body. The insulating value of duck down has been well known since ancient times, but in spite of recent advances in creating synthetic insulation, the insulating power of natural down has never been equaled.

6 PTERYLOSIS

In most modern birds the contour feathers of the wings and body are not uniformly distributed over the skin surface. This pattern of distribution is called **pterylosis**. The follicles of the contour feathers are concentrated in dense tracts called **pterylae** (singular, **pteryla**) and are separated by areas bare of contour feathers, called the **apteria** (singular, **apterium**). In some groups of birds, notably flightless penguins (Family Spheniscidae) and ratites like the Ostrich (F. Struthionidae) and the rheas (F. Rheidae), contour feathers are evenly distributed over the surface of the skin and are not organized into tracts. Toucans (F. Ramphastidae) also lack apteria, but this is probably a secondary adaptation and not an ancient or primitive character as it might be interpreted in other uniformly feathered groups (Harrison 1964).

Within the pterylae the contour feather follicles are arranged in evenly spaced, linear criss-crossing patterns across the skin surface (Lucas and Stettenheim 1972). The distribution of pterylae and apteria over the body surface is characteristic within taxonomic groups and can vary greatly depending on the family, genus, or even species of bird under examination. Although they are barren of contour feathers, the apteria are not entirely bare and will almost always show some down or semiplume quills. In highly aquatic groups like ducks (F. Anatidae) the apteria are considerably reduced and may be densely feathered with thick down to insulate the body from cold water.

Observe the following structures

Pterylosis is best observed in fresh specimens; birds preserved in formalin tend to have plumage so matted and disrupted that the normal configuration of the feathers is difficult to estimate. Begin your study by carefully parting the plumage over the breast, abdomen, back, and shoulders to note the general location of the **apteria** and major feather tracts (the **pterylae**). If you are observing a songbird you might try blowing softly on the plumage of the abdomen and chest to gently part the plumage in those areas. This technique is often used by bird banders and ornithologists to check the fat deposits of migrating birds. The amount of **subcutaneous fat** seen is a quick indication of the status of the bird in migration. If your specimen was particularly healthy and well fed when collected, masses of yellowish or whitish fat deposits may be seen just anterior to and lateral to the large muscles of the breast.

To continue examining the pterylosis of your specimen, use a pair of scissors to remove the contour feathers of the entire thorax, abdomen, back, and one of the wings (leave one wing intact for reference). Clip the feathers close to the body to reveal the pterylosis. Carefully clip or pluck the contour feathers of the wings from around the bases of the primary and secondary remiges. Try to leave the shafts of the major covert feathers intact, so you can see their relation to the bases of the flight feathers.

Use the diagrams on the following pages to locate the major feather tracts of the Rock Dove, or compare these diagrams with your particular specimen and make general comparisons. Many groups of birds have unique pteryloses that may vary somewhat from the Rock Dove or generalized passerine (songbird) illustrated here, but the major feather tracts and apteria should be similar.

Pigeon pterylosis redrawn after Lucas and Stettenheim 1972; generalized passerine after Miller 1928 and Stewart 1953.

Lateral View of Pterylosis
Rock Dove (*Columba livia*)
and Typical Passerine

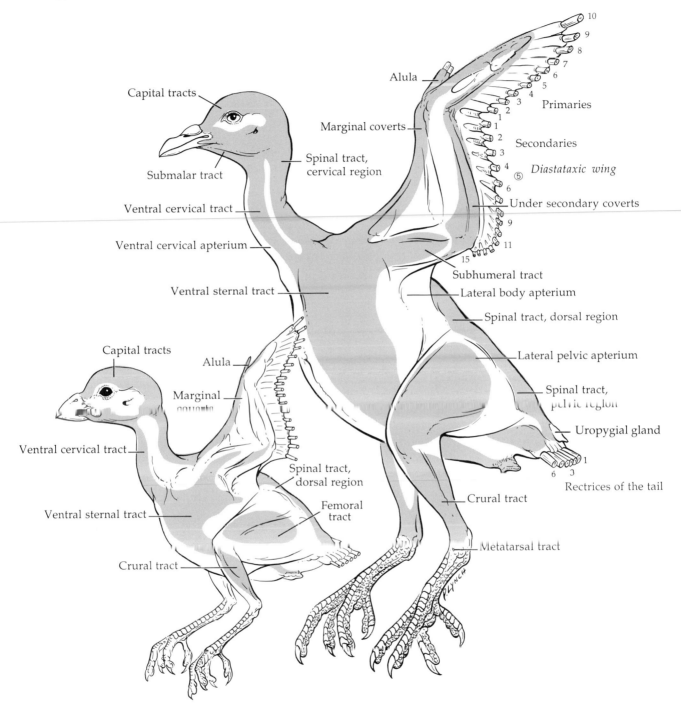

Capital tracts

Alula

Marginal coverts

Primaries

Secondaries

Diastataxic wing

Spinal tract,
cervical region

Submalar tract

Ventral cervical tract

Under secondary coverts

Ventral cervical apterium

Subhumeral tract

Ventral sternal tract

Lateral body apterium

Spinal tract, dorsal region

Capital tracts

Lateral pelvic apterium

Alula

Spinal tract,
pelvic region

Marginal
coverts

Uropygial gland

Ventral cervical tract

Rectrices of the tail

Spinal tract,
dorsal region

Ventral sternal tract

Femoral
tract

Crural tract

Crural tract

Metatarsal tract

Ventral View of Pterylosis
Rock Dove (*Columba livia*)
and Typical Passerine

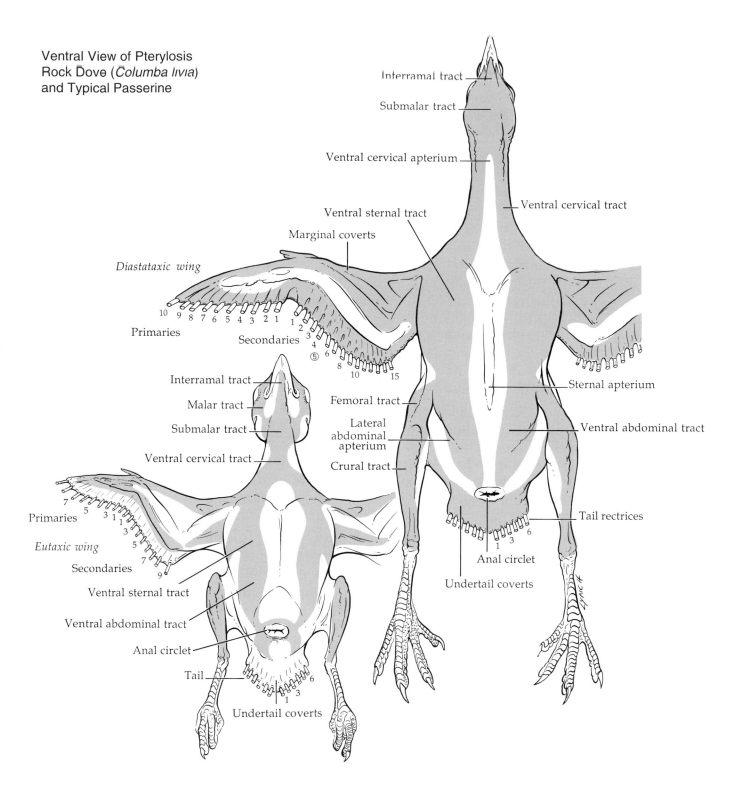

Interramal tract

Submalar tract

Ventral cervical apterium

Ventral cervical tract

Ventral sternal tract

Marginal coverts

Diastataxic wing

10 9 8 7 6 5 4 3 2 1

Primaries

Secondaries

1 2 3 4 ⑤ 6 8 10 15

Interramal tract

Malar tract

Submalar tract

Ventral cervical tract

Femoral tract

Lateral abdominal apterium

Crural tract

Sternal apterium

Ventral abdominal tract

Primaries

7 5 3 1 1 3 5 7 9

Eutaxic wing

Secondaries

Ventral sternal tract

Ventral abdominal tract

Anal circlet

Tail

1 3 6

Undertail coverts

Tail rectrices

1 3 6

Anal circlet

Undertail coverts

Note the number, location, and attachments of the major flight feathers of the alar and caudal tracts. In the Rock Dove (*Columba livia*) there should be ten primary remiges, fifteen secondary remiges, and twelve rectrices in the caudal tract. Be aware that your specimen may appear to be missing a flight feather or feathers from the alar or caudal tracts if it was molting when collected.

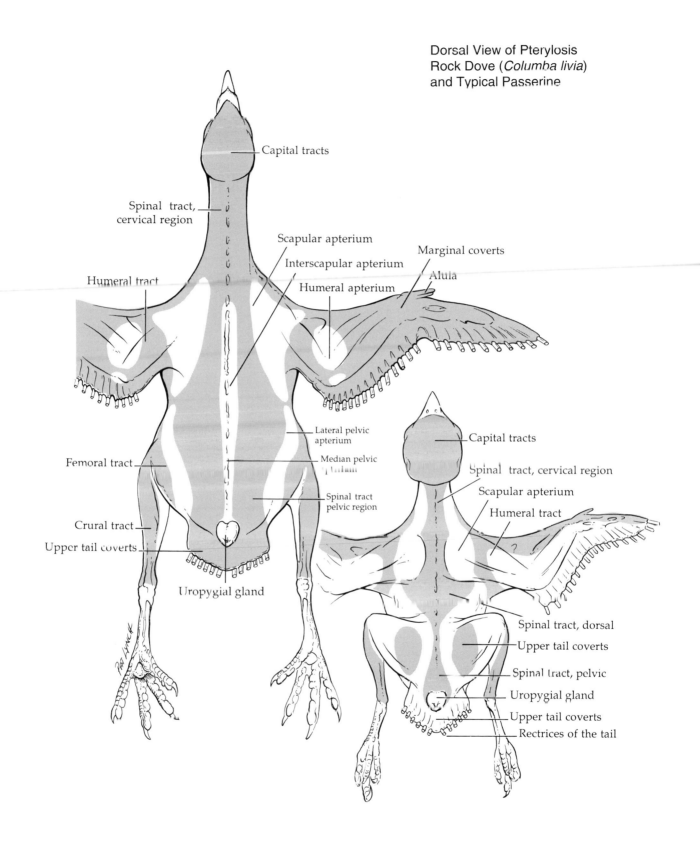

Dorsal View of Pterylosis
Rock Dove (*Columba livia*)
and Typical Passerine

Pigeon pterylosis redrawn after Lucas and Stettenheim 1972; generalized passerine after Miller 1928 and Stewart 1953.

The Major Pterylae, or Feather Tracts

Pteryla	Location and Boundaries
Capital tract	The capital tract extends over the entire dorsal surface of the head (the pileum) from the base of the maxilla (upper bill), over the forehead and crown areas (including **crest feathers,** where present), and down the back of the skull to the nape or occipital region, where the skull meets the spinal column. It is bounded laterally by an arbitrary line drawn along the lower edge of the mandible and passing posteriorly from the angle of the jaw (the articular bone) to the dorsal midline at the nape. This lateral boundary passes below the external ear opening, so the capital tract includes **auricular feathers, rimal feathers** around the eyelids, **rictal bristles,** and all other specialized feather regions of the head.
Spinal tract	The spinal tract—actually a complex and highly variable series of tracts and apteria lumped into one group—runs down the dorsal midline of the body from the base of the skull to the pygostyle at the posterior end of the vertebral column. In most birds it is a simple tract running dorsally along the midline of the back. In the Rock Dove, however, it is split, with a narrow apterium along the dorsal midline. The spinal tract is bordered along its length by the **cervical apteria** of the lateral neck (in most birds, but not the Rock Dove), the **scapular apteria** at the shoulders, and broad **lateral apteria** along the lateral walls of the abdomen. Note how the posterior end of the spinal tract broadens over the synsacrum and pelvis.
Ventral tract	The ventral tract covers the ventral neck, breast, and abdominal regions. It courses posteriorly along the ventral midline of the body as a single tract from the notch between the **mandibular rami** of the lower bill to the base of the neck just anterior to the breast musculature. The tract then bifurcates (divides) into two lateral bands (the **pecterosternal tracts**) that enclose a bare **sternal** or **mid-ventral apterium** along the ventral midline of the thorax and abdomen. The lateral ventral tracts reunite at a circle of feathers around the cloaca, the **cloacal circlet.**
Caudal tract	The caudal tract includes the major flight feathers of the tail, the **rectrices,** which vary in number depending on the taxonomic group. In the Rock Dove there are twelve rectrices, six on each side of the tail. Most passerines have twelve rectrices. The caudal tract also includes the upper tail coverts of the **dorsal caudal tract** and lower tail coverts of the **ventral caudal tract.**
Humeral tract	The humeral tract is a band of contour feathers that overlie the humerus bone (upper arm area) on the dorsal side of the wing, the "shoulder" area of the bird.
Alar tract	The alar tract comprises a series of smaller pterylae that cover both the dorsal and ventral surfaces of the outer wing; also included are the **primary, secondary,** and (if present) the **tertial remiges**—the major flight feathers of the wing—as well as the **alula feathers** of the second digit. The smaller alar tracts contain all of the rows of **greater, middle,** and **lesser covert feathers** that cover the wing both dorsally and ventrally, except those coverts within the humeral tracts.
Femoral tract **Crural tract**	The femoral tract covers the outer surface of the thigh in a diagonal strip from the knee joint upward toward the base of the tail. The rest of the leg contour feathers are included within the crural tract. In some large birds, and in birds with heavily feathered legs, an additional **metatarsal tract** is identified, covering the tarsometatarsal area of the lower leg.

References: Miller 1928; Lucas and Stettenheim 1972; Stewart 1953; Van Tyne and Berger 1976.

Brood Patches, a Special Form of Apteria

Brood patches are specialized temporary apteria that develop on the abdominal and breast feathers of female (and some male) birds during the incubation of eggs. Increasing levels of the hormone estrogen apparently cause some contour and down feathers of the ventral abdominal tracts to loosen and fall out, leaving one large apteria or a number of smaller patches on the brooding parent's abdomen (Bailey 1952; Drent 1970, 1975; Jones 1969). This bare skin contains many blood vessels, and when placed in contact with the incubating eggs it transfers heat from the parent to the eggs. In shorebirds and gulls (Order Ciconiiformes) the number of brood patches is matched to the number of eggs in a typical clutch, as shown in this view of the underside of a Herring Gull (*Larus argentatus*), which normally incubates two or three eggs.

Herring Gull (*Larus argentatus*)
View from below, with brood patches and eggs.

Muscovy Duck
(*Cairina moschata*)

Many birds have extensive barren areas on the head and neck. In the Turkey Vulture (*Cathartes aura*) and such other carrion feeders as the Marabou Stork (*Leptoptilos crumeniferus*), the barren skin of the head and neck obviously helps the birds avoid picking up debris during the messy business of scavenging carcasses. But in most birds that show large bare areas of the skin on the head, the apteria (featherless areas) expose colorful patches of skin that complement the plumage colors and patterns, distinguishing male from female in sexually dimorphic species. In many birds, such as the Great Blue Heron (*Ardea herodias*), the bare skin of the lores changes color markedly as the breeding season progresses. In domesticated varieties of the Muscovy Duck (*Cairina moschata*) aviculturists have selectively bred for males with bright red skin surrounding the eyes and base of the bill, a condition not normally seen in the drab wild Muscovy but very common in its domesticated variants.

Flight Feather Counts of Some Major Bird Groups

Group	Primary remiges	Secondary remiges*	Tail rectrices
PHASIANIDAE – Quail, pheasants	10	10–18 [e]	12–18
ANATIDAE – Ducks, geese, swans	11	15–24 [d]	12–24
PICIDAE – Woodpeckers	10	11 [e]	12
TROGONIDAE – Trogons	10	11 [e]	12
ALCEDINIDAE – Kingfishers	10–11	11–14 [de]	12
COCCYZIDAE – Cuckoos	10	11 [e]	8–10
PSITTACIDAE – Parrots, macaws	10	8–14 [d]	12
APODIDAE – Swifts	10	8–11 [de]	10
TROCHILIDAE – Hummingbirds	10	6–7 [de]	10
STRIGIDAE – Typical owls	10	11–19 [d]	12
CAPRIMULGIDAE – Nighthawks	10	12–15 [d]	10
COLUMBIDAE – Pigeons, doves	10	11–15 [de]	12
GRUIDAE – Cranes	11	~16 [d]	12
RALLIDAE – Rails, gallinules	10–11	~15 [d]	8–14
SCOLOPACIDAE – Sandpipers	10	11 [d, except *Philohela*]	12
CHARADRIIDAE – Plovers	10	11 [d]	12
LARIDAE – Gulls, terns, alcids	11	11 [d]	12
ACCIPITRIDAE – Hawks, eagles, osprey	10	15–20 [d]	12–14
FALCONIDAE – Falcons, caracaras	10	16 [d]	12–14
PODICIPEDIDAE – Grebes	12	~22 [d]	vestigial
PHALACROCORACIDAE – Cormorants	11	~20 [de]	12–14
ARDEIDAE – Herons	11	23 [d]	8–12
PELECANIDAE – Pelicans	11	20 [d]	22–24
CICONIIDAE – New World vultures	10	18–25 [d]	12–14
GAVIIDAE – Loons	11	22–23 [d]	16–20
PASSERINES – Songbirds	9–10	9–11 [e]	12

* e = eutaxic
d = diastataxic
de = mixed groups

References: Lucas and Stettenheim 1972; Van Tyne and Berger 1976

Feather Counts

The maximum number of major flight feathers of the wing (the remiges) and the tail (the rectrices) are a fairly constant characteristic of each bird taxonomic family. In general, smaller birds with short wings have fewer flight feathers than larger birds with long wings, and more ancient or primitive groups of birds (such as loons or ducks) have more flight feathers than more recently evolved groups (such as passerines). Most birds have ten primaries, but many passerines have nine primaries. The number of secondary feathers usually reflects the length of the wing. The longer the ulna bone is, the more secondary feathers there are, although here, too, more primitive groups usually have more secondaries than more modern groups. Albatrosses (Family Procellariidae) are both primitive and very long-winged, and can have as many as forty secondaries in the wing. Most birds have twelve rectrices, although primitive loons (F. Gaviidae) and pelicans (F. Pelecanidae) have about twice that many, with twenty-two to twenty-four flight feathers in their tails (Harrison 1964; Lucas and Stettenheim 1972; Welty and Baptista 1988).

A number of (very patient) Works Progress Administration workers assisting ornithologists have actually counted the number of contour feathers in various groups of birds (Ammann 1937; Brodkorb 1949; Wetmore 1936). Smaller birds, not surprisingly, have fewer contour feathers than larger birds, and the number of contour feathers in a species varies by season and molt. Winter birds have higher feather counts than birds collected in the summer, and aquatic birds generally have many more feathers than nonaquatic species (Wallace and Mahan 1975). Contour feather counts range from a low of 940 in the Ruby-throated Hummingbird (*Archilochus colubris*) to a high of 25,216 in the Tundra Swan (*Cygnus columbianus*). The typical songbird (about the size of an American Robin) has from 2,200 to 2,600 contour feathers, with higher counts in the larger species or in birds collected in fall or winter (Wetmore 1936).

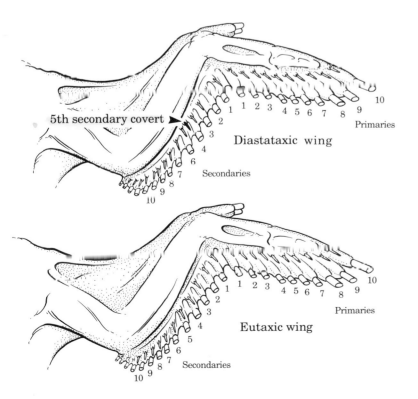

5th secondary covert ▶

Diastataxic wing

Primaries

Secondaries

Eutaxic wing

Primaries

Secondaries

Eutaxy and Diastataxy

Many groups of birds show a peculiarity in the series of secondary remiges along the ulna bone: the fifth remex is missing, leaving a slight gap between the fourth and sixth secondary remiges, called a **diastema** (see the figures at the left). Birds with a gap in the secondary series are said to be **diastataxic,** while birds with an even series of secondaries and no gap are said to be **eutaxic**. Although the reason for the gap is disputed, the trait can be a useful taxonomic character. In general, the older or more primitive groups of birds have diastataxic wings; some groups, such as pigeons and doves (Family Columbidae), include both eutaxic and diastataxic species, and most recently evolved groups, such as the passerines (songbirds), have uniformly eutaxic wings with no gap. If your dissection specimen is a Rock Dove, you may be able to discern a slight gap between the fourth secondary and the remex just proximal (in toward the body) to it. The gap is subtle, however, and is often detected only by carefully counting the number of major secondary coverts of the wing, for each secondary remex should have a matching greater covert. Passerine specimens are eutaxic and will not show a gap in the secondary series (Lucas and Stettenheim 1972).

7 PLUMAGES AND MOLTS

Feathers are not permanent structures; periodically they must be shed and replaced. This process of shedding and replacing worn feathers is called **molting,** and the feather coats worn between molts are called **plumages**. Most birds in the Northern Hemisphere molt all flight and body feathers late in summer and undergo a second, much more limited molt of body feathers in spring just before breeding season (Harrison 1964; Lucas and Stettenheim 1972; Welty and Baptista 1988).

Two methods for describing plumages and molts are in widespread use (see the chart on page 108). The traditional descriptive method was first proposed by Dwight in 1900 and remains in widespread use. In this scheme the two major plumages of the adult bird are the **winter** and the **breeding,** or **nuptial,** plumages. This scheme is fine for most species that breed in North America during the boreal spring and summer, but there are enough exceptions to cause some confusion. For example, in this terminology most bird species molt into a "nuptial" plumage after their first winter of life, but many birds do not actually breed until they are two or more years old, and most larger birds would still be wearing a distinctly sub-adult plumage in their second summer. In addition, many pelagic birds like shearwaters and fulmars (F. Procellariidae) breed in the Southern Hemisphere during the austral summer and are in their "winter" plumage when seen in the Northern Hemisphere during the boreal summer.

To avoid terminology based too closely on seasonal or descriptive language, Humphrey and Parkes (1959) proposed a different way of naming plumages. In this terminology adult birds molt into a basic plumage just after the breeding season. In the spring they molt some of their body and/or head feathers and then wear an alternate plumage (the "breeding" plumage) during the summer months. The terms **basic** and **alternate** help avoid some of the more confusing connotations of the traditional system. Both ways of referring to plumages and molts remain in widespread use, though the Humphrey and Parkes system seems to be gaining broader acceptance.

The exact schedule and number of molts varies for each species, depending on breeding cycle, habitat, and whether the species is migratory or sedentary. In general, birds that undertake long migrations have more frequent and complete molts than do sedentary species, and birds that live in such physically harsh conditions as grasslands and deserts undergo more molts than forest birds. Most birds go through a quick series of plumages in their first months of life, then cycle between a basic, or winter, plumage worn for most of the year and an alternate, or breeding, plumage worn only during spring and summer. The diagram to the right shows the Humphrey and Parkes terminology for a typical series of plumages and molts that would bring the average bird species through to adult plumage in its second winter. In later years the bird would cycle between the basic, or winter, plumage and its alternate, or nuptial, plumage. For most species the late summer prebasic, or post-nuptial, molt is the most important, because during this molt the bird replaces all of its body and flight feathers.

In most species young birds wear a series of distinct sub-adult plumages, marking them as nonbreeding members of the species. In a so-called four-

NATAL DOWN

|

Prejuvenal molt

|

JUVENAL PLUMAGE

|

Prebasic molt

|

BASIC PLUMAGE

|

Prealternate molt

|

ALTERNATE PLUMAGE

|

Prebasic molt

|

BASIC PLUMAGE

|

et cetera . . .

Plumage Development in the
Herring Gull (*Larus argentatus*)

Juvenile

First winter

Second winter

Third winter

Adult nuptial plumage

Plumage Sequence and Molt Terminology

		Plumage Name		Molt Replacing the Plumage		Extent of Molt
		Traditional (Dwight)	Humphrey & Parkes	Traditional (Dwight)	Humphrey & Parkes	
First Year of Life, or Hatching Year (HY)	**1**	Natal down	Natal down	Postnatal molt	Prejuvenal molt	Complete*
		The down coat precocial chicks are born with or altricial chicks develop in the first weeks of life.				
	2	Juvenal plumage	Juvenal plumage	Postjuvenal molt	First prebasic molt	Partial**
		Most birds wear their juvenal plumage until late in their first summer. The juvenal plumage includes the flight feathers the young bird will wear all winter.				
	3	First winter plumage	First basic plumage	First prenuptial molt	First prealternate molt	Partial
		In most species this is the "immature" plumage. The first prebasic molt is incomplete, usually involving only body feathers and rarely including flight feathers.				
	4	First nuptial plumage	First alternate plumage	First postnuptial molt	Second prebasic molt	Complete
		In most North American species this molt is completed by mid-spring, before birds reach their breeding grounds. Many birds do not breed in their first breeding season.				
Second Year	**5**	Winter plumage (2d)	Basic plumage (2d)	Prenuptial molt	Second prealternate molt	Partial
		In most North American species this is the first fully adult plumage and is indistinguishable from adult winter plumage, or mature basic plumage.				
		Most birds are now fully adult and are said to wear their **definitive plumage,** but larger or long-lived birds like gulls or eagles may take as many as five years to mature and show a distinct sub-adult plumage in each early year.				
	6	Nuptial plumage (2d)	Second alternate plumage	Second postnuptial molt	Second prebasic molt	Complete
		The first fully adult breeding, or alternate, plumage in most species. Most passerines are fully adult breeders at this stage.				
Third Year	**7**	Winter plumage (3d)	Basic plumage (3d)	Prenuptial molt	Prealternate molt	Partial
	8	Nuptial plumage (3d)	Alternate plumage	Postnuptial molt	Prebasic molt	Complete
	etc.	Winter plumage (4th)	Basic plumage (4th)	Prenuptial molt	Prealternate molt	Partial

*Complete molts into the basic, or winter, plumage replace all body as well as flight feathers of the body. Only a few body contour feathers remain from the previous plumage.

**Partial molts into the alternate, or breeding, plumage replace only a percentage of the body contour feathers and rarely involve the flight feathers. In many passerine species only some the feathers of the head and neck are molted.

References: Dwight 1900; Humphrey and Parkes 1959; Lucas and Stettenheim 1972; Cramp and Simmons 1977.

year gull like the Herring Gull (*Larus argentatus*), young birds show a series of four distinct immature plumages as they gradually attain the clear white head and gray mantle of the breeding adult. Even when finally wearing the fully adult plumage, the Herring Gull may not actually nest until the fifth year of life (Tinbergen 1960; Cramp and Simmons 1983). Many of the larger birds of prey also show a series of immature plumages. The Bald Eagle (*Haliaeetus leucocephalus*), for example, does not attain the clear white head and tail of the adult plumage until at least its fourth year (Clark and Wheeler 1987).

Form and Function
Feather wear in the Snow Bunting

The Snow Bunting (*Plectrophenax nivalis*) is typical of many passerines in that the breeding plumage worn by the male in summer is not a separate plumage but is actually just the worn feathers of the winter, or basic, plumage. This is often called a "wear plumage," and other common birds like the European Starling (*Sturnus vulgaris*) also exhibit this molt pattern. In the late-summer prebasic (post-nuptial) molt, the Snow Bunting male molts into a pale-brown-and-gray basic, or winter, plumage. The colors of the basic plumage blend well with the bunting's winter habitat of open shoreline areas and farm fields. Over the winter the pale white edges of the basic plumage wear off, revealing a bolder black-and white pattern that is fully developed by the time the male bunting is back on its tundra breeding grounds the following spring.

References: Welty and Baptista 1988 (after Chapman 1912).

10 AVIAN ECTOPARASITES

Parasites that live on and attack the outside of a host's body are called
ectoparasites, and birds suffer from five major groups of ectoparasitic
animals: parasitic flies, lice, mites, fleas, and ticks. Some avian parasites
feed directly on the feathers and on debris that naturally flakes off the skin
and feathers. Most ectoparasites feed by piercing the host bird's skin and
sucking blood and other fluids from the host. In either case the parasitic
infection reduces the bird's overall fitness and ability to maintain a warm,
healthy, and functional set of feathers. Although parasites rarely kill an
adult bird outright, heavy infestations can gradually degrade the condition
of the bird, making it more susceptible to predation or death by exposure
(Clay 1964; Terres 1980; Wheeler and Threlfall 1986).

Parasitic Flies

Flies (Order Diptera) are the most commonly encountered insect ectopara-
site of birds. The Finch Louse Fly (*Ornithomyia fringillina*) of the Family
Hippoboscidae is known to infest more than a hundred species of birds in
both hemispheres. Hippoboscid flies have an extremely dorsoventrally
flattened body shape (see side view at right), to better fit into the gaps
between feathers and to lie flat and secure against the host's skin. They also
have strong, sharp claws to cling to feathers and skin, as well as sucking
mouthparts cling to and extract blood from the host's skin. Hippoboscid
flies have an unusual and complex life cycle. Female flies retain a single
developing fly larva within their bodies and give birth to live, fully formed
young one at a time. The hippoboscids require no alternate or intermediate
hosts in their life cycle, and under favorable conditions several generations
of flies may live out their lives on a single bird (Bennett 1961). In a parasitic
phenomenon known as **phoresy,** hippoboscid flies often inadvertently
transport another parasitic group, the mallophagan lice. These lice attach
themselves to the abdomens of hippoboscid flies and are thereby trans-
ported when the flies leave one host and move to another.

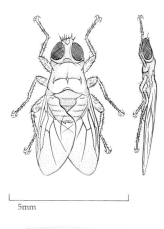

5mm

Finch Louse Fly
(*Ornithomyia fringillina*)

Bird Lice

Sucking lice of the Suborder Mallophaga attack both mammals and birds,
feeding on the host bird's feathers as well as sucking blood through its skin
(Borror and White 1970). The lice can live under and between feather shafts
throughout the bird's body, but one species of mallophagan is so specialized that it lives on blood, other fluids, and pith within the shaft of the
developing feather calamus. Biologists who handle birds often encounter
mallophagan lice, particularly in larger birds like hawks and eagles. The lice
seem to be a tolerable annoyance to the birds, but heavy infestations can
seriously damage the plumage of the host bird, rendering it susceptible to
exposure from the weather and degrading the performance of the flight
feathers.

Mallophagans attach their eggs directly to the feathers or lay them within
the feather shafts. No intermediate hosts are required, and most mallophagan
species are so closely adapted to their host species that the lice have been
used to determine taxonomic relations within closely related groups of
birds. Mallophagans have probably been associated with birds since early
in the evolutionary history of both groups. As bird families evolved and
differentiated, the lice associated with them followed a parallel develop-

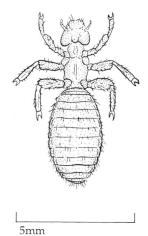

5mm

Mallophagan Louse
(Suborder Mallophaga)

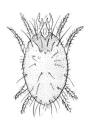

Itch Mite
(Order Acarina)

0.5mm

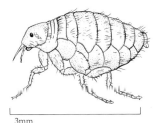

3mm

Sticktight Flea
(Family Tungidae)

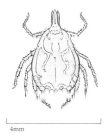

4mm

Lone Star Tick
(*Amblyomma americanum*)

ment. A close relation between lice found on two bird species can thus be used as a clue to how closely the birds are related. For example, ornithologists have long suspected that the large ground-dwelling birds the Ostrich (Family Struthionidae) and the rheas (F. Rhcidae) share a distant common ancestor. Interestingly, the mallophagan lice on both groups also show evolutionary affinities, suggesting how ancient the relationship between parasite and host can be in some groups.

Feather Mites, Itch Mites, Red Mites

Several groups of parasitic mites (Order Acarina) attack birds. Feather mites (Family Analgesidae) do not suck blood from their hosts but feed on the feathers and on debris that normally flakes off from skin and the plumage. Minor attacks of feather mites are of little consequence to the host bird, but heavy infestations can cause feather mange, a condition that can cause large patches of feathers to drop off the body. Itch mites (F. Sarcoptidae) often infest the legs and feet of birds, living under the scutes and scales of the skin. Serious infestations can cause swelling and debilitating infections of the legs. Nasal mites (F. Rhinonyssidae) infest the nasal cavities of birds, particularly birds that feed on nectar and visit many flowers each day. The mites use the birds for transport from flower to flower. Once at a new flower the mites infect the next bird that feeds there. When numerous, nasal mites can cause respiratory infections. Red mites (F. Dermannyssidae) infest the nests and roosts of birds, attacking resting adults and nestlings during the night. Large populations of red mites can kill nestlings and become permanent residents on the bodies of adult birds.

Bird Fleas

Fleas (Order Siphonaptera) are another blood-sucking group that attacks birds. This group tends to infest nestlings and birds that habitually use the same roosting area. Because fleas spend their early lives feeding on organic debris in and around nests, they have several disadvantages relative to parasites that live permanently on their hosts. The eggs and larvae are restricted to activity during the warmer months. Eggs are laid off the host animal in nest debris or ground litter, and so in each new generation the larvae must find a new host bird as they mature or they will die out (Borror and White 1970). Fleas that attack birds derive primarily from two families, the rodent fleas (F. Dolichopsyllidae) and the sticktight fleas (F. Tungidae).

Bird Ticks

Ticks (Family Ixodidae) attack the fleshy parts of the bird and are often found around the gape of the mouth, ears, and eyelids. In extreme cases ticks can cause blindness in birds. Ground-living birds and birds that roost communally are more susceptible to tick infection than more arboreal, mobile species. Most ticks have a complex life cycle and spend some of their lives living in debris around the host's nest or roost. Young tick larvae often attach to a host bird, obtain a blood meal, and then drop off again to complete the metamorphosis to adulthood. The adult tick then seeks a new host bird and may stay permanently attached if it finds a suitable host. Many of the factors that limit flea populations thus keep the numbers of ticks in and around nests from becoming overwhelming. These blood-sucking parasites have become famous in recent years as the vector for Lyme disease. Research has shown that birds may harbor the larval stages of the Deer Tick (*Ixodes dammini*) that actually transmits the Lyme disease spirochetes. In a study conducted in eastern Connecticut, twenty-seven bird species from eleven families were found to be carriers of *I. dammini* larvae (Anderson and Magnarelli 1984).

All avian parasites are quite dependent on their hosts and die unless they are able to find new hosts quickly. Predatory birds are often at risk of infection from parasites carried by prey birds. Any disruption of the host bird's feathers or body temperature causes lice, flies, and other parasites to move quickly to the surface of the feathers and prepare to transfer to a new host. Bird banders often see hippoboscid flies leave the host bird as it is handled in the process of banding and often encounter mallophagan lice and feather mites as well. These bird-specific parasites are no threat to humans, but with the recent rise of Lyme disease, banders and others who routinely handle wild birds should be aware of the (relatively minor) risk of infection from bird-borne Deer Tick larvae.

Common Avian Ectoparasites

Family	Type	Habits and Effects
Analgesidae	Feather mites	Feed on feathers and skin debris, causing feather destruction and skin infections.
Sarcoptidae	Itch mites	Feed on feather follicles, skin, and leg scutes, causing scaly leg, leg infections, and feather mange.
Rhinonyssidae	Nasal mites	Live in the nasal cavities of birds, causing nasal infections or restricting the flow of air through the nasal cavities.
Dermonyssidae	Red mites	Suck blood and irritate the skin of birds, sometimes causing itching, skin infections, or feather damage.
Ixodidae	Bird ticks	Suck blood from the skin and fleshy parts of birds, causing eye diseases and blindness and acting as vectors for a number of diseases, including Lyme disease.
Mallophagadae	Body lice	Attack the feathers and skin, sucking blood and fluids from the feathers. Some live within the feather shaft (calamus). Lice are vectors for a number of avian diseases and can cause feather loss and general loss of conditioning in heavily infected birds.
Hippoboscidae	Louse flies	Attach themselves to the skin under the plumage and suck blood from their avian hosts. Act as vectors for disease and sometimes kill young birds through loss of blood.
Siphonaptera	Fleas	Attack the skin and other fleshy areas by sucking blood. Fleas act as vectors for disease and can be so numerous around infected nests that they can kill young nestlings.

References: Bennett 1961; Clay 1964; Peters 1936; Terres 1980; Wheeler and Threlfall 1986.

Feathers

1. What are brood patches, and why do some birds have them and others lack them? Give examples.

2. Consider the Red-winged Blackbird (*Agelaius phoeniceus*) molt and plumage sequence. Chart the sequence, taking it from hatching to the second summer of its life.

3. What is a wear plumage? Name three species occurring in your area that have such plumage.

Chapter Worksheet, *continued*

4. How accurate is your memory for plumage patterns? Without review, describe the color patterns of a Blue Jay (*Cyanocitta cristata*) or Stellar's Jay (*Cyanocitta stelleri*), and then compare what you have drawn or written with a specimen or illustration of the jay.

5. Study the following skins: Northern Harrier (*Circus cyaneus*), Northern Flicker (*Colaptes auratus*), male Rose-breasted Grosbeak (*Pheucticus ludovicianus*). What color pattern is similar? Why? Compare patterns with male and female in each species. What are the differences? Why are there differences?

6. Define the following terms:

a. friction barbules

b. emarginated primaries

c. phoresy

d. eutaxy

References

Bakker, R. T. 1975. The dinosaur renaissance. *Sci. Amer.* 232(4):58–78.

Feduccia, A. 1980. *The age of birds.* Cambridge: Harvard University Press.

Gill, F. B. 1990. *Ornithology.* New York: W. H. Freeman.

Heilmann, G. 1927. *The origin of birds.* New York: D. Appleton. Reprint, 1972; New York: Dover.

Lucas, A. M., and P. R. Stettenheim. 1972. *Avian anatomy: integument.* Agric. Handb. 362:1–340. Washington, D.C.: U.S. Government Printing Office.

Martin, L. D. 1983. The origin and early radiation of birds. In A. H. Brush and G. A. Clark, eds., *Perspectives in ornithology: essays presented for the centennial of the American Ornithologists' Union.* Cambridge: Cambridge University Press, 291–337.

Ostrom, J. H. 1985. The meaning of *Archaeopteryx.* In M. K. Hecht, J. H. Ostrom, G. Viohl, and P. Wellnhofer, eds., *The beginnings of birds.* Eichstatt: Freunde des Jura–Museums, 161–76.

Parkes, K. C. 1966. Speculations on the origin of feathers. *Living Bird* 5:77–86.

Peters, D. S., and W. Fr. Gutmann. 1985. Construction and functional preconditions for the transition to powered flight in vertebrates. In M. K. Hecht, J. H. Ostrom, G. Viohl, and P. Wellnhofer, eds., *The beginnings of birds.* Eichstatt: Freunde des Jura-Museums, 233–42.

Regal, P. J. 1975. The evolutionary origin of feathers. *Q. Rev. Biol.* 50:35–66.

Regal, P. J. 1985. Common sense and reconstructions of the biology of fossils. *Archaeopteryx* and feathers. In M. K. Hecht, J. H. Ostrom, G. Viohl, and P. Wellnhofer, eds., *The beginnings of birds.* Eichstatt: Freunde des Jura-Museums.

Welty, J. C., and L. F. Baptista. 1988. *The life of birds.* 4th ed. New York: W. B. Saunders.

Golden Eagle
(*Aquila chrysaetos*)

A skeletal view is not just a look into the internal structure of an animal, it's also an X-ray into the ecology and behavior of that animal. The fierce head and beak of an eagle usually attract the most attention, but as you can see clearly here, the massive legs and feet are the most essential and characteristic anatomic developments in the Golden Eagle. Eagles hunt and kill with their legs and feet, using their bills only to dismember and eat their prey. No bird of prey would risk injury to its eyes by attacking a large animal with its beak.

The Skeleton

5 Although birds share a similar skeletal plan with all other advanced vertebrates, their skeletons have been heavily modified to meet the demands of flight. Birds have evolved both a strong skeletal framework characterized by extensive fusion of bones for rigidity and a unique system of hollow "pneumatic" bones interconnected with the respiratory system. Three principal adaptations for flight can be seen in the avian skeleton: rigidity, reduction and redistribution of mass, and specific modifications of the limbs for flight (Bellairs 1964, Dellalis et al. 1960; Bock 1974).

Rigidity

Compared with other modern vertebrates such as reptiles and mammals, avian skeletons show extensive fusion of individual bones into rigid structures that are both lightweight and strong enough to hold up to the tremendous forces generated by flight. The most distinct skeletal adaptations are in the thorax and pelvis. The bones of the pelvis (ilium, ischium, pubis, and sacral and some caudal vertebrae) have all fused into a strong yet lightweight platform for attachment of the leg, tail, and abdominal muscles (Bellairs 1964).

In the thorax (chest area) the thoracic vertebrae are tightly bound together, forming a rigid structure that resists the twisting and bending forces associated with wing flapping. The ribs of flying birds have a unique structure. In most bird species each rib has a small backward-pointing extension called the **uncinate process**. These small extensions are lateral braces that attach the ribs to each other in addition to the attachments to the spine and sternum, forming a strong supporting "basket" around the lungs and heart.

The bones of the avian hand area have also been extensively fused to make a rigid support (the carpometacarpus) for the primary flight feathers at the tip of the wings. Three digits are attached to the manus, supporting the alula feathers and primary remiges of the distal wing (Chamberlain 1943; Gill 1990; Welty and Baptista 1988).

Reduction and Redistribution of Mass

The most characteristic avian mass-reducing adaptation is the lack of teeth in modern birds. Although early birds like *Archaeopteryx* had a full complement of reptile-like teeth in their jaws, modern birds have lost both the teeth and the heavy jaw bones necessary to support them. Instead, birds have evolved a two-step system of digestion: the light but strong beak does the initial job of tearing or crushing the food, and then the strong muscles of the gizzard (part of the stomach) further "chew" the food into small bits suitable for digestion. The heavy "chewing" function has thus been moved back from the head toward the bird's center of gravity, the stomach. In the avian skull most of the cranial bones are so fused that their boundaries are usually invisible. Facial bones in most birds have been reduced to a network of struts that support the beak (Bellairs 1964; Bellairs and Jenkin 1960; Bock 1964).

Most of the major bones are also **pneumatic**—that is, they are hollow and filled with air spaces connected to the respiratory system. These hollow pneumatic bones are very strong relative to their mass, but they are not necessarily lighter than the bones of mammals of equivalent size. For example, the leg bones of many birds are considerably heavier than those of mammals of equivalent size, as are the avian sternum and humerus when compared to the equivalent bones in mammals (Prange et al. 1979).

The most obvious avian skeletal adaptation for flight is the huge blade of the sternum (breastbone), which supports both the muscles that draw the wing down and those that raise it up again in flapping flight. The pectoral girdle (shoulder) bones, too, have been extensively modified to support flapping flight. Birds have evolved a large pair of clavicles, together called the **furcula** but better known as the wishbone. This V-shaped wishbone allows flex during flight, and the large coracoid bones brace the thorax against the force of the flight muscles, which would otherwise collapse the chest in flight. The coracoid and scapula bones together form a unique "pulley" that allows the supracoracoideus muscle to come up over the shoulder girdle and onto the humerus, so that breast muscles below the wing can both raise and lower the wing in flight (see chapter 6, The Musculature). In the wing the humerus is short and stout, because it must bear the primary forces exerted by the flight muscles. Farther out on the wing, the bones of the wrist and hand have been extensively modified and fused to give the flight feathers a rigid support (Ostrom 1976b).

Birds have also lost most of the length of their ancestors' bony tail (*Archaeopteryx* had a long, bony reptilian tail). Instead, they have evolved a short stub of a tail to support a long, lightweight tail of feathers.

1 OVERVIEW OF THE SKELETON

The skeleton as a whole can be considered in two parts:

The Axial Skeleton
The skull, vertebral column, ribs, and sternum

The Appendicular Skeleton
The wings, legs, and pectoral and pelvic girdles

Three principal adaptations to flight can be seen in the avian skeleton:

Rigidity

Redistribution of mass

Adaptation of the limbs

References: Bellairs and Jenkin 1960; Bock 1974; Chamberlain 1943; Howard 1929; Lucas and Stettenheim 1972; Van Tyne and Berger 1976; Welty and Baptista 1988.

Lateral View of the Skeleton
Rock Dove (*Columba livia*)

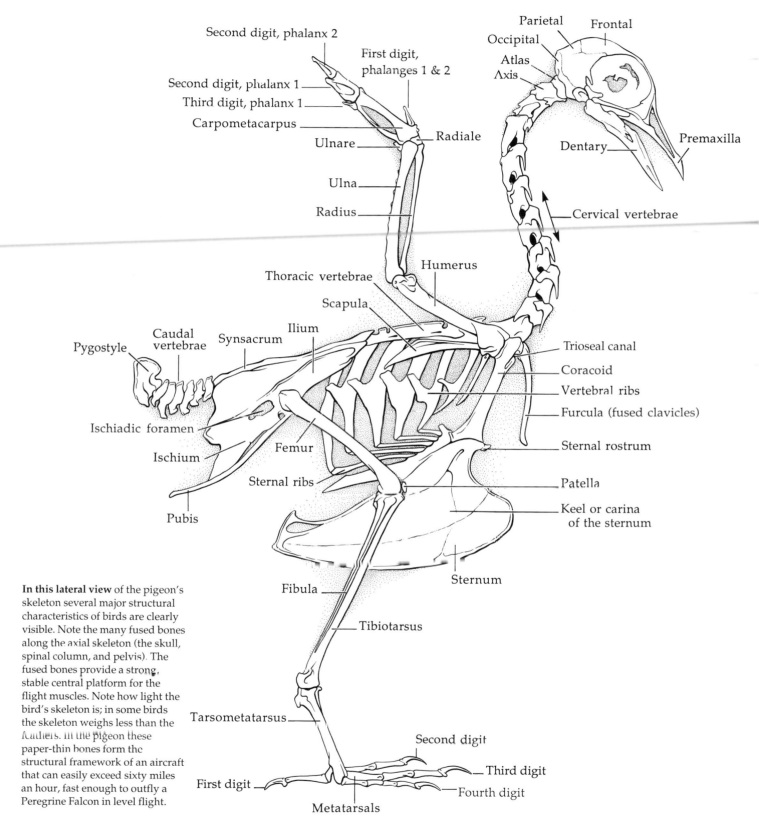

Second digit, phalanx 2

First digit,
phalanges 1 & 2

Second digit, phalanx 1

Third digit, phalanx 1

Carpometacarpus

Ulnare

Radiale

Ulna

Radius

Parietal   Frontal
Occipital
Atlas
Axis

Dentary

Premaxilla

Cervical vertebrae

Thoracic vertebrae

Humerus

Scapula

Ilium

Caudal
vertebrae

Synsacrum

Pygostyle

Trioseal canal

Coracoid

Vertebral ribs

Furcula (fused clavicles)

Sternal rostrum

Ischiadic foramen

Femur

Ischium

Patella

Sternal ribs

Keel or carina
of the sternum

Pubis

Sternum

In this lateral view of the pigeon's
skeleton several major structural
characteristics of birds are clearly
visible. Note the many fused bones
along the axial skeleton (the skull,
spinal column, and pelvis). The
fused bones provide a strong,
stable central platform for the
flight muscles. Note how light the
bird's skeleton is; in some birds
the skeleton weighs less than the
feathers. In the pigeon these
paper-thin bones form the
structural framework of an aircraft
that can easily exceed sixty miles
an hour, fast enough to outfly a
Peregrine Falcon in level flight.

Fibula

Tibiotarsus

Tarsometatarsus

Second digit

Third digit

First digit

Fourth digit

Metatarsals

Form and Function
The range of form in the avian skeleton

A side view of a ratite skeleton—an Emu of Australia. Note the relatively small sternum, which is typically shaped like a very shallow bowl, with no midline ridge or keel. In Emus the wing bones are almost vestigial and the whole pectoral girdle is much reduced in size. As in the other surviving flightless ratites—the African ostriches, the South American rheas, or the cassowaries of New Guinea—the Emu is a swift runner. Ratites usually defend themselves by rapid flight, but they can be very aggressive when necessary and are capable of inflicting dangerous wounds with their long, powerful legs.

Emu
(*Dromaius novaehollandiae*)

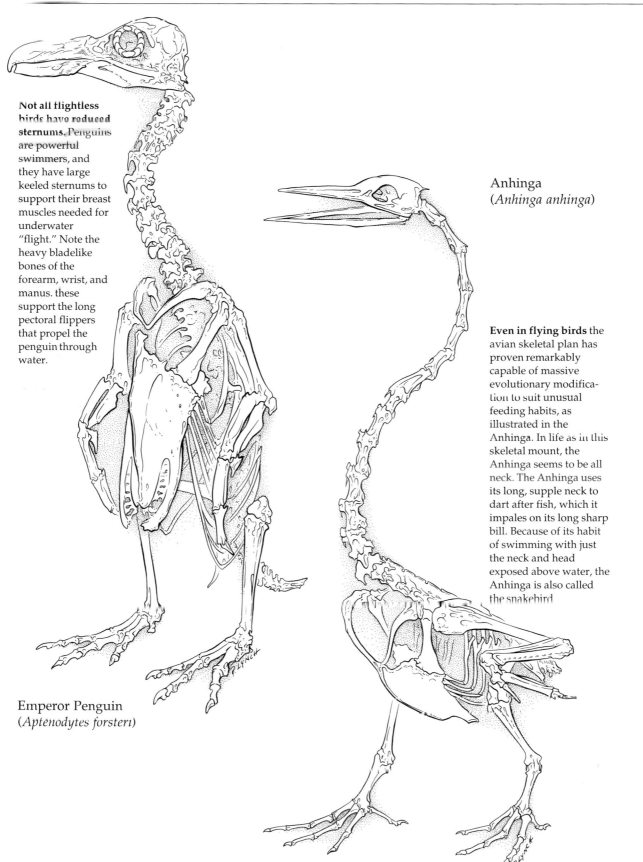

Not all flightless birds have reduced sternums. Penguins are powerful swimmers, and they have large keeled sternums to support their breast muscles needed for underwater "flight." Note the heavy bladelike bones of the forearm, wrist, and manus. these support the long pectoral flippers that propel the penguin through water.

Emperor Penguin
(*Aptenodytes forsteri*)

Anhinga
(*Anhinga anhinga*)

Even in flying birds the avian skeletal plan has proven remarkably capable of massive evolutionary modification to suit unusual feeding habits, as illustrated in the Anhinga. In life as in this skeletal mount, the Anhinga seems to be all neck. The Anhinga uses its long, supple neck to dart after fish, which it impales on its long sharp bill. Because of its habit of swimming with just the neck and head exposed above water, the Anhinga is also called the snakebird

2 BONES OF THE SKULL, TOP VIEW

The bones of the avian head are extensively fused to support and protect the head structures with a minimum of weight. The jaws are mere lightweight struts that support the beak and do not have to bear the weight and stress of chewing teeth. Note the huge spaces within the skull devoted to the eyes.

Although similar in overall plan to the skulls of other vertebrates, the avian skull shows several distinct differences:

1. The orbits are very large and are separated by a thin, bony membrane, the **mesethmoid,** or interorbital septum.

2. Teeth are absent. In their place, the horny bill sheath has evolved to take on various functions, sometimes shearing (as in parrots and hawks), crushing, or holding (as in the serrated edge of the merganser's bill).

3. The lower mandible of both birds and reptiles is composed of a complex fusion of five bones: the **prearticular, surangular, angular, splenial,** and **dentary.** The first four bones of the proximal lower mandible are so thoroughly fused that they are difficult to separate; the dentary bone forms most of the distal end of the lower mandible.

Observe the following structures

Frontals — Elongate, flat bones that form the major portion of the skull roof.

Nasals — Paired bones extending from the forward edge of the frontals.

Premaxillae — Two bones with processes that form the upper edge and tip of the upper mandible:
    a. The **nasal process** underlies the **culmen,** or upper ridge, of the beak.
    b. The **dentary processes** support the cutting edges of the beak.
    c. The **palatal process** forms part of the upper palate of the mouth.

The top view of the skull also shows the connecting portions of the quadrate bones.

Quadrates — Right and left bones connecting the lower mandible to the cranium; note how the quadrates articulate with the end of the zygomatic bar.

Zygomatic arch or **bar** — This distinctive lateral arch of the skull is made up of two bones:
    a. The **quadratojugal bones** form the posterior part of the zygomatic bar.
    b. The thin, toothpick-like **jugal bones** form most of the anterior portion of the zygomatic bar.

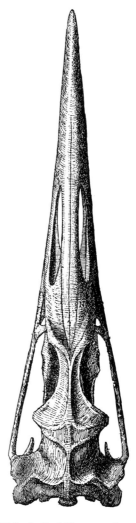

This skull of *Hesperornis regalis* **shows** how fast birds evolved into their present form. At a glance this skull looks strikingly similar to that of a present-day loon (Family Gaviidae). But the loon-like *Hesperornis* lived during the Cretaceous epoch, over ninety million years ago. A surprising number of modern bird groups were present at the dawn of the Cenozoic era: loons, grebes, pelicans, flamingos, and some shorebirds were among the groups present in remarkably modern form about sixty million years ago. After Heilmann 1927.

Superior View of the Skull
Rock Dove (*Columba livia*)

In looking at the skull you can see just how large the pigeon's eyes really are—they occupy most of the room within the skull. In life only a small portion of the eye is visible between the eyelids.

Dorsal View

Parietal region

Squamosal region

Frontal

Quadrate

Quadratojugal

Jugal
(zygomatic arch, or bar)

Frontal

Nasal

Maxilla

Frontal process
of the premaxilla

Maxillary process
of the premaxilla

The most prominent departure from the standard vertebrate skull plan is of course that birds have lost their teeth and the heavy jawbones needed to support teeth. The bill is used to reduce food to a size the bird can easily swallow. Once in the digestive tract, the food is "chewed" by the tough muscular gizzard section of the stomach.

Your skull specimen is likely to be from a mature bird, and thus the suture lines between the bones of the skull may be indistinct or even invisible. As birds age, such normally immobile joints as those in the skull gradually become calcified, fusing the bones in a continuous sheet. This gradual ossification of the joints between skull bones occurs in most vertebrates (including humans) and is often used as a means of determining the approximate age of animals. In birds the rate of ossification of the frontal bones is used to separate birds only several months old from older birds wearing similar plumage colors (see Skulling in chapter 12, Field Techniques).

References: Bellairs and Jenkin 1960;
Carroll 1988; Chamberlain 1943; Howard 1929.

3 | BONES OF THE SKULL, SIDE VIEW

Observe the following structures

Relocate the bones that were named in the previous side view of the skull.

Periotic capsule — A bulbous structure seen at the skull base near the articulation of the lower mandible with the skull; this structure encloses the inner ear.

Lacrimal — A small curved bone at the forward edge of the orbit.

Ethmoid — The thin bone at the anterior portion of the orbit; basically a continuation of the thin interorbital septum (the **mesethmoid**) that divides the two orbits.

Maxilla — Thin bones and processes that adjoin the processes of the nasal and premaxillary bones; the maxillas and premaxillas fuse to support the **tomia** (cutting edges) of the upper bill.

Note in this view the **quadrate bones** at the articulation point of the lower jaw and the skull, and the structure of the **jugal bone.**

The Lower Mandible

Each half of the lower mandible is formed by a fusion of five bones: prearticular, surangular, angular, splenial, and dentary.

Dentary — This forms most of the distal end of the lower mandible. The other four bones of the avian lower mandible are typically so fused that they are not easily distinguished. Note the **dentary foramen** at the midpoint of the lower mandible.

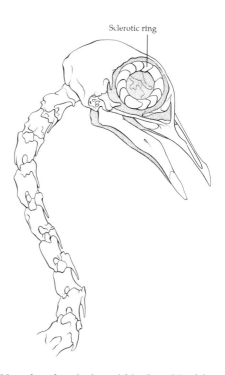

Sclerotic ring

Note the sclerotic ring within the orbit of the skull, supporting the globe of each eye. Birds share this unusual eye reinforcement with their dinosaurian ancestors. In birds the sclerotic ring reinforces the huge eyeball.

References: Bellairs and Jenkin, 1960; Carroll, 1988; Chamberlain, 1943.

Lateral View of the Skull
Rock Dove (*Columba livia*)

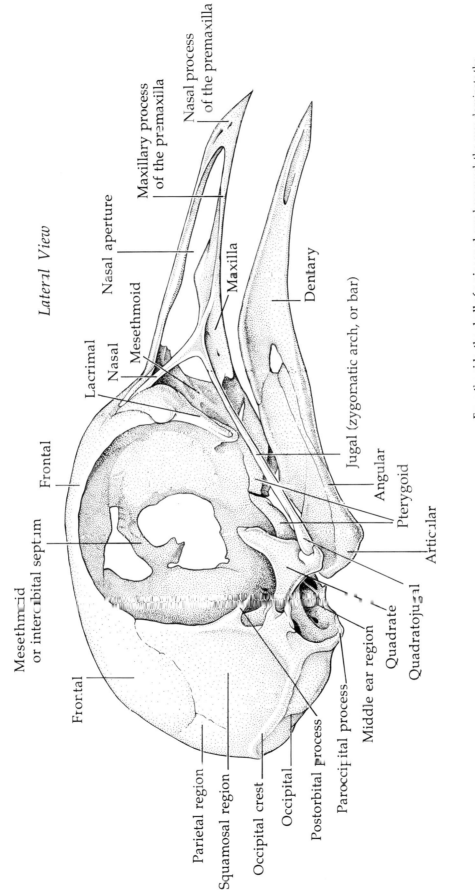

Lateral View

Nasal process
of the premaxilla

Maxillary process
of the premaxilla

Nasal aperture

Maxilla

Dentary

Lacrimal

Nasal

Mesethmoid

Frontal

Jugal (zygomatic arch, or bar)

Angular

Pterygoid

Articular

Mesethmoid
or interorbital septum

Quadratojugal

Quadrate

Middle ear region

Paroccipital process

Postorbital process

Occipital

Occipital crest

Squamosal region

Parietal region

Frontal

From the side, the skull of a pigeon shows how much the eyes dominate the avian head. To accommodate the large orbits the brain has been forced down and back into the occipital region and now tilts in the skull at an almost-45° angle. Note how open the bird's skull is. Composed largely of tiny struts and paper-thin sheets of bone, the structure is remarkably strong.

4 BONES OF THE SKULL, REAR VIEW

Observe the following structures

Foramen magnum — The large opening at the base of the skull. The spinal cord passes through this opening into the **cranium,** or brain case.

Occipital condyle — A rounded projection located just anterior to the foramen magnum. The first cervical vertebra of the neck (the **atlas**) articulates with the occipital condyle of the skull.

Occipital complex — The complex of four bones surrounding the foramen magnum.

Squamosals — The bones that wrap up and around the side of the head, forming much of the lower lateral walls of the skull and part of the posterior margins of the orbits.

Parietals — Square-shaped bones that form the posterior surface of the skull.

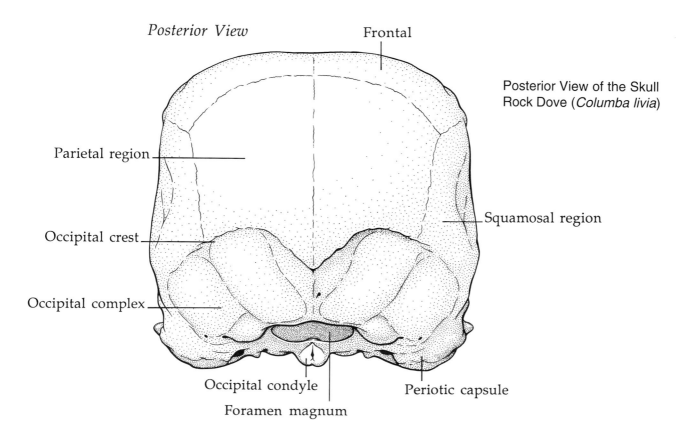

Posterior View

Frontal

Posterior View of the Skull
Rock Dove (*Columba livia*)

Parietal region

Squamosal region

Occipital crest

Occipital complex

Occipital condyle

Periotic capsule

Foramen magnum

References: Bellairs and Jenkin 1960; Carroll 1988; Chamberlain 1943.

5 BONES OF THE TONGUE

The bones of the tongue collectively are called the **hyoid apparatus**. The central portion of the hyoid apparatus is made up of three bones:

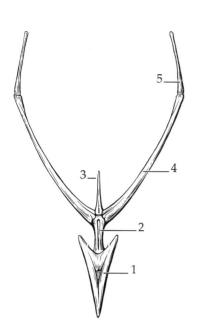

1. Paraglossale — An arrow-shaped bone projecting forward, forming the basic bone of the tongue.

2. Basihyal — The short central peg of bone on the midline of the hyoid apparatus.

3. Urohyal — The pointed bone that projects posteriorly from the basihyal bone.

Two long, thin horns of bone arise from the junction of the basihyal and basibranchial bones and project posteriorly. Each is composed of two bones:

4. Ceratobranchial — The heavier of the two bones composing each horn of the hyoid apparatus, articulating at the basihyal-basibranchial junction and projecting posteriorly.

5. Epibranchial — The thin spikes of bone that form the pointed tip of each horn of the hyoid apparatus.

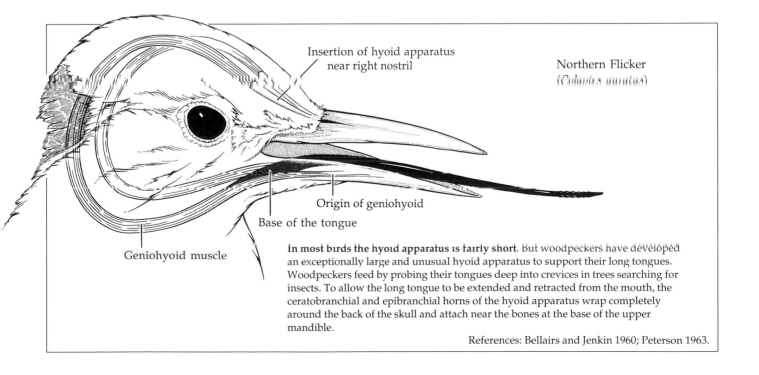

Insertion of hyoid apparatus near right nostril

Northern Flicker (*Colaptes auratus*)

Origin of geniohyoid

Base of the tongue

Geniohyoid muscle

In most birds the hyoid apparatus is fairly short. But woodpeckers have developed an exceptionally large and unusual hyoid apparatus to support their long tongues. Woodpeckers feed by probing their tongues deep into crevices in trees searching for insects. To allow the long tongue to be extended and retracted from the mouth, the ceratobranchial and epibranchial horns of the hyoid apparatus wrap completely around the back of the skull and attach near the bones at the base of the upper mandible.

References: Bellairs and Jenkin 1960; Peterson 1963.

6 THE BASE AND UNDERSIDE OF THE SKULL

Locate as many of the previously named bones as possible. In this view the **foramen magnum** and **occipital condyle** are clearly visible. Note the articulation points of the **quadrate** and **quadratojugal bones**.

Observe the following structures

Pterygoids — Short, thick T-shaped bones that connect the quadrate bones to the palatine bones.

Palatines — Long, thin, flat bones that make up the major portion of the hard palate.

Sphenoid complex — The flattened complex of four bones at the center of the skull base, just anterior to the occipital condyle.

Foramen magnum — The large opening on the inferior surface of the skull. The spinal cord passes through this opening into the cranium, or brain case.

Occipital condyle — Rounded projection located just anterior to the foramen magnum. The first cervical vertebra of the neck (the atlas) articulates with the occipital condyle of the skull.

Occipital complex — The complex of four bones surrounding the foramen magnum.

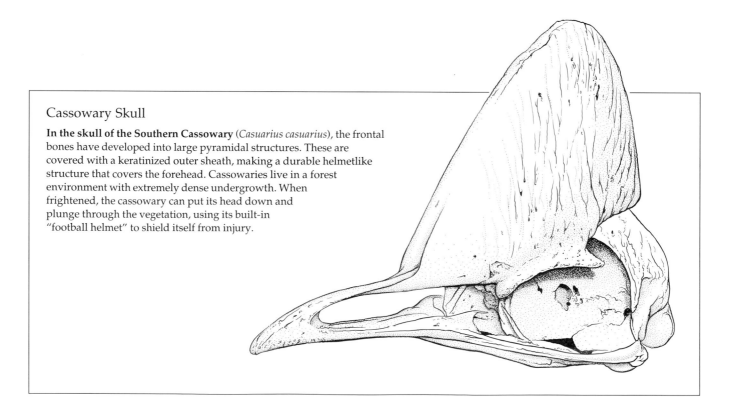

Cassowary Skull

In the skull of the Southern Cassowary (*Casuarius casuarius*), the frontal bones have developed into large pyramidal structures. These are covered with a keratinized outer sheath, making a durable helmetlike structure that covers the forehead. Cassowaries live in a forest environment with extremely dense undergrowth. When frightened, the cassowary can put its head down and plunge through the vegetation, using its built-in "football helmet" to shield itself from injury.

Inferior View of the Skull
Rock Dove (*Columba livia*)

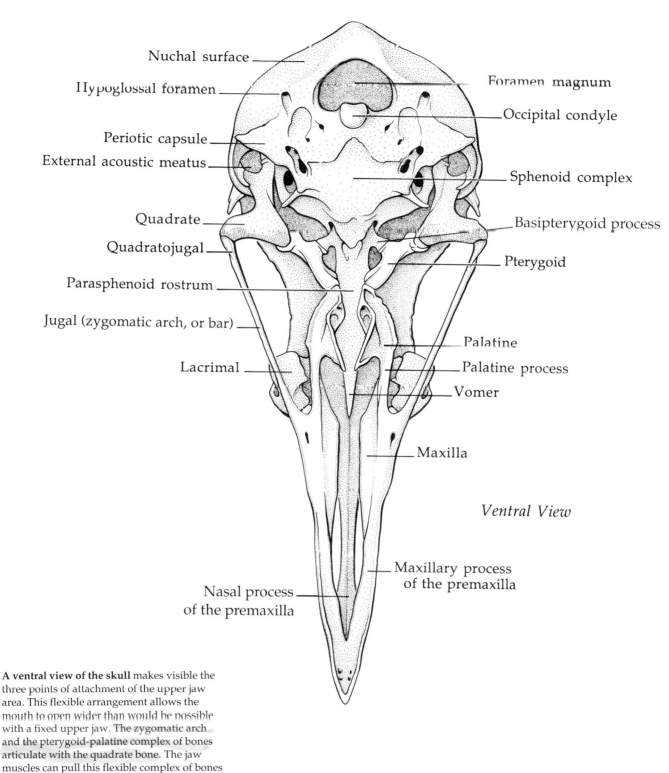

Nuchal surface

Hypoglossal foramen

Periotic capsule

External acoustic meatus

Quadrate

Quadratojugal

Parasphenoid rostrum

Jugal (zygomatic arch, or bar)

Lacrimal

Foramen magnum

Occipital condyle

Sphenoid complex

Basipterygoid process

Pterygoid

Palatine

Palatine process

Vomer

Maxilla

Ventral View

Maxillary process
of the premaxilla

Nasal process
of the premaxilla

A ventral view of the skull makes visible the three points of attachment of the upper jaw area. This flexible arrangement allows the mouth to open wider than would be possible with a fixed upper jaw. The zygomatic arch and the pterygoid-palatine complex of bones articulate with the quadrate bone. The jaw muscles can pull this flexible complex of bones upward to swallow large food items whole.

References: Bellairs and Jenkin 1960; Carroll 1988; Chamberlain 1943.

7 THE VERTEBRAL COLUMN AND RIB CAGE

The vertebrae are classified by their general position along the length of the spine, supporting the neck (**cervical region**), thorax (**thoracic region**), pelvic area (**sacral region**), and tail (**caudal region**). The avian vertebral column illustrates the widespread fusion of bones that strengthens the midline of the body and provides a stable platform for the pectoral and pelvic girdles.

Observe the following structures

Cervical vertebrae — Those vertebrae that lie between the skull and the first thoracic vertebra. (The first thoracic vertebra supports the first true vertebral rib.) The number of cervical vertebrae varies by species from thirteen to twenty-five, with longer-necked birds having more cervical vertebrae. Humans, by comparison, have seven cervical vertebrae. The first cervical vertebra articulates with the occipital condyle of the skull and is sometimes termed the **atlas**. The second cervical vertebra is sometimes referred to as the **axis**. In the pigeon several tiny **cervical ribs** may sometimes be seen near the base of the cervical spine articulating with the last two cervical vertebrae. Cervical ribs are very short and do not reach the sternum.

Thoracic vertebrae — Immediately posterior to the cervical spine are five fused thoracic vertebrae, each of which articulates with a large dorsal or **vertebral rib.**

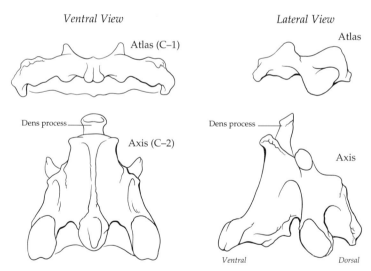

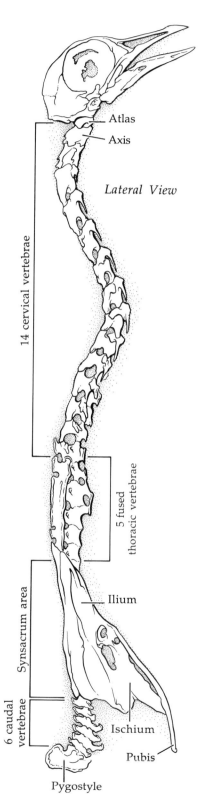

Lateral View

The Synsacrum

In birds the three lumbar (abdominal) vertebrae are fused with the sacral (pelvic area) and six of the caudal (tail) vertebrae to form a fused spinal column called the **synsacrum**. This synsacrum of vertebrae is extensively fused with the ilium and ischium of the pelvis to form one complete unit of bone.

Components of the Synsacrum
Sacral vertebrae — Seven vertebrae, three posterior to the thoracic vertebrae and four extensively fused within the synsacrum.
Caudal vertebrae — The six most anterior caudal vertebrae, fused within the synsacrum.

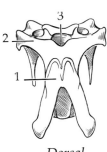

Dorsal

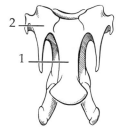

Ventral

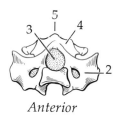

Anterior

The Tail

Caudal vertebrae — The six posterior caudal vertebrae are free of the synsacrum and extend posterior to it.

Pygostyle — A fusion of the final few caudal vertebrae; the pygostyle is a flat blade of bone that supports the musculature and connective tissues of the tail.

Vertebral Structure

Observe the following structures on one vertebra

1. Centrum — The main body of the vertebra; each centrum has a saddle-shaped articular surface on the cranial end of the centrum.

2. Transverse processes — The bony projections that surround the neural canal. In the cervical spine each transverse process is penetrated by the vertebral artery.

3. Neural canal — The open tube that runs through each vertebral body; it surrounds and protects the spinal cord.

4. Neural arch — The most dorsal part of the ring of bone that forms the neural canal.

5. Spinous process — The vertical ridge of bone projecting from the dorsal surface of each vertebra; the spinous processes are especially prominent in the thoracic spine, where they are fused. In the cervical spine they are reduced to a ridge line along the dorsal midline of each neural arch and do not project outward.

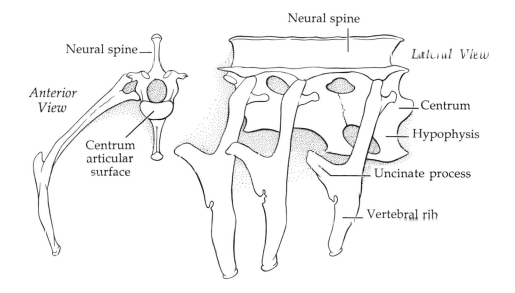

References: Bellairs and Jenkin 1960; Carroll 1988; Chamberlain 1943; Howard 1929.

8 THE THORAX AND STERNUM

Observe the following structures

Thoracic vertebrae — Immediately posterior to the cervical spine are five fused thoracic vertebrae, each of which articulates with a large dorsal or **vertebral rib.**

The Ribs

There are **seven ribs** on either side of the vertebral column. The first two **cervical ribs** are tiny and do not attach to the sternum. The next five ribs on each side are composed of two pieces: a **vertebral rib** articulating with the thoracic vertebrae of the spine and extending ventrally, and a **sternal rib** connecting the ventral end of the vertebral rib to the sternum. Note the **uncinate process** at the posterior margin of each vertebral rib. The uncinate processes form lateral braces between each vertebral rib, strengthening the rib cage against the forces of flapping flight.

The Sternum

The **sternum,** or breastbone, is one of the most specialized parts of the avian skeleton. There are two types of sterna:

Ratite — Certain flightless birds like the Ostrich and the rheas have a flat, bowl-like sternum without a distinctive keel or carina for attachment of enlarged breast muscles.

Carinate — In flying birds the sternum has been massively enlarged with a bony keel (the **carina**) to support the huge flight muscles of the breast. Most birds (even relatively reluctant flyers like the chicken) have carinate sterna.

Observe the following features of the sternum

Coracoid facets — Deep grooves that receive the base of the coracoid bones.

Rostrum — A small forked projection extending forward between the coracoidal facets.

Costal facets — Small depressions that receive the distal ends of the sternal ribs.

Lateral caudal process — Sometimes called the external lateral xiphoid process.

Sternal notch — The large opening just posterior to the lateral caudal process.

Medial caudal process — The posterior-pointing end of the sternum at the midline of the body.

Pneumatic foramen — A small opening in the superior surface of the sternum, on the midline just posterior to the rostrum, that communicates with the air sacs and respiratory system (see chapter 9, The Respiratory System).

Lateral View of the Thoracic Skeleton and Pelvis
Rock Dove (*Columba livia*)

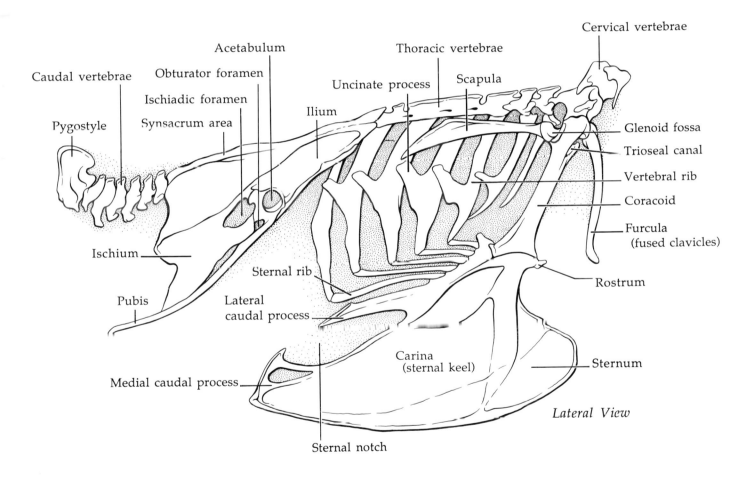

In flying birds the sternum has been massively enlarged with a bony keel to support the huge flight muscles of the breast. This flight-adapted sternum is called a **carinate sternum**. Flightless birds (ratites), such as the Ostrich, have no enlarged breast muscles and so have lost the keeled sternum. This reduced breast bone is called a **ratite sternum**.

9 THE PECTORAL GIRDLE

The pectoral girdle is perhaps the most distinctive region of the avian skeleton. This complex of bones provides the foundation for the mechanisms that direct force from the flight muscles to the wings, braces the body against the powerful action of the flight muscles, and anchors the wings to the body.

The pectoral girdle is formed from four major bones: sternum, coracoid, scapula, and clavicle. The **coracoid** forms the central pillar of the system, linking the **sternum** of the breast to the shoulder area and bracing the pectoral girdle against the downward pull of the breast musculature. The posteriorly pointing **scapula** forms the second leg of the pectoral "tripod." Its long, flat blade provides surface area for the attachment of the shoulder muscles. This allows the pectoral girdle some freedom to move over the surface of the rib cage while remaining firmly anchored to the body. The two **clavicles** (together forming the **furcula,** or "wishbone") provide a long surface area at the front of the chest for attachment of the breast muscles. In birds the furcula probably acts like a flexible spring suspension system, bending under the downward force of the flight muscles and springing back as the wing is raised (Jenkins et al. 1988).

Although the avian shoulder girdle shares structural and functional characteristics with the human shoulder, there is one crucial difference, which is the key to understanding how birds raise and lower the wing in powered flight. It is easy to see how birds pull the wing downward: the large pectoralis muscle of the breast inserts directly on the lower surface of the humerus bone of the wing, and pulls directly down. But there is no corresponding mass of muscle at the top of the shoulder to pull the wing up. A small but crucial flight adaptation is visible in the bones of the pectoral girdle. Look closely at the upper end of the coracoid for the **trioseal canal,** a tiny tunnel just anterior to the articulation of the scapula and the coracoid. The tendon of the supracoracoideus muscle (sometimes called the **pectoralis minor** or **subclavius**) of the breast goes up through this trioseal canal and over the top of the shoulder to insert on the upper surface of the humerus. The trioseal canal forms a pulley, allowing a muscle below the wing (the **supracoracoideus**) to pull the wing up. This unique arrangement keeps the muscle mass that powers flapping flight below the bird's center of gravity and relieves the shoulder muscles above the wing from the difficult task of raising the wing (George and Berger 1966; Ostrom 1976a). The supracoracoideus is most critical in the take-off phase of flight, being somewhat less important in recovering the wing from the downstroke once the bird is airborne. Experiments have shown that pigeons whose supracoracoideus muscle tendons were cut could not take off on their own but once launched were able to fly well without it (Sy 1936).

Observe the following structures

Coracoid — The largest bone of the shoulder joint, forming a near-vertical column between the coracoid facets of the sternum and the shoulder joint. Note the tiny **trioseal canal** at the superior end of the coracoid.

Scapula — A long, blade-like bone running horizontally back from the shoulder joint; on an articulated skeleton note how the forked anterior end of the scapula also forms part of the trioseal canal.

Trioseal canal — This is formed by a complex of three bones: coracoid, scapula, and clavicle. The canal starts as a small vertical groove along the medial surface of the coracoid. At the head of the coracoid, two projections of bone form a nearly complete tunnel. The forked anterior end of the scapula forms the upper end of the trioseal canal. This canal allows the tendon of the **supracoracoideus** (pectoralis minor) breast muscle to pass upward along the inner surface of the coracoid, over the top of the scapula and shoulder joint, and out onto the upper surface of the humerus.

Glenoid fossa — The cup-shaped depression formed by the scapula and coracoid. The rounded condyle at the proximal end of the humerus fits into the glenoid fossa, forming a ball-and-socket joint that permits the humerus to rotate freely around the shoulder joint.

Clavicles — The two long, thin bones that make up the V-shaped **furcula,** or wishbone, of the breast.

Humerus — Though not a part of the trioseal canal, the humerus is the arm bone upon which the supracoracoideus muscle attaches as it emerges from the trioseal canal. The tendon of the supracoracoideus passes up thorough the trioseal canal and out onto the upper surface of the humerus, allowing the supracoracoideus muscle *below* the wing to raise it in flight.

Pectoral Girdle
Rock Dove (*Columba livia*)

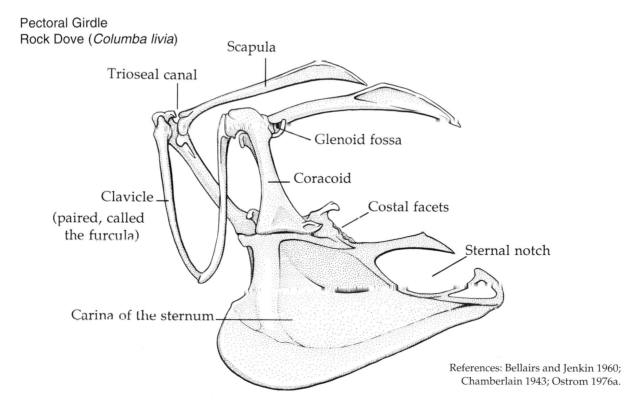

Scapula

Trioseal canal

Glenoid fossa

Coracoid

Clavicle
(paired, called
the furcula)

Costal facets

Sternal notch

Carina of the sternum

References: Bellairs and Jenkin 1960;
Chamberlain 1943; Ostrom 1976a.

10 THE BONES OF THE WING

The wing is built on the classic vertebrate plan of humerus (upper arm bone), radius and ulna (forearm bones), carpal (wrist) bones, and five digits. The avian humerus is fairly short and stout relative to the total length of the wing. This is because the main flight muscles of the breast attach only to the humerus, and it must bear large forces acting upon it as the wing moves up and down in flapping flight. Two large crests of bone at the near (proximal) end of the humerus show the attachment points of the flight muscles, the pectoralis and the supracoracoideus.

The radius and ulna form the support for the mid-wing (forearm). Note the series of small bumps along the trailing edge of the ulna (largest forearm bone). These bumps show where the secondary feathers of the wing attach directly to the ulna.

Most of the distal (outer) wing bones have been extensively fused and modified to strengthen the wing and form a support for the outer flight feathers (primaries). In particular, the bones of the manus (hand) have been fused so that they are hardly recognizable as the classic vertebrate five-fingered hand of humans.

In the avian wrist, two small carpal (wrist) bones lie between the radius and ulna (forearm bones) and the fused bones of the hand. These are the radiale (near the radius) and the ulnare (near the ulna). The rest of the carpal and metacarpal (hand) bones are fused into a blade-like structure called the **carpometacarpus**. In most birds the carpometacarpus has a small oval fenestra (window) in it, though in some birds the hole is closed by a paper-thin layer of bone. The fourth and fifth digits of the avian hand have been lost and are not visible in adult birds. At the distal end of the carpometacarpus the fused remnants of the second and third digits may be seen. The phalanges of the first digit (digital bones are called **phalanges**, singular, **phalanx**) can be seen at the wrist joint. This first digit is sometimes called the **pollex** or **thumb** and supports the feathers of the alula, a small "second wing" that helps the bird control airflow over the wing in flight.

Observe the following structures

Trioseal canal — A small notch in the upper end of the coracoid near the articulation of the humerus and the scapula. The trioseal canal is critical to the functioning of the wing in flight, as it forms the "pulley" that allows the supra-coracoideus muscle of the breast to pass a tendon up and over the shoulder joint and onto the upper surface of the humerus. Thus a muscle *below* the wing is able to *raise* the wing in flight (see sections on the Pectoral Girdle and Flight).

Humerus — The short, thick arm bone that articulates with the **glenoid fossa,** a cup-shaped cavity formed by the scapula and coracoid of the pectoral girdle. Note the **pneumatic foramen** near the proximal end of the humerus.

Radius — The slender, *straight* anterior bone of the forearm, articulating with distal end of the humerus and with the carpal bones of the manus.

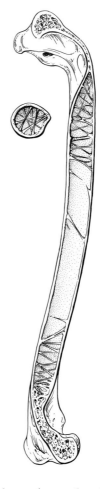

A balance of strength and rigidity, as seen in the sectioned humerus of a Herring Gull (*Larus argentatus*). Both long and short sections through the humerus are shown here. Note the internal struts of bone within the pneumatic cavity, further strengthening the shaft of the bone. Birds smaller than the Herring Gull tend to show less of a pneumatic cavity in the humerus, and some aquatic birds show few or no pneumatic cavities at all.

References: Bellairs 1964; Bellairs and Jenkin 1960; Chamberlain 1943; Howard 1929; Van Tyne and Berger 1976.

Superior View of the Left Wing Skeleton
Rock Dove (*Columba livia*)

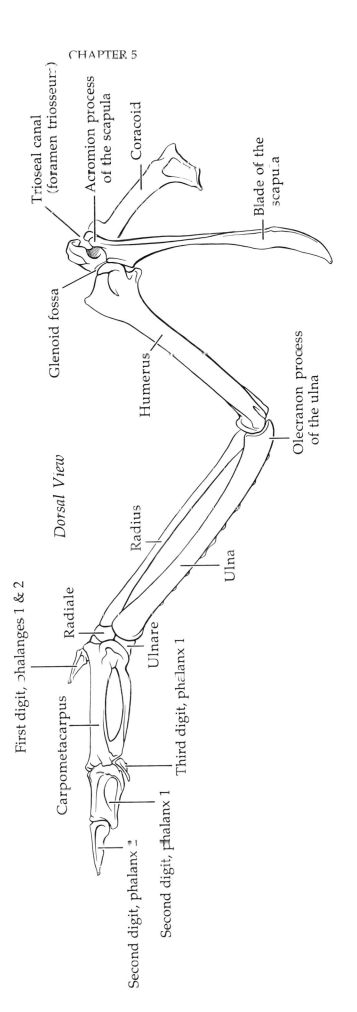

Trioseal canal
(foramen triosseum)

Acromion process
of the scapula

Coracoid

Blade of the
scapula

Glenoid fossa

Humerus

Olecranon process
of the ulna

Dorsal View

Radius

Ulna

Radiale

First digit, phalanges 1 & 2

Ulnare

Carpometacarpus

Third digit, phalanx 1

Second digit, phalanx 2

Second digit, phalanx 1

The wing plan seen from above is remarkably simple, yet it is structured for maximum efficiency and light mass. The flight muscles of the breast attach to the short, powerful humerus, which bears the main stresses during both the upstroke and the downstroke. The radius and the heavier ulna support the mid-wing area. As in the human forearm, the two bones are designed to allow some twisting (pronation and supination) of the wing during flight. Note the row of tiny dimples along the trailing edge of the ulna; these bony knobs show where the secondaries are fixed to the wing along the ulna. Farther out on the wing the primaries attach to the carpometacarpus and the phalanges of the second and third digits.

Ulna — The stout, curved posterior bone of the forearm, articulating with the distal end of the humerus and with the carpal bones of the manus. Note the small, bony **quill knobs** along the posterior margin of the ulna. They are the attachment points for the secondaries.

The Manus

Carpals — The two small, square bones called the **radiale** and **ulnare,** situated between the distal ends of the radius and ulna and the carpometacarpus complex.

Metacarpals — The first, second, and third metacarpals persist and are fused into a single blade-like structure called the **carpometacarpus**.

Digits and phalanges — Three digits (1, 2, 3) persist in the avian manus:

> **Digit 1** arises at the junction of the radiale and the carpometacarpus near the wrist, where it supports the feathers of the alula. It is composed of two fused phalanges.

> **Digit 2** arises at the distal end of the carpometacarpus. Its three phalanges form the flattened blade-like outer tip of the wing bones that project from the body of the carpometacarpus. The first two phalanges are fused, while the third phalanx projects distally from them.

> **Digit 3** is composed of one phalanx projecting from the posterior end of the junction between the carpometacarpus and the phalanges of digit 2.

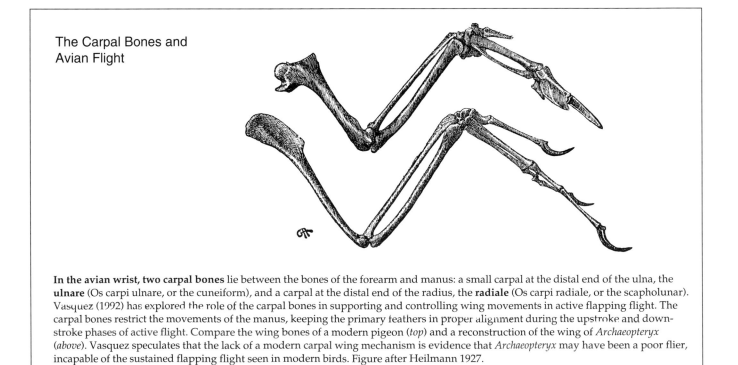

The Carpal Bones and
Avian Flight

In the avian wrist, two carpal bones lie between the bones of the forearm and manus: a small carpal at the distal end of the ulna, the **ulnare** (Os carpi ulnare, or the cuneiform), and a carpal at the distal end of the radius, the **radiale** (Os carpi radiale, or the scapholunar). Vasquez (1992) has explored the role of the carpal bones in supporting and controlling wing movements in active flapping flight. The carpal bones restrict the movements of the manus, keeping the primary feathers in proper alignment during the upstroke and down-stroke phases of active flight. Compare the wing bones of a modern pigeon (*top*) and a reconstruction of the wing of *Archaeopteryx* (*above*). Vasquez speculates that the lack of a modern carpal wing mechanism is evidence that *Archaeopteryx* may have been a poor flier, incapable of the sustained flapping flight seen in modern birds. Figure after Heilmann 1927.

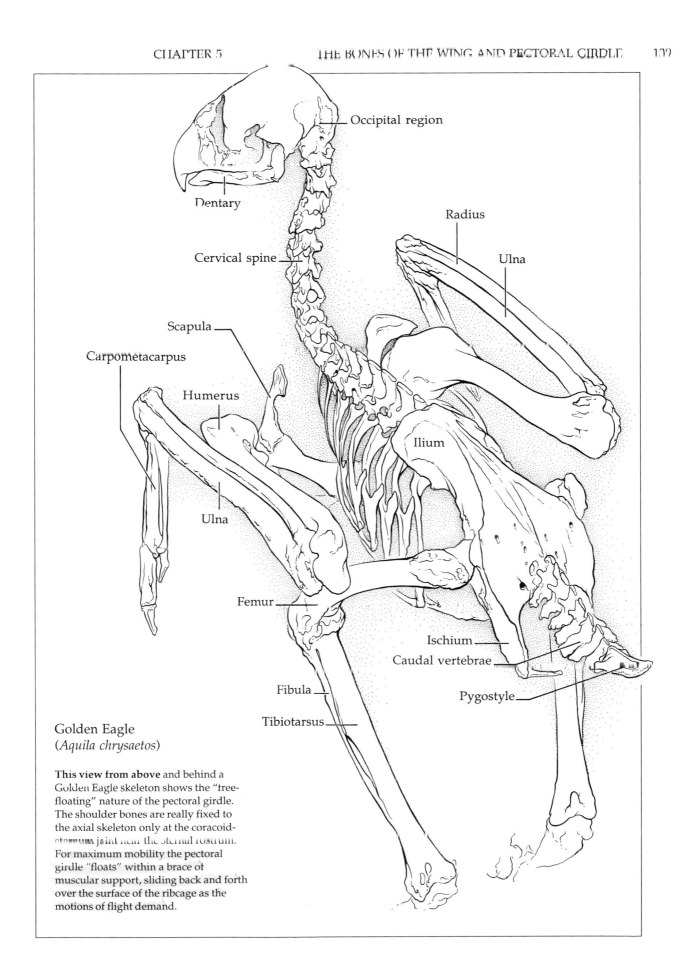

Occipital region

Dentary

Radius

Ulna

Cervical spine

Scapula

Carpometacarpus

Humerus

Ilium

Ulna

Femur

Ischium

Caudal vertebrae

Fibula

Pygostyle

Tibiotarsus

Golden Eagle
(*Aquila chrysaetos*)

This view from above and behind a
Golden Eagle skeleton shows the "free-
floating" nature of the pectoral girdle.
The shoulder bones are really fixed to
the axial skeleton only at the coracoid-
sternum joint near the sternal rostrum.
For maximum mobility the pectoral
girdle "floats" within a brace of
muscular support, sliding back and forth
over the surface of the ribcage as the
motions of flight demand.

11 | THE PELVIC GIRDLE

The three principal bones of the avian pelvis (ilium, ischium, and pubis) are extensively fused with the spinal vertebrae to form a strong, lightweight structure. In birds the last three thoracic (chest) vertebrae are fused with the sacral (pelvic) and most of the caudal (tail) vertebrae to form a fused spinal column called the **synsacrum**. This synsacrum of vertebrae is extensively fused with the ilium to form one complete unit of bone. These fused pelvic bones give the bird a strong central platform for the attachment of the leg and tail musculature. The pelvis also gives support and protection to the abdominal contents and large areas for attachment of the leg and abdominal musculature.

Observe the following structures

Synsacrum — The sacral (pelvic) and six of the caudal (tail) vertebrae together form a fused spinal column called the synsacrum. See also the section on the vertebral column.

Ilium — The largest pelvic bone, forming most of the anterior and lateral surface of the pelvis. Note the concave cup of the **acetabulum,** the articulation point of the femur bone of the leg. The concave surface of the acetabulum and the convex proximal head of the femur form a classic ball-and-socket joint.

Ischium — This bone forms the posterior and lateral corners of the pelvis. The ilium and ischium are completely fused, with no distinct borders in adult birds.

The fused ilium and ischium are sometimes called the **innominate bone.** The pelvis has two openings, or foramen: the **ischiadic foramen** is the larger opening just posterior to the acetabulum joint of the ilium. Just inferior to the ischiadic foramen is the smaller **obturator foramen**.

Pubis — A long, thin bone fused with the ilium just posterior to the acetabulum and running posteriorly along the inferior edge of the ischium.

Caudal vertebrae — Six of the twelve caudal vertebrae are free (not fused into the synsacrum) and project posteriorly from the pelvis and synsacrum.

Pygostyle — A fusion of the final few caudal vertebrae. The pygostyle is a flat blade of bone that supports the musculature and connective tissues of the tail.

Pelvic Girdle
Rock Dove (*Columba livia*)

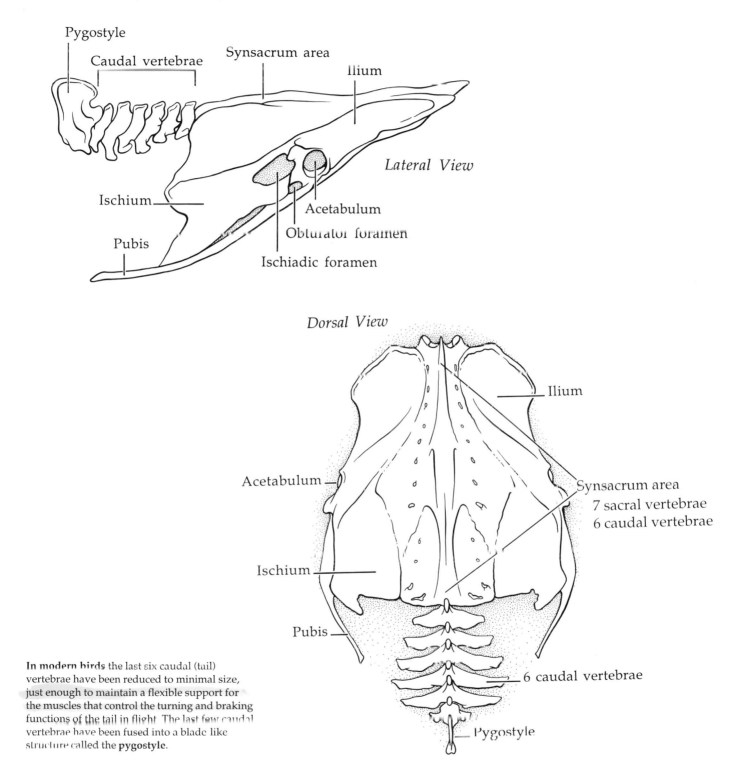

In modern birds the last six caudal (tail) vertebrae have been reduced to minimal size, just enough to maintain a flexible support for the muscles that control the turning and braking functions of the tail in flight. The last few caudal vertebrae have been fused into a blade like structure called the **pygostyle**.

References: Baumel 1988; Bellairs 1964; Bellairs and Jenkin 1960; Chamberlain 1943; Van Tyne and Berger 1976.

12 THE BONES OF THE LEG AND FOOT

The bones of the upper leg are laid out in a fairly conventional vertebrate pattern, with a femur (thighbone) and a tibia and fibula (leg bones), but the lower (distal) bones of the leg and foot have been extensively fused and modified to aid in absorbing the shock of takeoff, landing, and running. The bird's tibia (lower leg bone) has been fused with some of the upper bones of the foot to form the **tibiotarsus**. A small toothpick-like remnant of the fibula parallels the tibiotarsus along its length.

The lower bones of the foot are also fused and stretched to form the extended foot area known as the **tarsometatarsus**. Extending the bones of the foot in this way confers advantages to both flying and nonflying birds. In running animals and flying birds this extension of the foot bones gives the leg an extra lever length, useful to a limited extent for absorbing the force of footfalls and landings. The extended bones also give birds extra leverage from the length of the leg when they jump upward or push in the power stroke of running. (Think how much more effective a human runner's power "kick" would be if the foot were twice as long.)

In living birds the thigh region of the leg and the "knee" joint between the femur and tibia are covered by feathers and so are rarely visible, and this can lead to confusion over which way the bird's "knee" points. It points forward, of course, just as the human knee does. The potential for confusion arises because the only leg joint that is usually visible is the backward-pointing "heel" joint between the tibiotarsus and the fused bones of the tarsometatarsus. Birds have the same basic leg plan and joints as humans. Most birds, like the pigeon shown here, have four toes, three facing forward (digits 2, 3, and 4) and one pointing backward, digit 1. Digit 1 is homologous to the human big toe and in birds is often called the **hallux**. This classic three-in-front, one-behind "chicken foot" pattern of the toes is termed an anisodactyl foot. See chapter 3, Topography, for a review of avian foot structure.

Because they actually walk on their toes and not on all of the foot bones (as humans do), birds are called **digitigrade** (toe walkers). The bones and flesh of the foot are covered with a tough plating of scales and papillae that strengthen the foot and provide resistance from the wear and tear of walking and perching.

Observe the following structures

Femur — The largest leg bone. The femur articulates with the ilium of the pelvis at a cup-shaped structure on the ilium called the **acetabulum**.

Tibiotarsus — The main bone of the lower leg. The small **fibula** of the lower leg may be seen running parallel to the tibiotarsus on its lateral side.

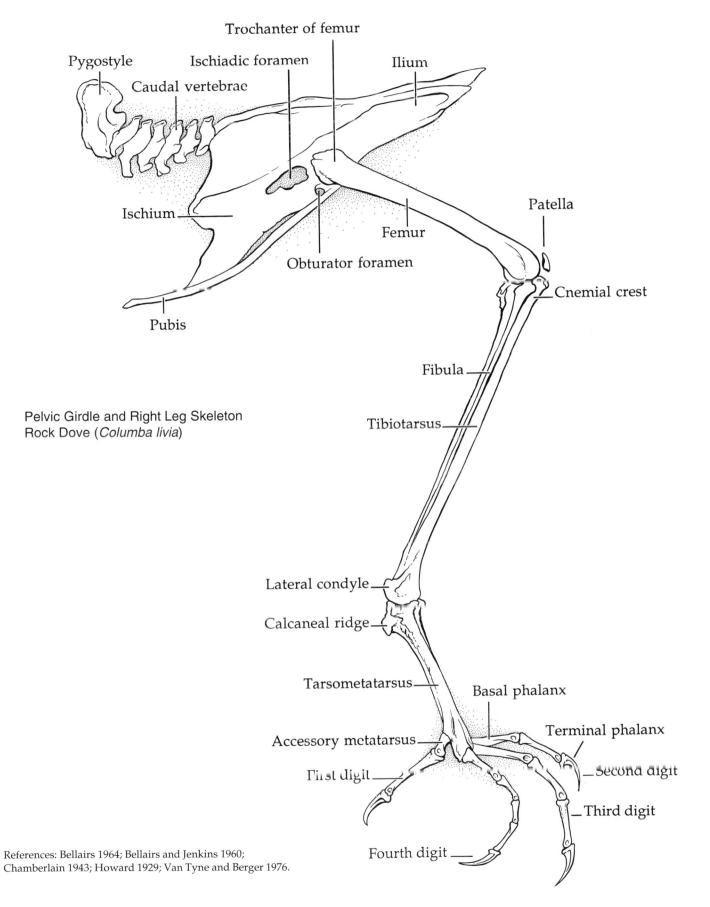

Pygostyle
Caudal vertebrae
Ischiadic foramen
Trochanter of femur
Ilium
Ischium
Obturator foramen
Pubis
Femur
Patella
Cnemial crest
Fibula
Tibiotarsus

Pelvic Girdle and Right Leg Skeleton
Rock Dove (*Columba livia*)

Lateral condyle
Calcaneal ridge
Tarsometatarsus
Accessory metatarsus
First digit
Basal phalanx
Terminal phalanx
Second digit
Third digit
Fourth digit

References: Bellairs 1964; Bellairs and Jenkins 1960;
Chamberlain 1943; Howard 1929; Van Tyne and Berger 1976.

Tarsometatarsus — The second, third, and fourth **metatarsals** fuse to form this bone. The first metatarsal is a rudimentary bone at the base of digit 1, and the fifth metatarsal is absent. This bone and region of the leg may also be called the **tarsus,** particularly in descriptions of external avian anatomy. See the leg and foot section of chapter 3, Topography.

Digits — Most birds have four toes, or digits, on the foot. The first digit, the **hallux,** projects to the rear in most birds. The forward toes are numbered in series from the innermost toe (2) through the outermost toe (4). The fifth digit is entirely absent in birds.

   Digit 1 — Note the small first metatarsal at the base of this digit. Beyond the metatarsal the first digit has two phalanges, the proximal phalanx and the terminal phalanx.

   Digit 2 — This digit has three phalanges.

   Digit 3 — This digit has four phalanges.

   Digit 4 — This digit has five phalanges.

The Skeleton   *Chapter Worksheet*

1. In what principal ways has the avian skeleton been adapted for flight?

2. How are the avian skeleton and respiratory system related?

3. View a drawing or cast of *Archaeopteryx*. List the differences in the skeletons of *Archaeopteryx* and modern birds.

Chapter Worksheet, continued

4. What are the principal regions of the axial skeleton? What are the principal regions of the appendicular skeleton?

5. Discuss the significance of the following structures:

   a. Trioseal canal

   b. Pygostyle

   c. Synsacrum

   d. Pollex

References

Bellairs, A. d'A. 1964. Skeleton. In A. L. Thompson, ed., *A new dictionary of birds*. New York: McGraw-Hill.

Bellairs, A. d'A., and C. R. Jenkin. 1960. The skeleton of birds. In A. J. Marshall, ed., *Biology and comparative physiology of birds*. 2 vols. New York: Academic Press, 1:241–300.

Bock, W. J. 1974. The avian skeletomusculature system. In D. S. Farner, J. R. King, and K. C. Parkes, eds., *Avian biology*. 8 vols. New York: Academic Press, 4:119–257.

Chamberlain, F. 1943. *Atlas of avian anatomy*. Mich. Agric. Exp. Sta. Mem. Bull. 5.

Feduccia, A. 1980. *The age of birds*. Cambridge: Harvard University Press.

Gould, S. J. 1982. The telltale wishbone. In S. J. Gould, *The panda's thumb*. New York: W. W. Norton, 267–77.

Heilmann, G. 1927. *The origin of birds*. New York: D. Appleton. Reprint, 1972; New York: Dover.

Howard, H. 1929. The avifauna of Emeryville shellmound. *Univ. Calif. Publ. Biol.* 32(2):301-94.

King, A. S., and J. McLelland, eds. 1970–89. *Form and function in birds*. 4 vols. New York: Academic Press.

Lucas, A. M., and P. R. Stettenheim. 1972. *Avian anatomy: integument*. Agric. Handb. 362:1–340. Washington, D.C.: U.S. Government Printing Office.

Nero, R. W. 1951. Pattern and rate of cranial ossification in the House Sparrow. *Wilson Bull.* 63(1):84-88.

Ostrom, J. H. 1976a. Some hypothetical stages in the evolution of avian flight. In S. L. Olsen, ed., *Collected papers in avian paleontology. Smithsonian Contr. Paleobiol.* 27:1–21.

Pough, F. H., J. B. Heiser, and W. N. McFarland. 1989. *Vertebrate life*. 3d ed. New York: Macmillan.

Van Tyne, J., and A. J. Berger. 1976. *Fundamentals of ornithology*. 2d ed. New York: John Wiley and Sons.

Belted Kingfisher
(*Megaceryle alcyon*)

Hovering flight is perhaps the single most strenuous flight maneuver birds can execute. Only a few other species like the American Kestrel (*Falco sparverius*) and hummingbirds are capable of true sustained hovering flight, although many birds of open tundra and grasslands like the Rough-legged Hawk (*Buteo lagopus*) also hover for short periods when hunting. Hovering Belted Kingfishers are a common sight in coastal, riverine, and lake environments throughout North America. Kingfishers prefer to hunt their fish from branches above the water surface, but in treeless areas they shift to hovering while looking for fish below the water surface. As soon as it spots a fish, the kingfisher collapses its wings and dives headfirst into the water.

The Musculature

6 Virtually every aspect of a bird's anatomy reflects the demands of flight, and analyzing the characteristics of the muscles can be a powerful aid in determining how the bird's metabolic energy has been invested through evolution to fit its ecological niche. The configuration of any animal's muscles is one of the clearest visual clues to its environmental adaptation. At a glance, even a casual observer can see that the most important muscular avian adaptations are the massive flight muscles of the breast. In a flying bird the breast muscles dominate the body so completely that they may account for almost a third of total body mass. Try to imagine an adult human with breast muscles weighing fifty pounds and you get some sense of the extraordinary metabolic and structural investment required to sustain powered flight. In such forest birds as grouse, woodcock, and tinamous that escape predators by using short, explosive bursts of flight, the breast muscles can account for 40 percent of body weight (Hartman 1961). Yet many important muscle groups seen in mammals are greatly reduced in birds. Most mammals (including humans) have an extensive network of dorsal muscles running along the length of the spine to support the spine and give the back flexible strength. But anyone who has ever carved a chicken or a turkey can tell you that there is almost no meat on a bird's back. In birds the axial skeleton of spine, ribcage, and pelvis is so tightly connected that no elaborate muscular support is necessary; the dorsal musculature has accordingly been reduced to a few thin muscle groups that stabilize the shoulder area and fix the wing to the body.

Birds exhibit a variety of adaptations in the structure of their body musculature. Flightless ratites such as the African Ostrich and the rheas of South America have huge leg muscles that allow them to run at speeds approaching those of the fastest four-legged mammals. Their breast muscles are much reduced and are not supported by a sternal carina. Conversely, in some flying birds such as the swifts (Family Apodidae) the legs and leg muscles have become almost vestigial. And not all flightless birds have minimal breast muscles. The penguins have retained large sternal keels and heavy breast muscles because they use their wings to "fly" in a medium even more demanding than the air—the ocean. The shape and configuration of a bird's body are under continuous dynamic pressure from evolution and the environment. The occasions when these pressures are suddenly removed illustrate this point. For example, rail species (Family Rallidae) that have reached isolated oceanic islands often lose the ability to fly after relatively few generations on the island: if no ground predators are present, the environmental pressure to fly is removed and the ability to fly quickly becomes a expensive waste of energy that the rails can do without.

Most birds have given over their pectoral limbs completely to flight and have thus lost the ability to use those limbs for grooming and holding food, though wings are used in display, thermoregulation, and as "flippers" by such aquatic birds as penguins. In compensation, birds have long, supple necks controlled by a marvelously complex neck musculature. Most birds, such as songbirds or pigeons, normally hold their necks in an S position except when stretching to groom themselves. The true length of the neck thus is usually not apparent due to the ruff of feathers surrounding the junction of the neck and the thorax. Without their long necks birds would be unable to perform the crucial job of grooming their feathers to maintain their flightworthiness and insulating ability. Neck mobility is crucial if the bird is to remain aware of its environment, and the neck muscles have become correspondingly complex (Berger 1960; Chamberlain 1943; George and Berger 1966; Hudson 1937; Hudson and Lanzillotti 1955, 1964).

1 OVERVIEW OF THE MUSCLES

Note the following major features of the pigeon's musculature:

Ventral concentration of the muscle mass

Note how the masses of the musculature are primarily ventral, below the wings and the center of gravity. This creates a particularly stable flight configuration in birds, not unlike the small, high-winged Piper Cub or Cessna aircraft popular in civil aviation. The rigid thoracic, lumbar, and synsacral spinal areas do not require extensive back musculature for support, so there are few muscles along the spine. In general, most muscles are concentrated near the center of gravity, sending long tendons out away from the body to control the movement of the outer wings, legs, and feet.

Emphasis on the pectoral girdle

Not surprisingly, the pectoral girdle shows the most distinctive muscular developments; it is so heavily modified for flight that the pectoral limbs can no longer be used for grooming or holding food. The breast muscles are huge, comprising as much as one-fifth or more of the pigeon's weight. Although the wing bones are extensively fused and reduced, however, the wing muscles are numerous and complex. This light but amazingly strong network of tendon slips and wire-thin muscles gives the avian wing far more subtle aerodynamic control than is available to human aircraft pilots.

Complexity of the neck musculature

Birds have such large eyes in proportion to their skulls that they have largely lost the ability to move the eyeball within the eye socket of the skull. To restore their ability to see the world around them, and to compensate for the loss of the pectoral limbs for grooming and other daily functions, birds have evolved long, flexible necks with complex muscular systems surrounding the cervical verte-brae.

References: Berger 1960; Chamberlain 1943; Lucas and Stettenheim 1972.

Lateral View of the Musculature
Rock Dove (*Columba livia*)

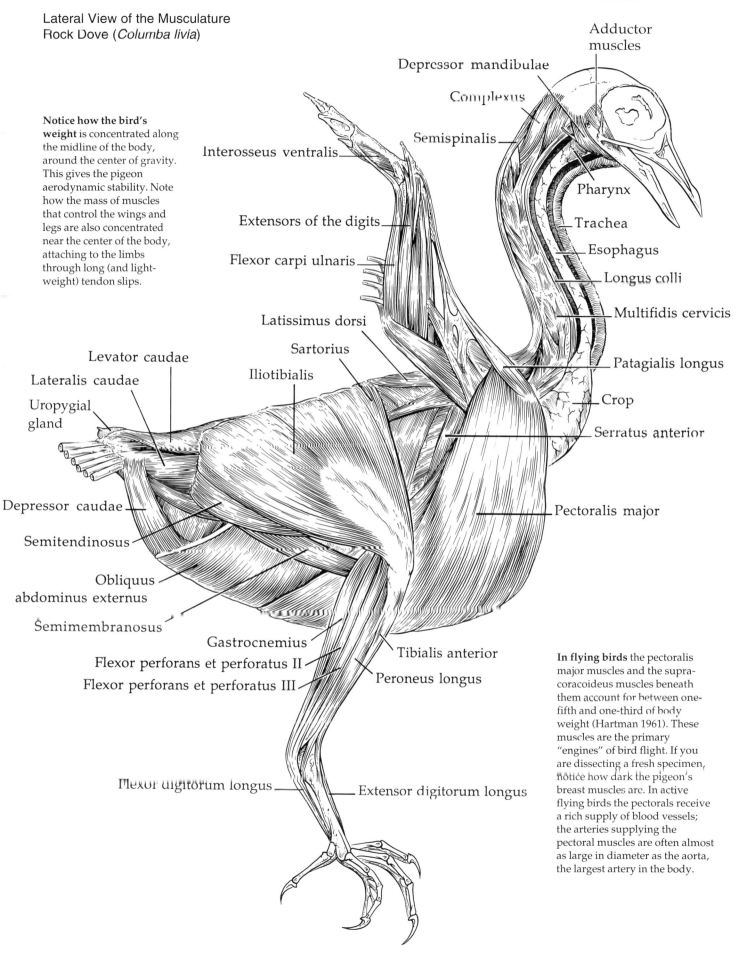

Notice how the bird's weight is concentrated along the midline of the body, around the center of gravity. This gives the pigeon aerodynamic stability. Note how the mass of muscles that control the wings and legs are also concentrated near the center of the body, attaching to the limbs through long (and lightweight) tendon slips.

Adductor muscles

Depressor mandibulae

Complexus

Semispinalis

Interosseus ventralis

Pharynx

Trachea

Extensors of the digits

Esophagus

Flexor carpi ulnaris

Longus colli

Multifidis cervicis

Latissimus dorsi

Sartorius

Patagialis longus

Iliotibialis

Crop

Levator caudae

Serratus anterior

Lateralis caudae

Uropygial gland

Depressor caudae

Pectoralis major

Semitendinosus

Obliquus abdominus externus

Semimembranosus

Gastrocnemius

Tibialis anterior

Flexor perforans et perforatus II

Peroneus longus

Flexor perforans et perforatus III

Flexor digitorum longus

Extensor digitorum longus

In flying birds the pectoralis major muscles and the supra-coracoideus muscles beneath them account for between one-fifth and one-third of body weight (Hartman 1961). These muscles are the primary "engines" of bird flight. If you are dissecting a fresh specimen, notice how dark the pigeon's breast muscles are. In active flying birds the pectorals receive a rich supply of blood vessels; the arteries supplying the pectoral muscles are often almost as large in diameter as the aorta, the largest artery in the body.

2 MUSCLES OF THE HEAD AND NECK

As in most animals, the configuration of a bird's head and jaw muscles often matches its preferred diet. Such birds as parrots and finches have a relatively heavy jaw musculature to enable them to crack the seeds that comprise most of their diet. More omnivorous birds such as pigeons and carnivorous birds such as hawks have less need for heavy jaw muscles. Birds do not chew their food and use their beaks only for tearing, cracking, and holding onto food. The chewing needs of the digestive system have been taken over by the stomach muscles of the gizzard. For this reason, even in larger birds the jaw and throat muscles tend to be relatively delicate.

Birds have evolved long, flexible necks to allow the head and eyes to be moved rapidly and effectively to survey their world. Birds also need their long necks to reach and groom their feathers, which require constant maintenance. In especially long-necked aquatic groups, such as cormorants, herons, and cranes, the neck is an essential element in hunting, functioning along with the beak as a lightning-fast system for catching prey.

The muscles that surround the cervical spine are numerous and complex, adapted both to support the cervical vertebrae and to provide precise control over movement. The borders between cervical muscle groups are often difficult to distinguish. There are many thin overlapping sheets of muscles, and most of the individual muscles, except the long dorsal and ventral midline muscles, are no longer than a third of an inch.

Observe any group of pigeons for a few minutes and you will see the delicate dermal muscles of the head and neck in action, as male birds puff their neck feathers up in display behavior. These paper-thin muscle sheets are usually dissected away from the neck as the skin is removed, so look carefully for them on the inner skin surface.

Observe the following structures

Adductor muscles — These muscles lie just below and posterior to the orbit.

Depressor mandibulae — This muscle lowers the mandible and opens the mouth.

Complexus — This broad sheet of muscle just below the occipital region helps extend the head when the bird looks upward.

Multifidis cervicis, Intertransversales — These tiny muscles run between each cervical vertebrae, providing lateral movement and dorsal flexion of the neck. They are often difficult to tell apart.

Longus colli — This long band of muscle covering the ventral surface of the cervical spine flexes the neck ventrally (forward and down).

Semispinalis capitis — This dorsal band of muscle along the back of the neck flexes the neck dorsally (upward and back).

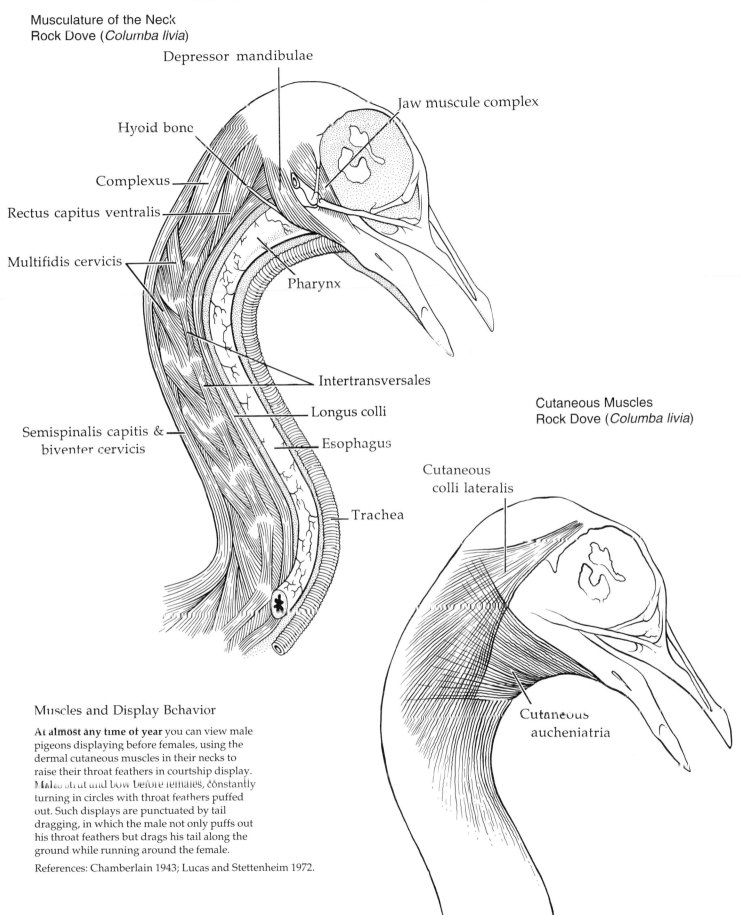

Musculature of the Neck
Rock Dove (*Columba livia*)

Depressor mandibulae

Jaw muscule complex

Hyoid bone

Complexus

Rectus capitus ventralis

Multifidis cervicis

Pharynx

Intertransversales

Cutaneous Muscles
Rock Dove (*Columba livia*)

Longus colli

Semispinalis capitis &
biventer cervicis

Esophagus

Cutaneous
colli lateralis

Trachea

Cutaneous
aucheniatria

Muscles and Display Behavior

At almost any time of year you can view male pigeons displaying before females, using the dermal cutaneous muscles in their necks to raise their throat feathers in courtship display. Males strut and bow before females, constantly turning in circles with throat feathers puffed out. Such displays are punctuated by tail dragging, in which the male not only puffs out his throat feathers but drags his tail along the ground while running around the female.

References: Chamberlain 1943; Lucas and Stettenheim 1972.

Muscles of the Head and Neck

Name	Origin	Insertion	Action
Jaw muscle complex (adductor muscles)	Temporal, parietal, and frontal regions of the skull.	Lateral surface of the mandible.	Raises the lower jaw, closes the mouth.
	Orbit wall, temporal area.	Lateral, medial, and dorsal surfaces of the mandible.	Raises the lower jaw, closes the mouth.
Depressor mandibulae	Occipital region of the skull.	Ventral and lateral surfaces of the mandible.	Lowers the lower jaw, opens the mouth.
Cutaneous colli lateralis	Supraorbital region of the frontal bone.	Skin and feathers of the neck.	Erects the feathers of the neck.
Cutaneous aucheniatria	Mastoid process of the temporal bone.	Ventral median raphe at midline of the throat.	Tenses the skin, erects the feathers of the throat area.
Complexus	Transverse processes of the 2d through 5th cervical vertebrae.	Occipital crest of the skull.	Extends the head dorsally and laterally.
Rectus capitus ventralis	Transverse processes of the 4th and 5th cervical vertebrae.	Lateral surface of the occipital region of the skull.	Flexes the head ventrally and laterally.
Multifidis cervicis	Transverse processes of the cervical vertebrae.	Spinous processes of the cervical vertebrae.	Flexes the neck dorsally and laterally.
Intertransversales	Transverse processes of the cervical vertebrae.	Transverse processes of the cervical vertebrae.	Flexes the neck dorsally.
Longus colli	Ventral surface of the thoracic vertebrae.	Ventral area of the 1st and 2d cervical vertebrae.	Flexes the head and neck ventrally.
Semispinalis capitis, or **Semispinalis cervicis**	Supraspinous ligaments above the 1st and 2d thoracic vertebrae.	Spinous processes of the cervical vertebrae.	Dorsal flexion of the neck.
Biventer cervicis	Supraspinous ligaments of the cervical vertebrae.	Nuchal surface of the occipital bone of the skull.	Dorsal flexion of the head.

Muscles of the Thorax and Abdomen

Name	Origin	Insertion	Action
Pectoralis major, or Pectoralis superficialis	Carina of the sternum, furcula, and sternal ribs.	Proximal ventral surface of the humerus.	Primary depressor of the wing in downstroke.
Supracoracoideus, or Pectoralis minor, Subclavius	Carina of the sternum, furcula, and sternal ribs.	Proximal dorsal surface of the humerus.	Primary elevator of the wing in upstroke.
External oblique	Ventral borders of the ilium and pubis, 7th rib.	Median raphe between the sternum and the anus.	Compresses the abdomen.
Internal oblique	Ventral borders of the ilium and pubis.	Caudal border of the last vertebral and sternal rib.	Compresses the abdomen.
Transversus abdominus	Ventral borders of the ilium and pubis, caudal vertebral ribs.	Median raphe between the sternum and the anus.	Compresses the abdomen.
Serratus anterior, or Serratus dorsalis	Lateral surfaces of the distal vertebral ribs.	Ventral surfaces of the scapula.	Aids in inspiration, expands the chest.
Serratus anterior, or Serratus dorsalis	Lateral surfaces of the distal vertebral ribs.	Ventral surfaces of the scapula.	In inspiration, expands the chest.
External intercostals	Posterior border of the vertebral ribs.	Anterior border of the vertebral ribs.	In expiration, contracts the chest.
Costosternalis	Sternal ribs.	Median surface of the sternum.	In inspiration, expands the chest.
Scaleneus	Transverse processes of the 11th and 12th cervical vertebrae.	Anterior lateral surface of the cervical ribs.	Draws cervical ribs forward.
Sternocoracoideus	Sternum and cervical ribs.	Medial surface of the coracoid	Draws the ribcage forward, and to the coracoid laterally.

References: Berger 1960; Chamberlain 1943; Lucas and Stettenheim 1972.

3 | MUSCLES OF THE THORAX AND ABDOMEN

The muscles of the thorax support the rib cage, power respiration, and fix the pectoral girdle to the body. In birds the relation of the pectoral girdle to the chest wall musculature is crucial, for the massive muscles of the breast power both the downstroke and the upstroke of flight (Berger 1960; George and Berger 1966). On your specimen note the lack of muscle mass along the dorsal surface of the thorax. The muscles here are thin sheets, playing relatively small roles in fixing the wing to the body and powering flight. The fusion of the thoracic and lumbar areas of the vertebral column make extensive dorsal musculature unnecessary. In fact, if birds carried more weight above the aerodynamic center of gravity, they would be less stable in flight (Burton 1990; Rüppell 1975; Terres 1987).

The muscles of the abdominal wall support and protect the viscera. Note the orientation of the muscle fibers in the thin sheets of oblique abdominal muscles. They cross each other at almost 90° angles, much like the fibers in a bias-ply automobile tire. This criss-crossing of muscle fibers for added strength is common in vertebrate anatomy, and in fact the same arrangement of abdominal muscle fibers is found in human external and internal oblique muscles (Chamberlain 1943; George and Berger 1966).

Observe the following structures

Pectoralis major — The largest muscle of the bird's body, it provides most of the force (downstroke) for flapping flight. The pectoralis is shown cut in the figure at the right, to reveal the antagonistic supracoracoideus muscle lying beneath it.

Supracoracoideus — This muscle is the principal muscle in raising the wing in flapping flight. The supracoracoideus sends its tendon up along the anterior-medial surface of the coracoid bone (note the groove in the coracoid), through the **trioseal canal** formed from the junction of the coracoid, scapula, and humerus at the shoulder joint, and onto the dorsal surface of the humerus. Through this natural "pulley" system, a muscle below the wing is able to raise the wing in flight. This arrangement avoids the need to have a large antagonistic muscle above the shoulder acting against the pectoralis, where its weight would render the body less stable in flight.

External and internal oblique — These thin sheets of muscle, and the transversus muscle lying below and dorsal to them, hold the abdominal viscera in place.

Intercostal muscles — These muscles, which lie between the vertebral ribs, strengthen the rib cage and aid in respiration.

Serratus anterior — These muscles also aid the bird in expanding and contracting its chest during respiration.

References: Berger 1960; Chamberlain 1943; George and Berger 1966; Lucas and Stettenheim 1972.

Thoracic and Abdominal Musculature
Rock Dove (*Columba livia*)

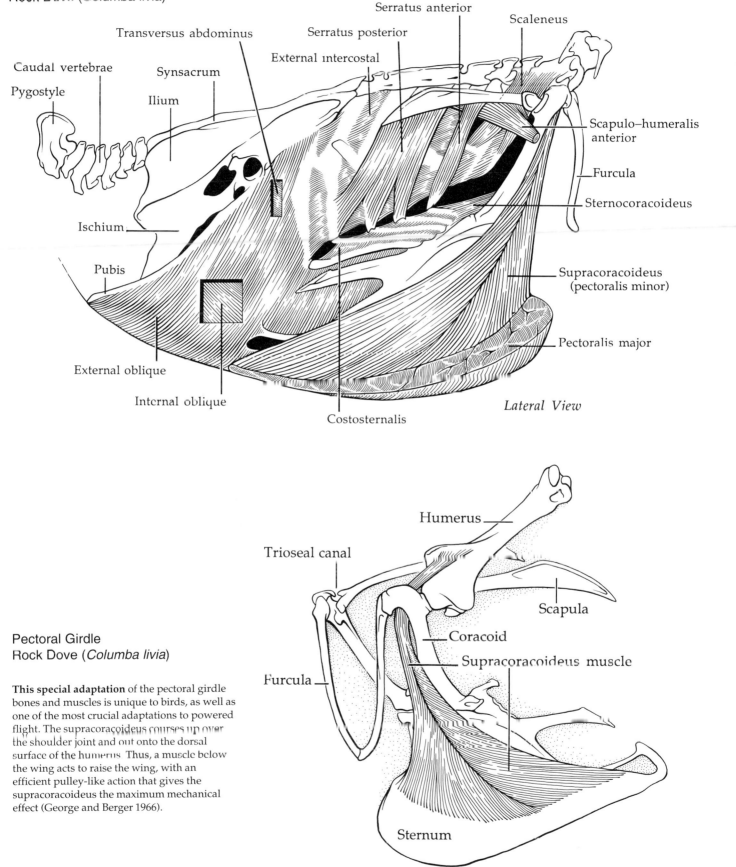

Lateral View

Pectoral Girdle
Rock Dove (*Columba livia*)

This special adaptation of the pectoral girdle bones and muscles is unique to birds, as well as one of the most crucial adaptations to powered flight. The supracoracoideus courses up over the shoulder joint and out onto the dorsal surface of the humerus. Thus, a muscle below the wing acts to raise the wing, with an efficient pulley-like action that gives the supracoracoideus the maximum mechanical effect (George and Berger 1966).

4 | VENTRAL MUSCLES OF THE WING

The avian wing is a marvel of biological engineering. Light yet amazingly strong, the wing is an exquisite example of nature's economy of form and function. After you have completed your dissection, hook a finger under the tendons of the mid-wing of a fresh specimen and pull on them. Those wiry tendons can sustain great loads in flight—in some cases hundreds of times their weight—without breaking. We take the humble pigeon for granted, but pigeons are remarkably accomplished fliers, capable of out-flying most birds of prey in a level chase.

The wing muscles—by nature thin, tough bands of similar-looking tissue—can be difficult to dissect. Do your best to locate the major muscle groups illustrated here, but note especially how the muscle groups relate to the overall wing plan. Try expanding and collapsing a fresh wing specimen, noting how the muscles and bone form a strong, stable wing core. Note the actions of the flexor and extensor groups, and see how each type of muscle is located along the wing. The details of the locations and actions of these muscles are not simply nature's way of torturing ornithology students. Take a few minutes to study the motions of the wing, and the details of the musculature will be much easier to learn and understand.

Observe the following structures

Patagialis longus — A tough band of fiber that actually forms the leading edge of much of the inner wing.

Patagialis brevis, Patagialis accessorius — These muscles support the anterior patagium of the wing, tensing it under flight stresses. Many authors refer to the patagialis accessorius as the **biceps slip**.

Biceps brachii — A flexor of the wing running just ventral to the humerus along its length. The muscle originates from two heads, the first from the area of the glenoid fossa (shoulder joint), the second from the proximal end of the humerus.

Triceps brachii — This muscle flexes the shoulder joint and extends the forearm. The muscle originates from two separate heads, the first (**scapulotriceps**) from the scapula just posterior to the glenoid fossa, the second (**humerotriceps**) from the proximal end of the humerus. Both heads insert on the olecranon process of the proximal ulna.

Expansor secondariorum — This muscle is unique to birds, running from the axilla to the forearm to expand the secondary feathers. Unlike other striated skeletal muscles of the wing, the expansor secondariorum is composed of smooth muscle fibers. Be very careful when dissecting the area around this muscle. Use blunt tools to avoid destroying it. The muscle appears as a thin band of silvery tendon-like tissue, running along the fibers of the postpatagium toward the secondary feather quills. If you are working on a fresh specimen, grab the muscle with a clamp and pull it to see the expanding action on the secondary feather quills of the wing.

References: Berger 1960; Chamberlain 1943; George and Berger 1966; Hudson and Lanzillotti 1955, 1964.

Ventral Musculature of the Breast and Wing
Rock Dove (*Columba livia*)

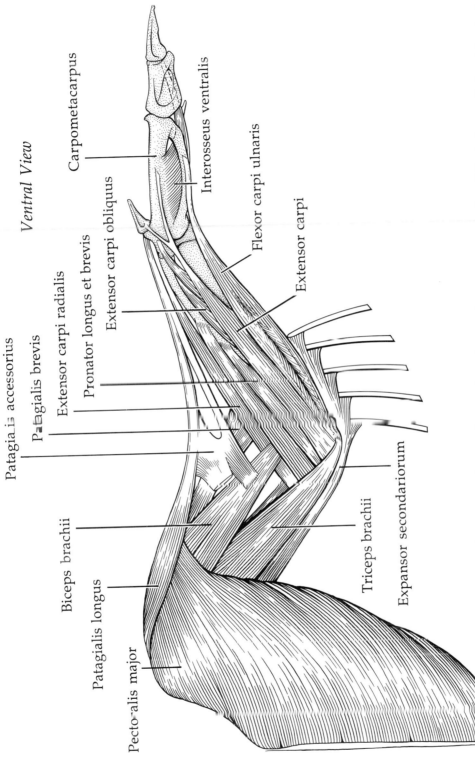

Ventral View

Carpometacarpus

Interosseus ventralis

Flexor carpi ulnaris

Extensor carpi

Patagia is accessorius

Patagialis brevis

Extensor carpi radialis

Pronator longus et brevis

Extensor carpi obliquus

Biceps brachii

Patagialis longus

Pectoralis major

Triceps brachii

Expansor secondariorum

The muscles of the wing can be intimidating to dissect because they are numerous and small. Study the overall plan of the wing before you begin, and note the actions of the muscles as you move, expand, and collapse the wing. This is especially easy to see in a fresh specimen, such as a fresh whole chicken wing.

There is an overall logic to the wing musculature, which can be considered in three parts: arm, forearm, and manus. The humerus (arm) area fixes the wing to the body and transmits the actions of the flight muscles to the rest of the wing. The muscles on the humerus primarily control the gross position of the radius and ulna (forearm); the anterior muscles around the humerus flex the wing; and the muscles posterior to the humerus extend the forearm. The muscles anterior to the forearm extend the manus and flight feathers, while those posterior to the forearm flex the manus and flight feathers. The forearm muscles also twist the whole outer wing in pronation and supination, which are essential motions in powered flight. The tiny muscles around the manus help flex and extend the primary remiges.

Muscles of the Pectoral Girdle and Wing

Name	Origin	Insertion	Action
Pectoralis major, or **Pectoralis superficialis**	Carina of the sternum, furcula, and sternal ribs.	Proximal ventral surface of the humerus.	Primary depressor of the wing in downstroke.
Supracoracoideus, or **Pectoralis minor, Subclavius**	Carina of the sternum, furcula, and sternal ribs.	Proximal dorsal surface of the humerus.	Primary elevator of the wing in upstroke.
Patagialis longus	Distal coracoid bone and pectoralis muscle.	Cranial (forward) surface of the carpal bones.	Flexes elbow, extends the carpus, tenses patagium of wing.
Biceps brachii	Near the glenoid fossa and proximal head of the humerus.	Proximal anterior surface of the radius.	Flexes elbow, extends carpus, tenses patagium of wing.
Patagialis accessorius, or **Biceps slip**	Biceps and axillary border of the pectoralis muscle.	Patagialis longus.	Tenses the patagium.
Triceps brachii, in two parts—long and short heads	From the neck of the scapula and proximal humerus.	Olecranon process of the proximal ulna.	Flexes the shoulder, extends the elbow.
Expansor secondariorum	Tendon from the coracobrachialis.	Follicles of the secondary feathers.	Expands the secondary feathers.
Deltoideus	Proximal clavicle.	Deltoid tuberosity of the (proximal) humerus.	Flexes shoulder, rotates wing outward.
Teres major	Lateral caudal surfaces of scapula bone.	Proximal ventral surface of the humerus.	Elevates and adducts the wing.
Latissimus dorsi	Supraspinous ligament of the thoracic spine.	Proximal surface of the humerus.	Adducts and flexes the wing, moving it backward and dorsally.
Coracobrachialis	Coracoid and sternal ribs.	Proximal dorsal surface of the humerus.	Depresses the wing.
Extensor carpi radialis	Lateral distal humerus.	Dorsal surface of the 2d metacarpal and carpometacarpus.	Flexes the elbow, extends the wing.

References: Berger 1960; Chamberlain 1943; George and Berger 1966; Hudson and Lanzillotti 1955, 1964.

More Muscles of the Pectoral Girdle and Wing

Name	Origin	Insertion	Action
Extensor carpi ulnaris	Ventral surface of the distal ulna.	Ventral surface of the 2d metacarpal and carpometacarpus.	Extends the carpus (outer) area of the wing.
Extensor carpi obliquus	Ventral surface of the medial ulna.	Dorsal surface of the 2d metacarpal.	Extends and rotates the wing inward.
Pronator longus et brevis	Medial epicondyle of the (distal) humerus.	Medial surfaces of the radius.	Pronates the manus and outer wing.
Flexor carpi radialis	Medial epicondyle of the (distal) humerus.	Carpometacarpus.	Supinates the manus, flexes the elbow, helps erect the secondaries.
Flexor carpi ulnaris	Medial epicondyle of the (distal) humerus.	Ventral surface of the carpometacarpus.	Flexes the manus and outer wing.
Anconeus	Lateral epicondyle of the humerus.	Proximal ventral surface of the ulna.	Extends the elbow.
Ulnaris lateralis	Lateral epicondyle of the humerus, olecranon process of the ulna.	Proximal ventral surface of the carpometacarpus.	Flexes and adducts the manus and outer wing.
Extensor of digits 2 and 3, or **Extensor communis**	Lateral epicondyle of the humerus.	1st phalanx of digits 2 and 3.	Flexes the elbow, extends digits 2 and 3.
Interosseus ventralis	Ventral surfaces of the carpometacarpus bone.	1st and 2d phalanges of digit 2.	Flexes digit 2.

Ventral Musculature of the Breast and Wing
Rock Dove (*Columba livia*)

5 DORSAL MUSCLES OF THE WING

Note the overall plan of the wing musculature and tendons. The muscle masses in birds are generally located as close as possible to the center of gravity. The muscles send long, tough tendons out along the bones of the wing, and these tendons transfer the action of the muscle and help hold the wing together under the stresses of flight. The more massive breast and shoulder muscles fix the wing to the body and provide power to the wing through the humerus. Beyond the elbow the muscles are reduced to tiny slips, for the laws of leverage dictate that every ounce of mass out away from the shoulder translates into many ounces of lever force on the humerus as the bird flies. The fine muscles of the outer wing act to expand the manus of the wing and the primary feathers out to full flight position. In flight they also twist (pronate or supinate) the wing to control the exact attitude of the wing in the airstream to control turning and gliding.

Observe the following structures

Deltoideus — This sheet of muscle covers the dorsal area of the shoulder and acts much as the human deltoid does, pulling the arm up and to the rear. Note that the deltoid is much too small a muscle to raise the wing alone. It is not massive enough to be a proper antagonist to the large pectoralis muscle of the breast. Without the unique trioseal canal–supracoracoideus arrangement, sustained powered flight would be impossible. *Archaeopteryx* did not have a trioseal canal and must have relied on the dorsal muscles of the shoulder (such as the deltoideus) to recover the wing in flapping flight. *Archaeopteryx* also lacked a sternal carina, and its pectoral muscles were probably small. These observations have led some researchers to propose that *Archaeopteryx* must have been at best a modest flier.

Latissimus dorsi — In quadrupedal mammals and bipedal primates, this muscle covers much of the dorsal thorax and abdomen; running mammals and primates have flexible spines that need a great deal of muscular bracing. But the avian spinal column is so extensively fused and stiffened that birds do not need heavy back muscles for support. In birds, the latissimus dorsi and other dorsal muscles are reduced to small muscles that support the shoulder joint from above. Follow the latissimus dorsi from its origin at the dorsal midline lateral to its insertion on the posterior surface of the humerus.

Extensor carpi ulnaris, Extensor carpi radialis — These two muscles of the forearm extend the manus area of the wing, and thereby the primary flight feathers attached to the manus. Follow the tendons of these extensors out along the anterior edge of the wing and onto the carpometacarpus and phalanges of the wingtip. If you have a fresh specimen, pull on either muscle to see how each draws the carpometacarpus out away from the body into a fully extended position.

Flexor carpi ulnaris — This is the main flexor muscle of the mid-wing. Follow the belly of this muscle along the trailing edge of the wing and onto the surface of the carpometacarpus. Pull it to see how the flexor pulls in the outer wing and helps fold the wing into the resting position.

References: Berger 1960; Chamberlain 1943; George and Berger 1966; Hudson and Lanzillotti 1955, 1964.

Dorsal Musculature of the Wing
Rock Dove (*Columba livia*)

Dorsal View

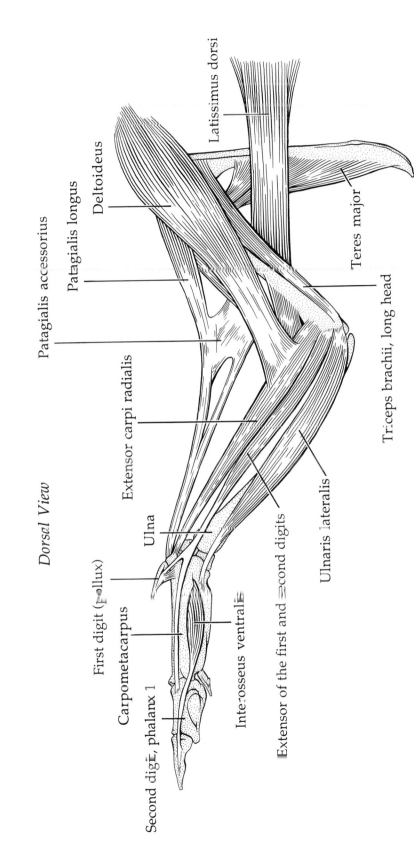

Latissimus dorsi

Deltoideus

Patagialis longus

Patagialis accessorius

Teres major

Triceps brachii, long head

Extensor carpi radialis

Ulna

First digit (pollux)

Carpometacarpus

Second digit, phalanx 1

Interosseus ventralis

Extensor of the first and second digits

Ulnaris lateralis

6 PELVIC AND LEG MUSCLES, LATERAL VIEW

The muscles of the avian thigh and leg are much less unusual in arrangement than the muscles of the wing. Flight makes few demands on the leg other than as a shock absorber for landing, and the avian pelvic appendage has probably remained quite similar to those of the bipedal dinosaurs from which birds are thought by many to have evolved. As the most likely immediate ancestors of birds were small bipedal dinosaurs similar to the tiny Jurassic coelurosaur *Compsognathus*, flightless birds are especially interesting to observe because they are the only models we have of how these small running dinosaurs must have lived and behaved. Some of the most swift and agile of these dynamic carnivorous dinosaurs—for example, the Cretaceous coelurosaur *Struthiomimus* (meaning "ostrich-like")—bear a startling resemblance to modern flightless birds.

As birds have radiated throughout the world into numerous ecological roles, however, a striking variety of form and function in bird legs has developed. The huge legs and almost hooflike claws of such flightless ratites as the Ostrich (*Struthio camelus*) and Greater Rhea (*Rhea americana*) are the most notable anatomic features of these birds. In contrast, many of the most aerial birds, such as the Chimney Swift (*Chaetura pelagica*) and Barn Swallow (*Hirundo rustica*), have almost vestigial legs and feet. Birds use their legs to land, take off, and capture food, as well as in other ways that reflect their adaptation to particular habitats and ways of life.

Observe the following structures

Sartorius — The long, thin band of muscle that forms the anterior border of the thigh area. The sartorius inserts in a tough **aponeurosis** (band of connective tissue) at the knee, the **patellar tendon**. The patellar tendon covers the knee joint and encloses the patella.

Iliotibialis — This broad, thin muscle lies over most of the thigh musculature and is composed of muscular tissue and tough connective tissue. The layer of connective tissue that covers the lateral surface of the thigh is sometimes referred to separately as the **tensor fascia latae**. The iliotibialis is sometimes called the **gluteus maximus**.

Semitendinosus — This muscle originates on the lateral surface of the ilium and ischium and inserts into a short tendon (the accessory semitendinosus) that attaches it to the caudal surface of the distal femur.

Gastrocnemius — This is the largest leg muscle, by mass, covering the posterior surface of the leg and extending from the knee joint down to the Achilles tendon to flex the digits of the foot.

Flexor perforans et perforatus II and III — These long, thin muscles insert on the dorsal surface of the phalanges to flex the second and third digits.

References: Berger 1960; Chamberlain 1943; George and Berger 1966; Hudson 1937; Hudson et al. 1959.

Lateral View of the Leg Musculature
Rock Dove (*Columba livia*)

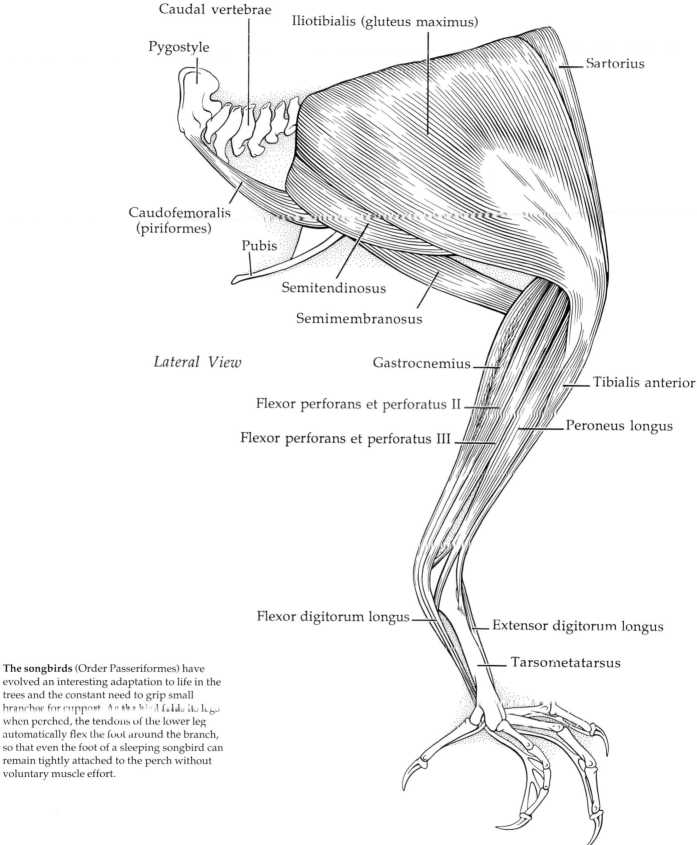

Caudal vertebrae

Pygostyle

Iliotibialis (gluteus maximus)

Sartorius

Caudofemoralis
(piriformes)

Pubis

Semitendinosus

Semimembranosus

Lateral View

Gastrocnemius

Tibialis anterior

Flexor perforans et perforatus II

Peroneus longus

Flexor perforans et perforatus III

Flexor digitorum longus

Extensor digitorum longus

Tarsometatarsus

The songbirds (Order Passeriformes) have
evolved an interesting adaptation to life in the
trees and the constant need to grip small
branches for support. As the bird folds its legs
when perched, the tendons of the lower leg
automatically flex the foot around the branch,
so that even the foot of a sleeping songbird can
remain tightly attached to the perch without
voluntary muscle effort.

7 PELVIC AND LEG MUSCLES, MEDIAL VIEW

In this medial view of the leg musculature note the overall structure of the pigeon leg. The mass of the muscles controlling the thigh, leg, and foot is concentrated up near the center of gravity of the bird, and only thin slips of muscle are present farther out on the leg and foot. The inner parts of the lower leg and foot are thus almost entirely composed of bone, ligaments, and tendons, forming a tough, light lower leg structure that resists routine wear and tear.

Observe the following structures

Quadriceps femoris — In the medial view this muscle lies just posterior to the sartorius, near the anterior border of the thigh.

Ambiens — Trace the ambiens from its origin on the ilium and pubis down through the knee tendons to its insertion at the top of the tibiotarsus. The ambiens is a muscle birds share with some reptiles but not with mammals.

Tibialis anterior — This muscle covers the anterior surface of the leg along the tibiotarsus, originating on the distal end of the femur and inserting on the proximal tarsometatarsus. The tibialis flexes the tarsometatarsus forward, helping the bird lift its foot off the ground.

Flexor hallucis, extensor hallucis — These two tiny muscles are the only muscular tissue along the length of the tarsometatarsus bone. Try to follow the tendons onto the medial phalanx of the hallux (digit 1) to see the antagonistic actions of the flexor and extensor hallucis.

Hooded Merganser
(*Lophodytes cucullatus*)

Heat Exchange and the Legs

Because the lower leg has so little soft tissue, an extensive network of blood vessels is not needed, reducing heat loss through the walls of veins and arteries. Some birds have a counter-current heat exchange mechanism whereby heat in the arterial blood going out into the leg is transferred to cooler venous blood returning from the legs and feet to reduce heat loss due to cold wind or water (Mitård 1980, 1984).

Medial View of the Leg Musculature
Rock Dove (*Columba livia*)

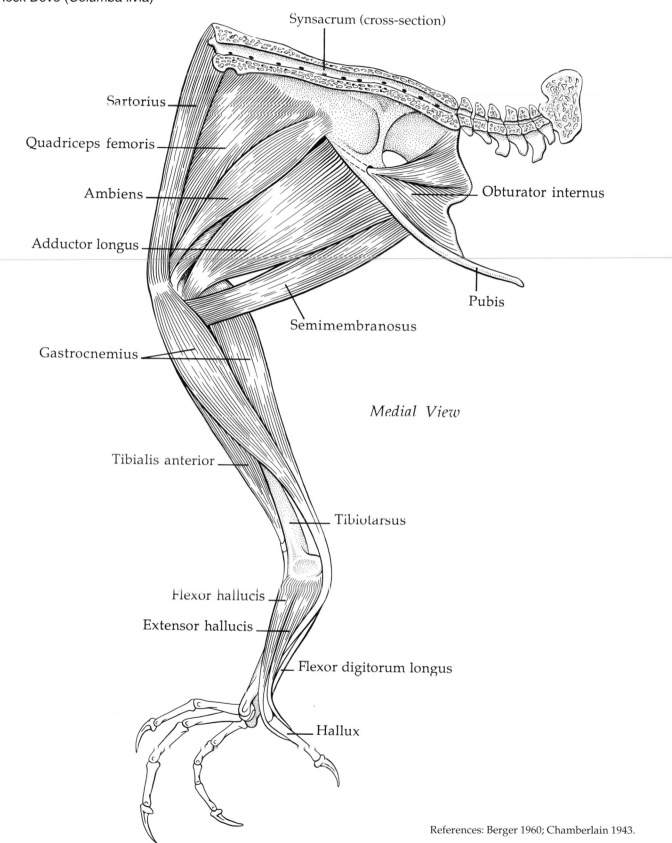

Synsacrum (cross-section)

Sartorius

Quadriceps femoris

Ambiens

Adductor longus

Obturator internus

Pubis

Semimembranosus

Gastrocnemius

Medial View

Tibialis anterior

Tibiotarsus

Flexor hallucis

Extensor hallucis

Flexor digitorum longus

Hallux

References: Berger 1960; Chamberlain 1943.

8 PELVIC AND TAIL MUSCLES

Although skeletal support for the tail in birds has been reduced to a few small caudal vertebrae and the blade-like pygostyle, the tail and rectrix feathers perform many functions. In flight the tail works as a rudder, elevator, and airbrake. As a rudder the lateral muscles of the tail twist the tail to cause rapid turns in flight; as an airbrake the tail feathers are fanned to cause drag as the bird slows down to land. The tail can also be bent downward or upward to act like the elevator of an airplane. Soaring birds like the *Buteo* hawks fan their broad, triangular tails to create a slot effect between the tail and the trailing edge of the wing—the high-speed flow of air between the tail and the wing increases the lifting efficiency of the wing airfoil (Burton 1990; Rüppell 1977). In long-tailed forest birds like the Blue Jay (*Cyanocitta cristata*) and Sharp-shinned Hawk (*Accipiter striatus*) the tail becomes a dynamic counterweight, enabling the birds to make rapid changes in direction when flying through foliage and branches. In many diving birds, such as penguins and sea ducks, the tail is used as a rudder to "fly" underwater. Some ducks even use their tails as a rudder when swimming at the water's surface, submerging the tail and twisting it from side to side to turn more rapidly than they could from kicking the feet alone.

Note how the pelvis and the synsacrum combine to provide a broad, stable platform that anchors the thigh and tail muscles. Without such extensive fusion of the spine and pelvis, the aerodynamic and muscular forces acting on a bird in flapping flight might bend and twist the thorax and pelvis, dissipating muscular energy that could otherwise be channeled to the wings. Without fused central skeletons birds would need heavy back muscles to keep their bodies rigid and would constantly waste energy by tensing these back muscles to counteract the twisting forces generated by the powerful flight muscles of the breast. In the pelvis and tail only the last few caudal vertebrae of the spine are unfused, to allow the tail to be twisted and bent down or up in flight.

The quills of the rectrices attach to the tail in a band of tough connective tissue extending from left to right and incorporating most of the pygostyle and posterior tail muscles. The broad lateral processes of the caudal vertebrae further reinforce the tail area, which must bear all the lever forces generated by the long tail feathers (Baumel 1988).

Observe the following structures

Levator coccygis, Levator caudae — These two muscles lying on the dorsal midline of the tail bend the tail dorsally (upward). There are corresponding **Depressor caudae** muscles on the ventral side of the tail.

Lateralis caudae — These lateral muscle masses allow the bird to spread out its tail rectrices into a broad fan, and also help to twist the tail to facilitate turns in flight.

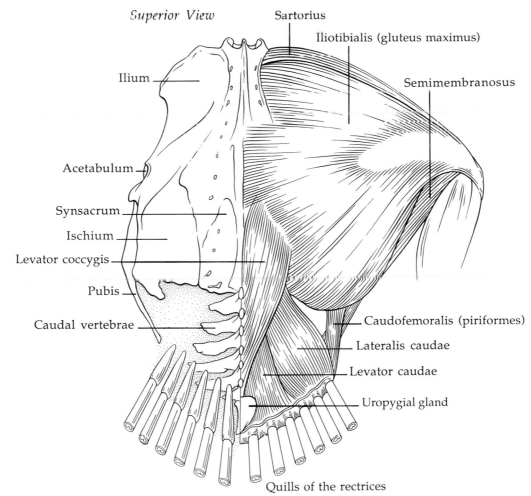

Superior View

Sartorius

Iliotibialis (gluteus maximus)

Ilium

Semimembranosus

Acetabulum

Synsacrum

Ischium

Levator coccygis

Pubis

Caudal vertebrae

Caudofemoralis (piriformes)

Lateralis caudae

Levator caudae

Uropygial gland

Quills of the rectrices

References: Baumel 1988; Berger 1960; Chamberlain 1943.

The Uropygial Gland

Also called the **preen gland** or **oil gland,** the uropygial gland sits at the base of the spine on the dorsal surface of the bird's fleshy stub. This heart-shaped structure is composed of many microscopic glandular tubules that exude an viscous liquid of fatty acids, waxes, and fats. The tubules drain from each lobe of the gland into a central duct, which has a midline nipple opening onto the dorsal surface of the gland (Jacob and Ziswiler 1982).

If you observe a bird preening, you will see that the bird often rubs its bill over the rump, collecting oil from the uropygial gland to spread onto its feathers. Ornithologists have long believed that the uropygial gland is somehow related to helping the bird maintain the insulating and waterproofing properties of its feathers. The oils from the uropygial gland apparently both waterproof the feathers and act as a grooming aid and lubricant when smoothing feathers and "zipping up" split feather vanes (Elder 1954). But a number of studies over the past century have shown that even aquatic birds like the Tufted Duck (*Aythya fuligula*) can apparently maintain waterproofed, well-groomed plumage if their uropygial glands are removed when they are hatchlings (Fabricius 1959). Oil on the plumage is not always helpful. One of the worst effects of an oil spill on aquatic birds is that the heavy crude oil destroys the insulating and waterproofing quality of the feathers, soaking the plumage and exposing the bird's thin skin to cold water (Wallace 1963).

Muscles of the Pelvic Girdle and Leg

Name	Origin	Insertion	Action
Iliotibialis, or Gluteus maximus, External iliotibialis lateralis	Fascia over the synsacrum at the midline and the ilium.	Fascia over proximal tibia at the tibial crest (patellar ligament).	Flexes the hip, extends the knee and lower leg.
Sartorius, or External iliotibialis anterior	Spinous processes of the lumbo-sacral area.	Fascia over proximal tibia at the tibial crest (patellar ligament).	Flexes the hip and extends the knee.
Semitendinosus	Crest and lateral surface of the ilium and ischium.	By ligament to the caudal surface of the femur.	Extends the thigh.
Semimembranosus	Lateral surface of the ischium.	Caudal surface of the tibiotarsus.	Extends the thigh, flexes the knee.
Femorotibialis medius	Medial and inferior borders of the ilium.	Fascia over proximal tibia at the tibial crest (patellar ligament).	Extends the thigh.
Ambiens	Ilium and proximal pubis.	Fascia over proximal tibia at the tibial crest (patellar ligament).	Flexes the thigh.
Adductor longus	Ventral border of the ilium and pubis.	Medial condyle of the distal femur.	Adducts and extends the thigh.
Caudofemoralis, or Piriformis, in part	Ventral surface of the pygostyle area.	Caudal surface of the tibiotarsus.	Flexes the thigh, moves the tail laterally.
Tibialis anterior, or Tibialis anticus	Anterior distal femur and proximal tibiotarsus.	Anterior surface of the tarsometatarsus.	Flexes tarsometatarsus forward toward the leg.
Gastrocnemius	By 3 heads—posterior, lateral, and medial—from the distal femur and proximal tibiotarsus.	Tarsometatarsus and phalanges of the foot via the Achilles tendon.	Flexes the knee, extends the foot.
Peroneus longus	Proximal tibiotarsus.	Tendon of the flexor perforans et perforatus III.	Flexes the digits.
Flexor perforans et perforatus II, III	Medial condyle of the proximal tibiotarsus.	Phalanges of digits 2 and 3.	Flexes digits 2 and 3.
Flexor digitorum longus	Medial condyle of the proximal tibiotarsus.	Distal phalanges of the digits.	Flexes the digits.
Extensor digitorum longus	Anterior surface of the proximal tibiotarsus.	Distal phalanges of the digits.	Extends the digits.

References: Berger 1960; Chamberlain 1943; Lucas and Stettenheim 1972.

The Musculature   *Chapter Worksheet*

1. List the primary characteristics of the avian musculature. What major muscular adaptations have birds made?

2. Explain the importance of the supracoracoideus muscle.

Chapter Worksheet, continued

3. Make a simple diagram below of the wing and its muscles, noting the general location of the muscles that extend and flex the outer wing.

4. Discuss the significance of the following muscles:

   a. Deltoideus

   b. Patagialis longus

   c. Expansor secondariorum

References

Berger, A. 1960. The musculature. In A. J. Marshall, ed., *Biology and comparative physiology of birds*. 2 vols. New York: Academic Press, 1:301–44.

Carroll, R. 1988. *Vertebrate paleontology and evolution*. New York: W. H. Freeman.

Chamberlain, F. 1943. *Atlas of avian anatomy*. Mich. Agric. Exp. Sta. Mem. Bull. 5.

Elder, W. 1954. The oil glands of birds. *Wilson Bull.* 66:6–31.

Fabricius, E. 1959. What makes plumage waterproof? *Rep. Waterfowl Trust* 10:105–13.

George, J., and A. Berger. 1966. *Avian myology*. New York: Academic Press.

Gill, F. B. 1990. *Ornithology*. New York: W. H. Freeman.

Heilmann, G. 1927. *The origin of birds*. New York: D. Appleton. Reprint, 1972; New York: Dover.

Hudson, G. E. 1937. Studies on the muscles of the pelvic appendage in birds. *Amer. Midl. Nat.* 18:1–108.

Hudson, G. E., and P. J. Lanzillotti. 1955. Gross anatomy of the wing muscles in the family Corvidae. *Amer. Midl. Nat.* 53:1–44.

Hudson, G. E., and P. J. Lanzillotti. 1964. Muscles of the pectoral limb in galliform birds. *Amer. Midl. Nat.* 71:1–113.

Hudson, G. E., P. J. Lanzillotti, and G. D. Edwards. 1959. Muscles of the pelvic limb in galliform birds. *Amer. Midl. Nat.* 61:1–67.

Terres, J. K. 1980. *The Audubon Society encyclopedia of North American birds*. New York: Alfred A. Knopf.

Wallace, G. J. 1963. *An introduction to ornithology*. 2d ed. New York: Macmillan.

Welty, J. C., and L. F. Baptista. 1988. *The life of birds*. 4th ed. New York: W. B. Saunders.

Raggiana Bird-of-Paradise
(*Paradisaea raggiana*)

The Digestive System

7

As small, active homoiotherms birds depend heavily on their digestive systems to remain nourished and healthy. Many birds operate on extremely thin margins of metabolic safety and can starve to death in mere hours if deprived of food or subjected to harsh weather that causes them to burn more metabolic "fuel" than they can quickly replace. Of necessity, then, the digestive system in birds is faster and more efficient than those in other vertebrate groups. Birds cannot afford to store heavy food materials within their bodies for long periods and usually need a constant supply of nutrients to sustain activity. Most birds digest their food quite quickly: shrikes have been reported to digest a mouse in about three hours, and many fruit-eating birds pass wastes and seeds in less than half an hour. Birds are uniquely able to put on body fat rapidly in preparation for long periods of exertion in migration. Some songbirds can increase their body weight by as much as 40 percent in ten days or less of intensive feeding. This avian version of the "carbohydrate loading" practiced by human endurance athletes shows why it is crucial to have adequate wildlife refuges and feeding areas along migratory flyways. If migrating birds are forced to fly too far between stopovers, they will become stressed and starved.

The avian mouth is simple and relatively unimportant in eating and digesting food in comparison with, for example, the mammal's mouth. Birds bolt their food in gulps and swallow without chewing. Most bird tongues are short, narrow, and triangular, with few taste buds. The bill is used to crush, tear, or simply hold the food before swallowing. But birds do have a complex system of touch sensors on their tongues, so although the taste of food may be relatively unimportant to birds, they are capable of making fine distinctions in food items based on the feel of the food to the tongue and hard palate. Most aquatic birds have reduced or absent salivary glands (ducks are the only prominent exception), but birds that feed on less moist prey usually have well-developed salivary glands to lubricate the food on its way down the esophagus (Worden 1964).

The esophagus is simple in structure and is basically a long, thin tube of muscle connecting the mouth with the **proventriculus**, or glandular stomach (McLelland 1989; Ziswiler and Farner 1972). In the pigeon and many other birds the esophagus has been widened at its midpoint to form a temporary food storage area called the **crop**. The crop allows birds like the pigeon to "load up" large amounts of food quickly and then fly off to digest the meal in safe cover. Birds are much more tolerant of food lodged in the esophagus than are mammals. Parent Common Terns (*Sterna hirundo*) will often feed chicks on fish so large that the tails stick out of the chicks' mouths for most of a day, yet the chicks appear to suffer no ill effects from the experience. Many other species without prominent crops can store large amounts of food in the esophagus temporarily when needed.

Diet and Sexual Selection

It is no great surprise that the diet of a species can profoundly affect the form and general habits of the species, but diet as a determining factor in sexual selection is somewhat surprising, even in such unusual groups as the New Guinea birds-of-paradise (Family Corvidae, Tribe Paradisaeini). Zoologist Bruce Beehler (1989) postulates that in such species as the Raggiana Bird-of-Paradise whose diets force females to establish large, overlapping feeding territories, the males are more likely to evolve display groups called **leks**, where each male competes with his fellows for the opportunity to mate with wandering females. Females of more omnivorous species such as the Magnificent Bird-of-Paradise (*Cicinnurus magnificus*) establish discrete, nonoverlapping feeding territories, and males of this species visit and display to females singly, not in groups. In both cases the competition for females causes intense selective pressure on the males to evolve distinctive, sexually dimorphic plumages, but it seems that the female's diet determines which display behaviors emerge in the male.

Birds have a much more complex and efficient stomach than their reptilian ancestors, but the avian stomach differs markedly in structure from the typical mammalian stomach. Birds have evolved a complex, two-part stomach that is a considerable improvement over the simple reptilian gizzard-like stomach. The first stomach area, the **proventriculus,** is a soft-walled glandular tube. The epithelial cells of the gastric mucosa in the proventriculus secrete strong hydrochloric acids (pH 0.7 to 2.3), digestive peptic enzymes, and mucus and start the process of breaking down the structure of the food material. The food then passes to the second part of the stomach, the **gizzard,** a disk-shaped section posterior to the proventriculus with thick muscular walls and a hard, sandpaper-like inner surface. The gizzard performs the function of mammalian teeth, grinding and disassembling the food to allow the digestive enzymes a maximum of surface area to attack. In most birds the gizzard contains sand grains or small rocks to aid the grinding process. The gizzard is amazingly strong: it has been reported that the gizzard of turkeys can completely crush twenty-four walnuts (in the shell) in under four hours and turn surgical lancet blades into grit in less than sixteen hours (Streseman 1927–34). The gizzards of carnivorous birds can crush large bones, although most birds of prey regurgitate the bones, hair, scales, and feathers of their prey in the form of pellets that are often found under or near their roosts (Craighead and Craighead 1956).

The **small intestine** is where food is digested and absorbed, and it varies in length and structure depending on the preferred diet of the species. This

Most predatory birds consume their smaller prey whole, later regurgitating the indigestible hair, bones, feathers, scales, or insect parts as pellets. The pellet shown here at life-size is from a Short-eared Owl (*Asio flammeus*) and was collected in a coastal salt marsh. The pellet contains the remains of a small rodent, probably *Microtus pennsylvanicus,* the Meadow Vole.

1 THE THORAX AND ABDOMEN, VENTRAL VIEW

The sternum has been removed in the illustration at the right to show the locations of the major organs of the thorax and abdomen.

The thoracic cavity — Note the general configuration of the heart and lungs within the thoracic cavity. The **heart** is large and easily identified. It lies within the **pericardial cavity** and is surrounded by a thin membrane, the **pericardium**. The **lungs** are located lateral and dorsal to the heart, on either side of the thoracic spine. They lie in left and right **pleural cavities**. Two **anterior thoracic air sacs** to either side of the pericardium may be identifiable, depending on how (or if) your specimen was preserved and how the sternum was removed. The **trachea** descends from the neck, passes dorsal to the heart and then bifurcates into **left and right bronchi**. The **syrinx** is located at the bifurcation. Note the tiny inverted V of the **bronchotracheal muscles** connecting the trachea to each bronchus. The major vessels anterior to the heart are studied in greater detail in chapter 8, The Circulatory System.

Note that birds lack the muscular diaphragm found in mammals. Instead, the avian thoracic cavity is divided from the abdominal cavity by a thin, double-walled membrane, the **oblique septum**. The airspace between the membranes of the oblique septum is called the **posterior thoracic air sac**.

References: King and Custance 1982; Petrak 1969; Pettingill 1985; Van Tyne and Berger 1976.

Thoracic and Abdominal Organs
Rock Dove (*Columba livia*)

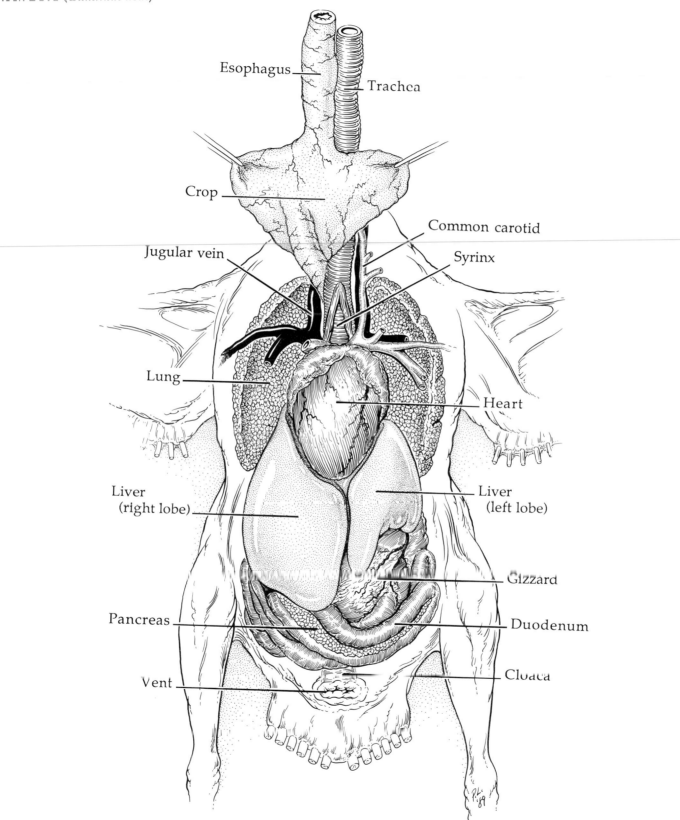

Esophagus

Trachea

Crop

Common carotid

Jugular vein

Syrinx

Lung

Heart

Liver
(right lobe)

Liver
(left lobe)

Gizzard

Pancreas

Duodenum

Vent

Cloaca

intestine is relatively featureless compared to mammalian small intestines and is not readily divided into duodenum, jejunum, and ileum, as it can be in mammals. Carnivorous birds tend to have shorter, less complex small intestines with small or absent intestinal caeca. Herbivorous birds have longer small intestines and well-developed intestinal caeca. The **intestinal caeca** are blind pouches off the end of the small intestine, where bacteria are thought to aid in the breakdown of cellulose plant material, much the way the complex stomachs of mammalian ruminants function. In birds like the Willow Ptarmigan (*Lagopus lagopus*) that feed on especially tough plant materials, the combined length of the paired caeca may equal the small intestine.

The liver and pancreas are connected by ducts to the **duodenum** (near the stomach) of the small intestine and contribute bile and pancreatic secretions that further attack starches, emulsify fats, and neutralize strong gastric acids. The avian liver is proportionately larger than that of mammals and is bigger in fish- and insect-eating birds than it is in grain- and meat-eating species. The **liver** stores fats and carbohydrates, synthesizes many proteins, and helps filter and neutralize metabolic waste products in the bloodstream. The **pancreas** is the chief regulator of carbohydrate metabolism, secreting insulin and other compounds to regulate the concentration of blood sugars. The pancreas also produces enzymes that break down proteins and fats in the small intestine.

Nutrients are then absorbed through the intestinal membranes and into the bloodstream, principally into the vessels of the superior and inferior mesenteric veins. These vessels then bring the nutrients into the portal hepatic system of circulation through the liver, where many of the nutrients are processed and stored before being distributed to the tissues of the body. The avian **large intestine** is reduced to a short, featureless connection between the end of the small intestine and cloaca. The **cloaca** is the final holding area for the waste products of digestion until they are voided through the vent (Farner 1960; Gill 1990; McLelland 1989; Pough et al. 1989; Welty and Baptista 1988; Ziswiler and Farner 1972).

2   THE THORAX AND ABDOMEN, LATERAL VIEW

The illustration depicts a left lateral view of the thoracic and abdominal cavities, with the sternum and breast muscles in place.

The abdominal cavity — Note the prominent right and left lobes of the **liver**. The liver is proportionately larger in birds than in mammals. A thin membrane, the **peritoneum**, surrounds the abdominal viscera and adheres to the abdominal wall. Within the peritoneum, the **proventriculus** and **gizzard** lie on the left side of the cavity. Posterior to them lies a loop of the **duodenum** enclosing the body of the **pancreas**. The coiled **small intestine** fills the posterior area of the abdominal cavity, ending in a short **large intestine** and the **cloaca** and **anus**. Intestinal **mesenteries** hold the coils of the small intestine in place and support the mesenteric blood vessels. Lateral to the abdominal viscera on each side of the cavity, large **abdominal air sacs** extend from the lungs posterior to the cloaca.

Lateral View of the
Thoracic and Abdominal Organs
Rock Dove (*Columba livia*)

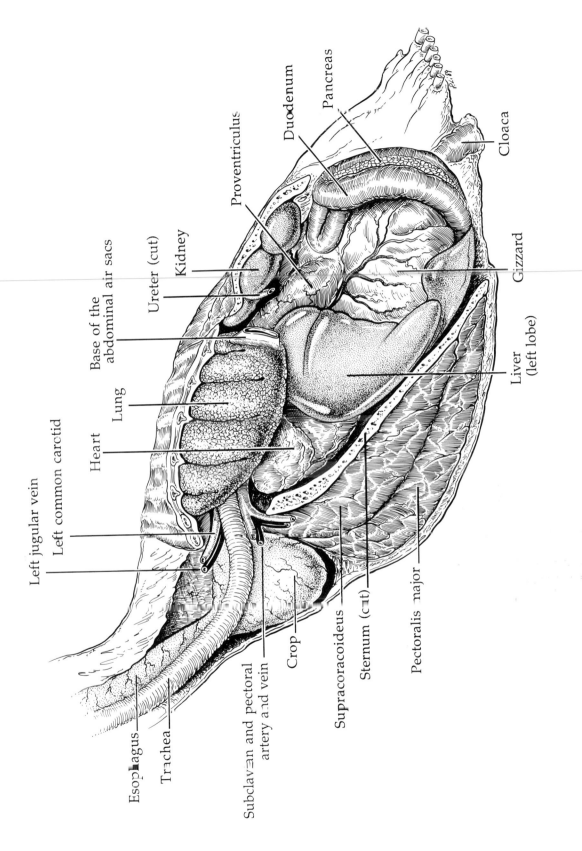

Proventriculus

Duodenum

Pancreas

Cloaca

Ureter (cut)

Kidney

Base of the
abdominal air sacs

Gizzard

Lung

Liver
(left lobe)

Heart

Left jugular vein

Left common carotid

Esophagus

Trachea

Subclavian and pectoral
artery and vein

Crop

Supracoracoideus

Sternum (cut)

Pectoralis major

References: King and Custance 1982; Petrak 1969; Pettingill 1985; Van Tyne and Berger 1976.

3 | THE DIGESTIVE TRACT

Observe the following structures

Esophagus — Follow the esophagus from the oral cavity down the neck. Note the **crop**, an enlarged segment of the esophagus at the base of the neck.

Liver — This largest of the glands dominates the abdomen in all vertebrates but is particularly large in birds. This may be due to birds' need for quick storage and retrieval of large amounts of carbohydrates to fuel flight and maintain homoiothermy. Two large left and right lobes enclose the lower **pericardium** and occupy most of the anterior and ventral space within the abdomen. Try to locate the **coronary ligament** fixing the anterior liver to the pericardium and the **falciform ligament** attaching between the ventral body wall and the ventral surface of the liver lobes.

Proventriculus — Retract (or cut away) the left lobe of the liver to expose the proventriculus, or glandular area of the stomach.

Gizzard — A tough, muscular organ just posterior and lateral to the proventriculus. The stomach is discussed further on the following page. The **mesogastric ligament** connects the gizzard to the body wall.

Spleen — A small, reddish bean-shaped organ found within the mesentery between the gizzard and the liver. The spleen is not always present in specimens (it is absent in pigeons, for example).

Duodenum — The area of the small intestine immediately following the gizzard.

Pancreas — Locate the relatively large body of this gland surrounded by a long loop of duodenum. The **pancreatic ducts** are normally hidden by the tissue of the gland. Note the **bile ducts** from the liver entering the duodenum near the pancreas. Many birds do not have a gall bladder to collect bile.

Small intestine — The **jejunum** and **ileum** areas of the small intestine follow the duodenum but are virtually indistinguishable from it. The small intestine is anchored by the intestinal **mesentery** to the dorsal abdominal wall. Note the superior and inferior **mesenteric arteries and veins** within the mesentery tissue.

Colic caeca — In pigeons these blind pouches are tiny, but in birds that must digest fibrous plant foods the combined length of the caeca may match that of the small intestine. The caeca harbor bacteria that break down plant cellulose into simple digestible carbohydrates. The featureless **large intestine** follows immediately posterior to the colic caeca.

Cloaca — Here fecal material from the intestines are collected before ejection from the body through the **vent,** or anus.

References: Farner 1960; Sturkie 1976; Welty and Baptista 1988; Ziswiler and Farner 1972.

Digestive System
Rock Dove (*Columba livia*)

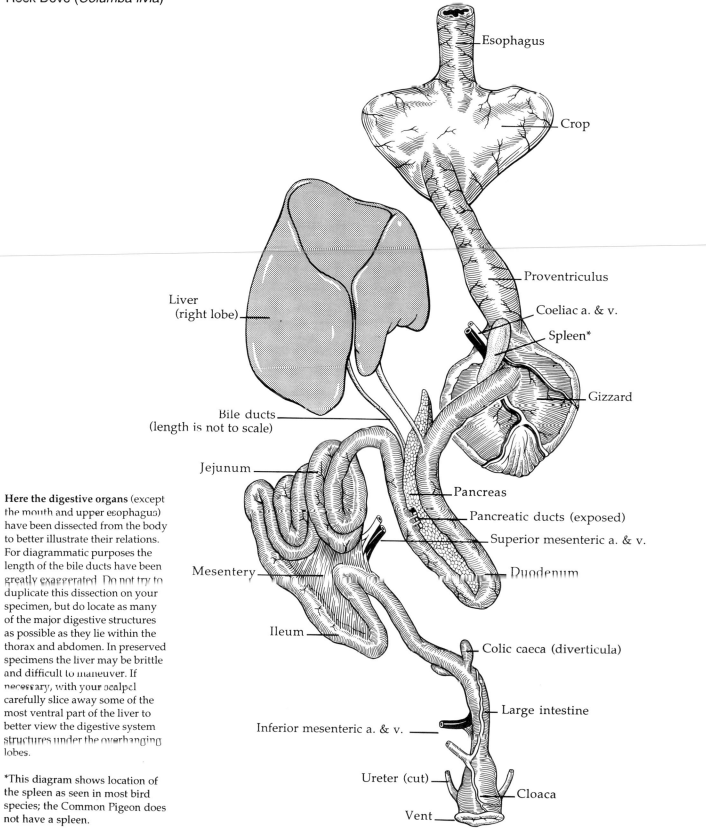

Esophagus

Crop

Liver
(right lobe)

Proventriculus

Coeliac a. & v.

Spleen*

Gizzard

Bile ducts
(length is not to scale)

Jejunum

Pancreas

Pancreatic ducts (exposed)

Superior mesenteric a. & v.

Mesentery

Duodenum

Ileum

Colic caeca (diverticula)

Large intestine

Inferior mesenteric a. & v.

Ureter (cut)

Cloaca

Vent

Here the digestive organs (except the mouth and upper esophagus) have been dissected from the body to better illustrate their relations. For diagrammatic purposes the length of the bile ducts have been greatly exaggerated. Do not try to duplicate this dissection on your specimen, but do locate as many of the major digestive structures as possible as they lie within the thorax and abdomen. In preserved specimens the liver may be brittle and difficult to maneuver. If necessary, with your scalpel carefully slice away some of the most ventral part of the liver to better view the digestive system structures under the overhanging lobes.

*This diagram shows location of the spleen as seen in most bird species; the Common Pigeon does not have a spleen.

4 | THE STOMACH

Observe the following structures

Proventriculus — This glandular area of the stomach produces the mucus, pepsin, and hydrochloric acid necessary for digestion. Note the ridges of the **papilla proventriculus**.

Isthmus gastris — A narrowing of the proventriculus as it enters the gizzard.

Gizzard — The tough, muscular section of the stomach that acts as the chewing "teeth" for toothless birds, grinding food material into a soft pulp and acting as a safety valve to block the passage of bones and other objects that might injure the soft intestinal tissues.

Pylorus — The muscular valve between the **lumen** (interior space) of the gizzard and the duodenum.

Lateral muscle — When you note how remarkably tough and hard the muscle masses are, you can easily understand how the gizzard can crack walnuts and reduce small bones to pulp.

Cutica gastrica — The hard, sandpaper-like lining of the gizzard lumen that both protects the muscle mass of the gizzard and helps break down food. At the base of the gizzard lumen the **saccus caudalis** forms a fibrous connection between the lateral muscle masses.

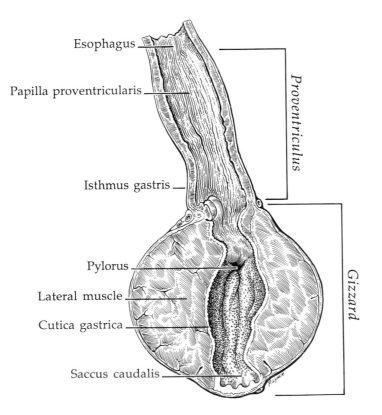

Esophagus

Papilla proventricularis

Isthmus gastris

Pylorus

Lateral muscle

Cutica gastrica

Saccus caudalis

Proventriculus

Gizzard

To study the structure of the stomach further you will need to remove it from the abdominal cavity. Dissect away or move the left liver lobe from the proventriculus, and follow the proventriculus forward toward the esophagus and neck for about 1.0 inch (2.5 cm). Divide the esophagus at this point. Next, relocate the gizzard and the duodenum exiting from the gizzard's dorso-medial surface. Divide the duodenum about 0.4 inch (1 cm) posterior to the gizzard, and remove the proventriculus and gizzard from the abdominal cavity. With a scalpel, divide the esophagus, proventriculus, and gizzard as in the diagram below. Do not try to divide tough muscle tissue in a few quick swipes; you will risk destroying some of the structures within the gizzard or perhaps accidentally dissecting a finger.

Evening Grosbeaks
(*Coccothraustes
vespertinus*)

Form and Function
Songbirds and their diets

Differences in diet give birds considerable variety in their digestive systems. The Rock Dove (*Columbia livia*) eats a range of plants and invertebrates but is primarily granivorous (eating grains or small seeds), and its diet requires the well-developed proventriculus and gizzard illustrated opposite. Evening Grosbeaks, pictured above, feed primarily on heavy nuts and seeds. They, too, need a substantial gizzard to crush food items into particles fine enough for their intestines to digest. But not all birds need a powerful grinding gizzard, and birds that specialize in soft, easily digestible foods such as fruits and berries may have a gizzard that is substantially reduced in size and importance. Fruits and berries are rich in simple sugars, require little digestion, and pass rapidly through the avian alimentary canal. In fruit-eating euphonias, such as the Blue-hooded Euphonia depicted below, the gizzard may be reduced to a tiny nub of flesh on a stomach composed almost entirely of glandular proventriculus (Van Tyne and Berger 1959). Other fruit-eating birds such as the Blackcap (*Sylvia atricapilla*), an Old World warbler (Family Sylviidae), have been reported to begin passing parts of a meal of berries as soon as twelve minutes after eating (Groebbels 1932).

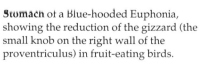

Stomach of a Blue-hooded Euphonia, showing the reduction of the gizzard (the small knob on the right wall of the proventriculus) in fruit-eating birds.

Blue-hooded Euphonia
(*Euphonia elegantissima*)

Form and Function
The Hoatzin's unique stomach

Hoatzins (pronounced "WAT-sins") are large, bizarre birds of the South American rain forest that have long attracted the attention and wonder of ornithologists. In addition to having striking features, including a spiked crest, a bare blue face, and flame-red eyes, Hoatzins are unique among modern birds in that their chicks have clawed wings. Flightless Hoatzin chicks use these claws to grip branches as they clamber about the rain forest canopy. Because of these unique wing claws the Hoatzin was once considered to be a primitive bird, intermediate in form between modern birds and such ancient ancestors as *Archaeopteryx*, which also had clawed wings. Recent genetic analysis (Sibley and Ahlquist 1990) suggests that Hoatzins may be related to the cuckoos (Order Cuculiformes) and that any resemblance between Hoatzins and ancient birds results from convergent evolution of the wing claws.

Recent studies on the natural history of Hoatzins have yielded an even more interesting fact about this unusual bird: it is a true ruminant, feeding almost entirely on leafy foliage and fermenting the ingested leaves with a stomach that functions much as does that of a cow (Grajal 1991; Strahl et al. 1989). Like cows, Hoatzins are foregut fermenters. A series of chambers near the anterior of the intestinal tract house bacteria that break down indigestible cellulose fibers into simpler digestible carbohydrates. The Hoatzin's large crop consists of two fermentation chambers in which bacteria break down the leaves. The fermented mash of leaves then passes through an enlarged lower esophagus area, where it is further fermented. The crop and esophagus are both quite muscular in the Hoatzin, contracting to grind the leaf mash much as a cow "chews its cud" to aid fermentation. In the Hoatzin the proventriculus and gizzard areas of the true stomach are relatively small and of secondary importance in digestion.

The Hoatzin's digestive system is unique among birds. Although large herbivores such as cows and deer commonly are foregut-fermenting ruminants, small warm-blooded animals rarely digest food in this way. Leaves are a relatively poor source of nutrients; digesting them takes a lot of time and fermentation space within the body. The Hoatzin's unusual feeding strategy probably depends on its warm, food-rich rain forest environment and its sedentary habits. Hoatzins are comically poor fliers (most landings seem to be barely controlled crashes) and spend most of the day roosting to conserve energy.

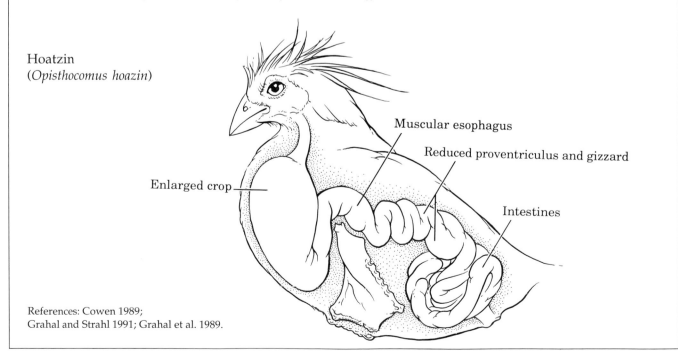

Hoatzin
(*Opisthocomus hoazin*)

Muscular esophagus

Reduced proventriculus and gizzard

Enlarged crop

Intestines

References: Cowen 1989;
Grahal and Strahl 1991; Grahal et al. 1989.

Digestive System   *Chapter Worksheet*

1. What adaptations does the avian digestive system show to the demands of homoiothermy and flight?

2. What is the function of the gizzard? How big a gizzard might you expect to find in a granivorous species?

3. How are the colic caeca analogous in function to the chambered stomachs of mammalian ruminants?

Chapter Worksheet, continued

4. Discuss the function of the following structures:

  a. Crop

  b. Proventriculus

  c. Pylorus

  d. Duodenum and ileum

  e. Pancreas

References

Beehler, B. 1989. The birds of paradise. *Sci. Amer.* 261(6):116–23.

Berger, A. J. 1961. *Bird study.* New York: John Wiley. Reprint, 1971; New York: Dover.

Farner, D. S. 1960. Digestion and the digestive system. In A. J. Marshall, ed., *Biology and comparative physiology of birds.* 2 vols. New York: Academic Press, 2:411–68.

Grajal, A., and S. D. Strahl. 1991. A bird with the guts to eat leaves. *Nat. Hist.* 100(8):48–54.

Groebbels, F. 1932. *Der vogel.* Vol. 1: *Atmungswelt und nahrungswelt.* Berlin: Gebruder Borntraeger.

McLelland, J. 1979. Digestive system. In A. S. King and J. McLelland, eds., *Form and function in birds.* 4 vols. New York: Academic Press, 1:69–181.

Miller, M. R. 1975. Gut morphology of Mallards in relation to diet quantity. *J. Wildl. Mgmt.* 39:168–73.

Pough, F. H., J. B. Heiser, and W. N. McFarland. 1989. *Vertebrate life.* 3d ed. New York: Macmillan.

Streseman, E. 1927–34. In W. Kukenthal and T. Krumbach, eds., *Handbuch der zoologie. Sauropsida: Aves.* Berlin: W. de Gruyter.

Sturkie, P. D., ed. 1976. *Avian physiology.* 3d ed. New York: Springer-Verlag.

Van Tyne, J., and A. J. Berger. 1959. *Fundamentals of ornithology.* New York: John Wiley.

Wallace, G. J., and H. D. Mahan. 1975. *An introduction to ornithology.* 3d ed. New York: Macmillan.

Welty, J. C., and L. F. Baptista. 1988. *The life of birds.* 4th ed. New York: W. B. Saunders.

Ziswiler, V., and D. S. Farner. 1972. Digestion and the digestive system. In D. S. Farner, J. R. King, and K. C. Parkes, eds., *Avian biology.* 8 vols. New York: Academic Press, 2:343–430.

Long-billed Starthroat
(*Heliomaster longirostris*)

Hummingbirds live on the ragged
edge of what is physiologically
possible for such small warm-blooded
animals. Their metabolism burns so
fiercely that they must find and eat
nearly their own weight in flower
nectar each day or risk starving to
death overnight. To support the
hovering flight needed to feed from
tubular flowers the hummingbird's
heart must beat almost 1,300 times a
minute; even at rest its heart rate is
almost 500 beats a minute. To save
energy during especially cool nights
hummingbirds may enter a state of
physiologic torpor in which their
bodies cool down to nearly ambient
air temperature. In spite of their small
size and intense metabolism,
hummingbirds may live surprisingly
long lives without "burning out."
Captive hummingbirds may live as
many as fourteen years, and banded
wild birds have survived nine years
(Skutch 1973).

The Circulatory System

8

Sustained flapping flight may be the most strenuous activity in which any vertebrate engages. It taxes the circulatory system more than running, swimming, or climbing. To support the intense demand for oxygen and nutrients produced by active flight, birds have developed the largest hearts relative to their body mass of all the vertebrates. As in mammals, in birds there is a complete separation between the **pulmonary** (circulation to the lungs) and **somatic** (circulation to the body) portions of the circulatory system. Only a true four-chambered heart is efficient enough to satisfy the metabolic oxygen demand produced by homoiothermy. Yet a few features of the avian circulatory system are more akin to birds' reptilian ancestors than to mammals, such as the renal portal network of veins in the abdomen. Birds retain this network because, like reptiles, they excrete nitrogenous wastes as urates (uric acid) and not as urea, as do mammals.

The four-chambered avian heart is similar to mammalian hearts in most structural details. The most prominent exception is the structure of the aortic arch, which arches to the right in birds and to the left in mammals. In birds the aortic arch derives from the right half of the fourth gill arch of the embryo, whereas in mammals it develops from the left half of the fourth gill arch. The largest hearts relative to body mass are found in shorebirds, kingfishers, swallows, hummingbirds, and other groups that regularly hover in flight or undertake long migrations over water. Birds of northern latitudes and alpine environments also tend have larger hearts than those living nearer the equator (Hartman 1961, Norris and Williamson 1955).

Bird hearts pump blood more efficiently than do the hearts of mammals of comparable size. The avian heart is proportionately as much as 40 percent larger than the mammal heart, and it moves more blood per beat than does the mammal heart (Lasiewski and Calder 1971). The resting heart rates of most small songbirds range from 350 beats per minute (bpm) to 480 bpm, or almost 10 beats a second. During short periods of extreme exertion the heart rate in small birds can reach 1,000 bpm. But heart rates are inversely proportional to body mass. An Ostrich has a resting heart rate of just 40 bpm, rising to 175 bpm during heavy exertion (Schmidt-Nielsen et al. 1969).

Birds have a less conspicuous lymphatic circulatory system than do reptiles and mammals (Welty and Baptista 1988). Prominent lymph nodes are rarely seen in birds. There is lymphatic tissue within the thymus gland, in the spleen, near the intestines and mesenteric tissues, and in the bursa of Fabricius. In some specimens paired thoracic ducts may be observed on either side of the spine. These ducts are the major collecting vessels of the lymphatic system and deliver lymph into the jugular veins at the base of the neck.

1 THE MAJOR ARTERIES

The avian arterial system is remarkably similar to the arterial system found in mammals. One unique feature of the avian system, however, is the right-hand arch of the aorta. Avian and mammalian aortic arches both develop from the fourth gill arch, which is bilaterally symmetrical in the developing embryos of both groups. In all mammals, the left side of the fourth gill arch forms the aortic arch, and in the mammalian adult (including humans), the aortic arch has a prominent leftward orientation as it passes up, over, and behind the heart and down along the left side of the spinal column. In birds, the reverse occurs; the embryonic aortic arch derives from the right half of the fourth gill arch, and a mirror image of the mammalian aorta develops along the right side of the spine (Jones and Johansen 1972; Simons 1960; Sturkie 1976).

In the illustration of the avian arterial system, opposite, the abdominal viscera have been retracted to expose the underlying abdominal aorta and its major branches. Take great care when retracting the liver to preserve the hepatic portal vessels, which are studied in following pages. At times you may wish to refer to the detailed drawings of the heart that appear later in this chapter. The chart on the following two pages also details the origin and major branches of the central blood vessels.

Observe the following structures

Brachiocephalic, carotid, and subclavian arteries — The prominent paired vessels just anterior to the heart (at the top of the heart). The brachiocephalic arteries branch off the **ascending aorta** almost immediately above the aortic valve. The first major arteries to branch off each brachiocephalic artery are the paired **common carotid arteries** that run anteriorly from the brachiocephalics along each side of the cervical spine. The carotid arteries carry the main blood supply to the head and anterior neck. As the brachiocephalic arteries give off the common carotids they pass under the clavicular region of the anterior thorax and are then called the **subclavian arteries**.

Pectoral arteries — Follow the subclavian arteries as they quickly branch yet again into **pectoral and axillary arteries**. Note the relatively large size of the pectoral arteries. In most mammals these vessels are tiny and quickly dissipate into the mass of the pectoral muscles. In active flying birds such as pigeons the pectoral vessels are very large to ensure an adequate supply of oxygenated blood to the massive pectoral musculature during flight.

Aortic arch and abdominal aorta — Just dorsal to the brachiocephalic vessels at the anterior of the heart lies the aortic arch. The aortic arch passes dorsal to the heart along its right dorsal surface and runs posteriorly along the thoracic and lumber spine to supply the posterior regions of the body with oxygenated blood. In most major details the posterior arterial systems of birds and mammals are alike. See the charts on the following pages for details on the major branches of the abdominal aorta.

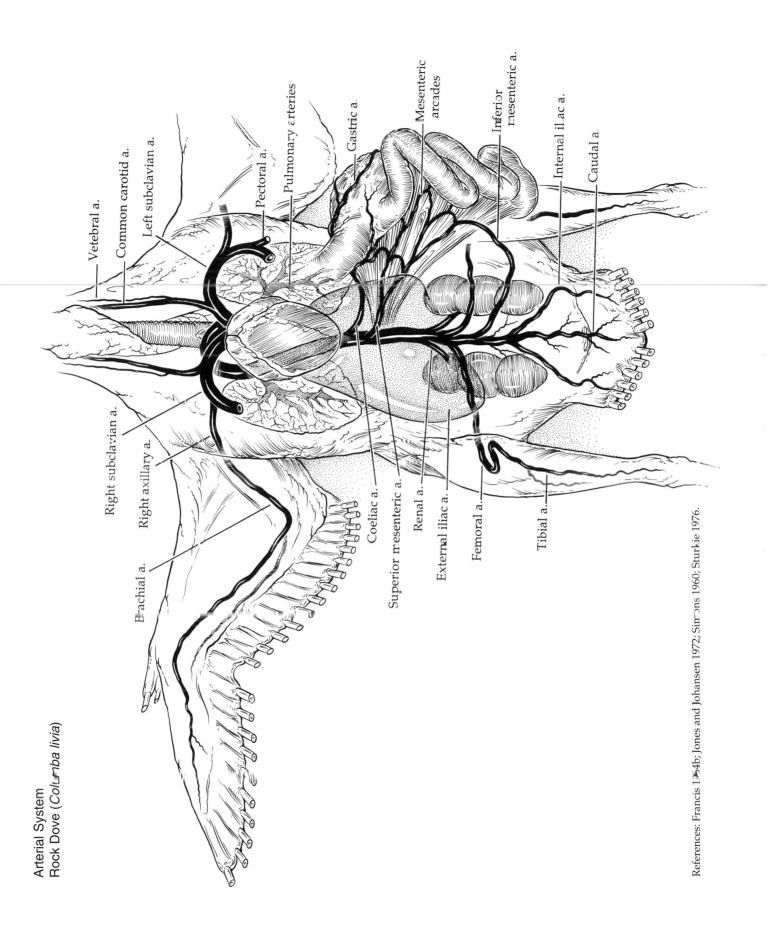

Arterial System
Rock Dove (*Columba livia*)

Vetebral a.
Common carotid a.
Left subclavian a.
Pectoral a.
Pulmonary arteries
Gastric a.
Mesenteric arcades
Inferior mesenteric a.
Internal iliac a.
Caudal a.

Right subclavian a.
Right axillary a.
Brachial a.
Coeliac a.
Superior mesenteric a.
Renal a.
External iliac a.
Femoral a.
Tibial a.

References: Francis 1964b; Jones and Johansen 1972; Simons 1960; Sturkie 1976.

The Major Arteries

Name	Originates from	Supplies	Major Branches
Brachiocephalic artery	Ascending aorta.	Carotid and subclavian arteries.	Carotid and subclavian arteries.
Aorta	Aortic valve of heart (left ventricle).	Whole body, including heart.	Brachiocephalic artery, coronary arteries, descending aorta.
Common carotid artery	Brachiocephalic artery.	Neck and head.	External and internal carotid arteries.
Pulmonary artery	Right ventricle.	Carries unoxygenated blood to the lungs, the body's only "blue" vein.	Left and right pulmonary arteries.
Subclavian artery	Brachiocephalic, after carotid branches off.	Pectoral muscles, wing, axillary region.	Pectoral, axillary, and brachial arteries.
Pectoral artery	Subclavian.	Pectoral musculature.	Capillaries of the pectoral muscles.
Branchial artery	Axillary artery (extension of subclavian artery).	Wing.	Ulnar and radial arteries.
Descending aorta	Aortic arch.	Posterior body.	See below.
Coeliac artery	Abdominal aorta.	Stomach, spleen, liver, pancreas, stomach region.	Gastric and hepatic arteries.
Superior mesenteric artery	Abdominal aorta.	Small intestine, pancreas.	Mesenteric arcades.
Renal artery	Abdominal aorta.	Kidneys, adrenal glands.	Capillaries of the kidney.
Ovarian or spermatic artery	Abdominal aorta.	Ovaries or testes.	Capillaries of the testes or ovaries.
External iliac artery	Abdominal aorta.	Thigh and leg.	Femoral, tibial, popliteal arteries of the leg.
Inferior mesenteric artery	Abdominal aorta.	Large intestine, end of small intestine.	Mesenteric arteries.
Internal iliac artery	Abdominal aorta.	Posterior abdomen, tail musculature, body wall.	Femoral, tibial, popliteal arteries of the leg.
Caudal artery	Terminal end of the abdominal aorta.	Tail region, muscles that control the tail.	Capillaries within the tail muscles.
Femoral artery	External iliac.	Thigh and leg.	Popliteal and tibial arteries of the lower leg.

The Major Veins

Name	Drains region	Empties into	Major source vessels
Precava, left and right	Anterior regions of the body—head, neck, pectoral limb, and anterior thorax.	Right atrium of the heart	Jugular and subclavian veins.
Posterior vena cava, or **Postcava**	Posterior regions of the body—abdomen, viscera, tail, and leg.	Right atrium of the heart.	Hepatic portal system, common iliac veins, and femoral veins.
Jugular veins	Head and neck.	Precavas, right and left.	Internal and external jugulars.
Pulmonary veins	Carries oxygenated blood from the lungs; the body's only "red" veins.	Left atrium.	Capillaries within the lungs.
Subclavian veins	Wing and pectoral regions, anterior of thorax.	Precavas, right and left.	Axillary and pectoral veins.
Pectoral veins	Pectoral musculature, anterior thorax region.	Subclavian vein, right and left.	Capillaries within the pectoral muscles.
Brachial veins	Wing region.	Axillary vein (becomes subclavian vein).	Radial and ulnar veins of the wing.
Hepatic veins	Liver tissues.	Inferior vena cava.	Capillaries of the liver.
Portal vein, or **Hepatic portal vein**	Small and large intestines, stomach.	Capillaries within the liver tissue.	Capillaries of the intestines and mesenteries.
Common iliac veins	Posterior abdomen, kidneys, leg and tail regions.	Posterior vena cava.	External iliacs, femorals, and renal portals.
Common iliac veins	Posterior abdomen, kidneys, leg and tail regions.	Posterior vena cava.	External iliacs, femorals, and renal portals.
External iliac veins	Thigh and lower leg.	Common iliac veins.	Femoral vein from the leg.
Renal portal veins	Kidneys.	Common iliac veins.	Nephrons within the tissue of the kidneys.
Renal veins	Kidneys.	Common iliac veins.	Nephrons within the tissue of the kidneys.
Caudal vein	Tail region and muscles controlling the tail.	Common iliac veins.	Capillaries within the tail muscles.
Femoral veins	Thigh and leg.	External iliac vein.	Nephrons within the tissue of the kidneys.

References: Francis 1964b; Jones and Johansen 1972; Simons 1960; Sturkie 1976.

2 THE VENOUS SYSTEM

The avian venous system is similar to that of birds' reptilian ancestors. In particular, birds have retained a renal portal vein. This reptile-like renal portal system, however, is much reduced in birds, and large venous valves at the junction of the renal veins and the common iliac veins permit birds to bypass the renal portal circulation during heavy exertion.

Birds have a **hepatic portal system** similar in structure and function to that of mammals. A portal vein system is one that begins and ends with a capillary network. The hepatic portal vein system begins at capillaries in the walls of the intestines and ends with a capillary network within the liver tissue. The portal vein carries nutrients and minerals directly from the walls of the intestines to the liver, where excess carbohydrates and fats are stored before being distributed throughout the body. Within the liver the renal portal vein breaks down into a capillary network to supply the liver tissues, and a new set of capillaries then collects the blood and aggregates into the major vein draining the liver, the renal vein (Jones and Johansen 1972; Simons 1960; Sturkie 1976). Think of the liver as a filter both supplied and drained by veins: the renal portal vein brings in food from the intestines, and the renal vein drains the liver and sends the blood out of the liver and into the inferior vena cava.

Now that you have seen both arteries and veins, look carefully at their differences. Note that **arteries** are relatively thick-walled vessels and generally lighter in color than veins. The whitish color of arteries is due to the white collagen fibers of the outer tunica adventitia layer of the artery walls, which toughens the arteries against the forces of blood pressure and pulse. Larger arteries are composed of multiple layers of fibrous tissue, smooth muscle, and an internal epithelial lining layer that prevents blood components from adhering to the artery wall. **Veins** (except for the pulmonary vein) have much thinner walls, and the walls are much simpler in structure than arterial walls. But only veins contain flow valves within their lumens; these venous valves prevent blood from pooling in the lower legs and keep it flowing back toward the heart.

Observe the following structures

Use the figure opposite and the venous system chart on the previous page to identify the major veins in the bird's thorax and abdomen. Note in particular:

Hepatic portal vein and hepatic veins — Try to find and follow the hepatic portal vein as it collects blood from the intestines and stomach and delivers the nutrient-rich blood to the liver. Note the several large hepatic veins coming out of the liver and into the inferior vena cava just below the heart. In preserved latex-injected pigeons the hepatic portal system is often injected with yellow latex.

Renal veins and renal portal vein — The venous arrangements around the avian kidney are quite unlike the mammalian renal vein system. Note the renal portal veins lying just ventral to each of the three-lobed kidneys.

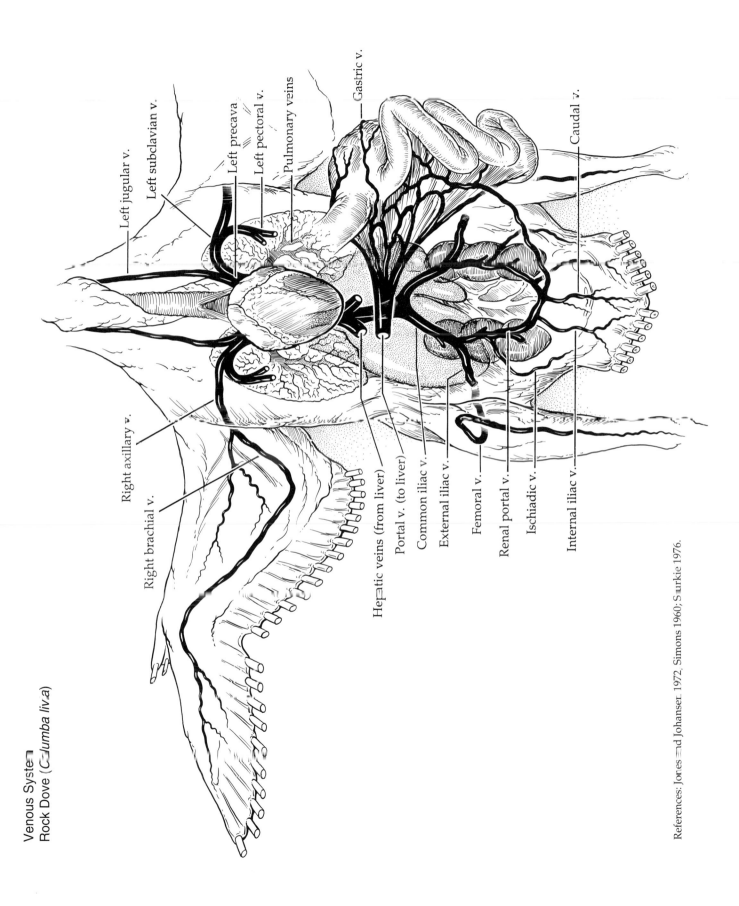

Venous System
Rock Dove (*Columba livia*)

Left jugular v.

Left subclavian v.

Left precava

Left pectoral v.

Pulmonary veins

Gastric v.

Caudal v.

Right axillary v.

Right brachial v.

Hepatic veins (from liver)

Portal v. (to liver)

Common iliac v.

External iliac v.

Femoral v.

Renal portal v.

Ischiadic v.

Internal iliac v.

References: Jones and Johansen 1972, Simons 1960; Sturkie 1976.

3 EXTERNAL ANATOMY OF THE HEART

The avian heart is quite similar in its major features to the hearts of mammals. Both birds and mammals have true four-chambered hearts that completely separate pulmonary and somatic circulation. But the avian heart differs in several ways from the mammalian heart. The most distinctly avian feature, mentioned above, is the rightward orientation of the aortic arch (rather than leftward, as in mammals). In addition, in birds the brachiocephalic vessels branch off the aorta at the base of the ascending aortic arch. The aortic arch is thus relatively smaller than it appears in mammals, where the brachiocephalic vessels branch off the top of the aortic arch. Because they supply the massive pectoral arteries, each brachiocephalic artery is nearly equal in diameter to the aortic arch.

The avian arrangement of the anterior veins entering the heart is similar to that found in reptiles. Two large precava veins meet in a sinus venosus just before entering the right atrium. There is no single superior vena cava, as found in mammals. As in the mammalian heart, however, the venous blood that flows from the posterior regions of the body collects into a single inferior vena cava (postcava) that enters the right atrium (Francis 1964a; Jones and Johansen 1972; Simons 1960; Sturkie 1976).

Before you attempt to remove the heart for dissection, carefully study the arrangements of the major blood vessels entering and leaving the heart, as some may be damaged in the attempt to remove the heart from the body. After the major blood vessels have been identified, cut them as far from the heart as possible to aid in identifying the vessels later. If all that remain on the heart are mysterious stumps of arteries and veins, the anatomy will be much harder to discern. Be careful when you dissect the dorsal surface of the heart to free it from the body; you do not want to destroy the delicate pulmonary veins, precavas, and postcava.

Observe the following structures

Identify as many of the external structures of the heart as you can, using the two diagrams opposite and other depictions of heart anatomy on following pages.

Note how large the heart is in relation to the thoracic cavity. Relative to their size and mass birds have the largest and strongest hearts of all the vertebrates. The heart is surrounded in the thoracic cavity by a thin membrane, the **pericardium**. Identify the major external structures of the heart, including the **right and left atria** and **right and left ventricles**. Most of the **apex of the heart** is composed of the muscular walls of the left ventricle. In both birds and mammals the right ventricle is much smaller and less muscular than the left ventricle. Note the **coronary arteries** visible on the dorsal and ventral surfaces of the ventricles.

References: Francis 1964a; Jones and Johansen 1972; Simons 1960; Sturkie 1976.

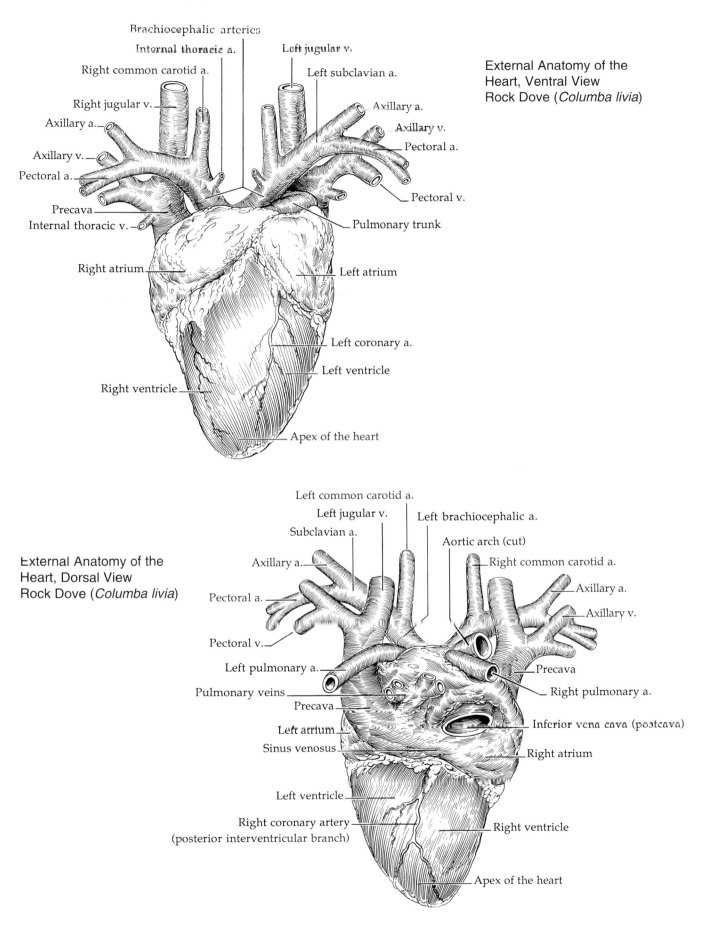

Brachiocephalic arteries
Internal thoracic a.
Right common carotid a.
Right jugular v.
Axillary a.
Axillary v.
Pectoral a.
Precava
Internal thoracic v.
Right atrium
Right ventricle

Left jugular v.
Left subclavian a.
Axillary a.
Axillary v.
Pectoral a.
Pectoral v.
Pulmonary trunk
Left atrium
Left coronary a.
Left ventricle
Apex of the heart

External Anatomy of the
Heart, Ventral View
Rock Dove (*Columba livia*)

External Anatomy of the
Heart, Dorsal View
Rock Dove (*Columba livia*)

Left common carotid a.
Left jugular v.
Subclavian a.
Axillary a.
Pectoral a.
Pectoral v.
Left pulmonary a.
Pulmonary veins
Precava
Left atrium
Sinus venosus
Left ventricle
Right coronary artery
(posterior interventricular branch)

Left brachiocephalic a.
Aortic arch (cut)
Right common carotid a.
Axillary a.
Axillary v.
Precava
Right pulmonary a.
Inferior vena cava (postcava)
Right atrium
Right ventricle
Apex of the heart

4 | INTERNAL ANATOMY OF THE HEART

Because both birds and mammals are homoiothermic, their hearts must work efficiently to supply the body tissues with enough oxygen to support a high metabolism. Hence, both have developed a four-chambered heart that separates the circulation of oxygenated and unoxygenated blood. Even in such advanced reptilian groups as the crocodilians the heart mixes some arterial and venous blood and is thus less effective in keeping the oxygen content of arterial blood high (Webster and Webster 1974).

To observe the internal details of the heart, try dissecting the heart by making incisions down the right and left sides of the ventricles, joining the two cuts at the apex of the heart to separate the dorsal and ventral surfaces of the heart as in the figure opposite. Preserving the structure of the pulmonary artery and aortic arch is more problematic, but do the best you can to open up the anterior heart without destroying these delicate structures. In injected specimens be very careful when removing the latex within the chambers of the heart to avoid ripping out such delicate structures as the valve leaflets and papillary muscles. Do try, however, to leave the latex intact in the major blood vessels; this will keep them from collapsing and make it easier to locate them later.

Observe the following structures

Note the semilunar structure of the **pulmonary and aortic valves**, with their tiny three-part valve cusps. These valves are identical in structure and function to the pulmonary and aortic valves of the human heart. The **mitral and tricuspid valves** between the **atria** and **ventricles** are controlled by the **papillary muscles** within each ventricle. Note the delicate **chordae tendineae** fibers that run from the papillary muscles to the valve leaflets. As the heart contracts in systole the chordae tendineae limit the travel of the valve leaflets and keep the valve from leaking blood back into the atria. As the ventricle expands in diastole the myocardial muscles relax and the valve leaflets open freely to allow the ventricles to fill again.

Cross-section of the Ventricles
Rock Dove (*Columba livia*)

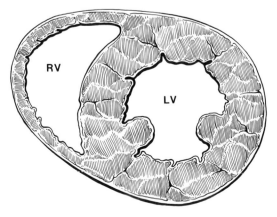

Most of the muscle mass of the ventricular portion of the heart is actually the left ventricular wall. Note how much thicker the left ventricular wall is relative to the right ventricular wall. Once you have dissected the ventricle area, try slicing off the apex of the heart, cutting horizontally across both ventricles to expose their walls in another plane. In this view it is easier to see that the left ventricle forms most of the heart's mass.

Internal Anatomy of the
Heart, Ventral View
Rock Dove (*Columba livia*)

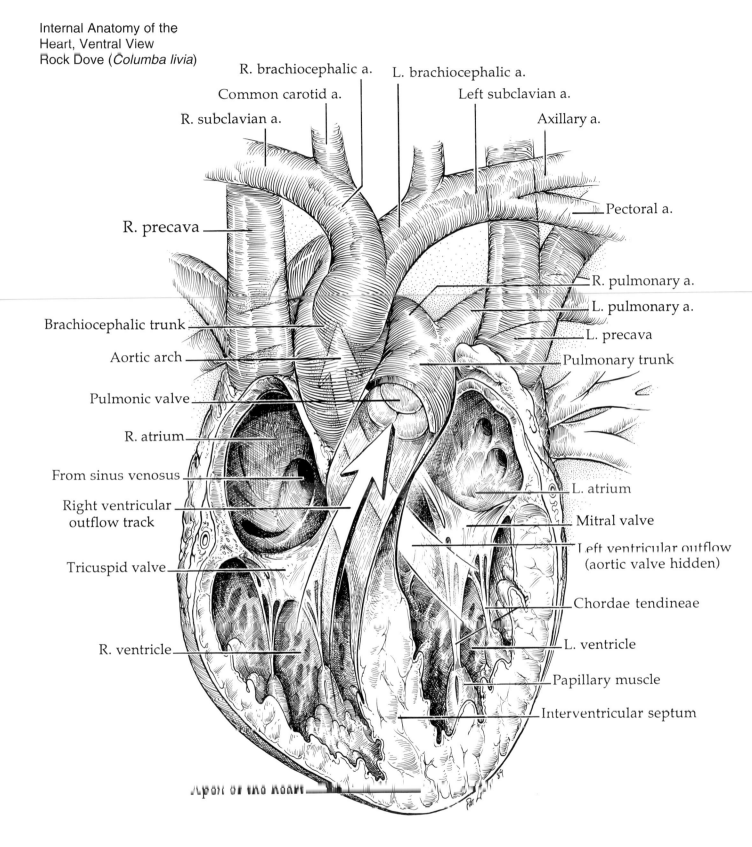

R. brachiocephalic a.

Common carotid a.

L. brachiocephalic a.

R. subclavian a.

Left subclavian a.

Axillary a.

R. precava

Pectoral a.

R. pulmonary a.

L. pulmonary a.

Brachiocephalic trunk

L. precava

Aortic arch

Pulmonary trunk

Pulmonic valve

R. atrium

L. atrium

From sinus venosus

Mitral valve

Right ventricular
outflow track

Left ventricular outflow
(aortic valve hidden)

Tricuspid valve

Chordae tendineae

L. ventricle

R. ventricle

Papillary muscle

Interventricular septum

References: Francis 1964a;
Jones and Johansen 1972; Simons 1960; Sturkie 1976.

4 | INTERNAL ANATOMY OF THE HEART, *continued*

The internal details of the avian heart are remarkably similar to these found in mammals, with the prominent exception of the right-oriented aortic arch. Blood from the anterior and posterior body flows into the right atrium via the precava and postcava veins. From there it moves though the tricuspid valve and into the right ventricle. In both birds and mammals the right ventricle is typically much less muscular than the left ventricle, although it may equal the left ventricle in volume. This is because blood flowing from the right ventricle into the pulmonary arteries must not damage the lungs by entering them at too high a pressure. Oxygenated blood from the lungs returns to the heart via the pulmonary veins, the only "red" veins of the body. The pulmonary veins empty into the left atrium. From there blood flows through the mitral valve and into the left ventricle. The left ventricle is the major engine of blood circulation and does most of the real work of moving blood throughout the body and maintaining blood pressure. The walls of the left ventricle are very muscular and contain small papillary muscles that control the opening and closing of the mitral valve. Blood flows out of the left ventricle via the aortic valve, a complex valve with three semilunar cusps. Just above the aortic valve two small coronary arteries branch from the aorta to supply the heart muscle with oxygen and nutrients.

In this anterior (top) view of the heart note how the **brachiocephalic arteries** branch from the **aortic arch** near its base. The drawing is oriented with the ventral side of the heart on top and the dorsal side on the bottom. Because the brachiocephalic arteries supply the large pectoral and brachial arteries of the breast and wing, the brachiocephalics are almost as large as the aorta itself. This dominance of the brachiocephalic arteries is yet another sign of how much avian anatomy has been shaped by the requirements of flight. In most mammals the brachiocephalic arteries are less than half as large as the aorta; in birds the brachiocephalics are so large that the heart seems to have three aortas coming off the left ventricle instead of one.

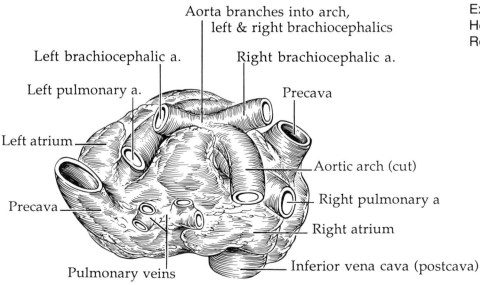

Aorta branches into arch,
left & right brachiocephalics

Left brachiocephalic a.

Right brachiocephalic a.

Left pulmonary a.

Precava

Left atrium

Aortic arch (cut)

Right pulmonary a

Precava

Right atrium

Pulmonary veins

Inferior vena cava (postcava)

External Anatomy of the Heart, Anterior View
Rock Dove (*Columba livia*)

Circulatory System *Chapter Worksheet*

1. How does the avian circulatory system differ from the reptilian circulatory system? How is it similar?

2. List the major arteries branching from the aorta and the principal organs or regions of the body they supply.

Chapter Worksheet, continued

3. Outline the major adaptations of the avian circulatory system to homoiothermy and flight.

5. Discuss the significance of the following structures:

   a. Papillary muscles

   b. Coronary arteries

   c. Left ventricle

   d. Brachiocephalic arteries

References

Akester, A. R. 1971. The blood vascular system. In D. J. Bell, ed., *Physiology and biochemistry of the domestic fowl*. 5 vols. New York: Academic Press, 2:783–839.

Calder, W. 1968. Respiratory and heart rates of birds at rest. *Condor* 70:358–65.

Francis, E. T. B. 1964a. The heart. In A. L. Thomson, ed., *A new dictionary of birds*. New York: McGraw-Hill.

Francis, E. T. B. 1964b. The vascular system. In A. L. Thomson, ed., *A new dictionary of birds*. New York: McGraw-Hill.

Hartman, F. A. 1955. Heart weight in birds. *Condor* 57:221–38.

Jones, D. R., and K. Johansen. 1972. The blood vascular system of birds. In D. S. Farner, J. R. King, and K. C. Parkes, eds., *Avian biology*. 8 vols. New York: Academic Press, 2:157–285.

Lasiewski, R. C., and W. A. Calder. 1971. A preliminary allometric analysis of respiratory variables in resting birds. *Resp. Physiol.* 11:152–66.

Norris, R., and F. Williamson. 1955. Variation in the relative heart size of certain passerines with increase in altitude. *Wilson Bull.* 67:78–83.

Simons, J. R. 1960. The blood vascular system. In A. J. Marshall, ed., *Biology and comparative physiology of birds*. 2 vols. New York: Academic Press, 1:345–62.

Skutch, A. F. 1973. *The life of the hummingbird*. New York: Crown.

Sturkie, P. D., ed. 1976. *Avian physiology*. 3d ed. New York: Springer-Verlag.

Van Tyne, J., and A. J. Berger, 1959. *Fundamentals of ornithology*. New York: John Wiley.

Webster, D., and M. Webster. 1974. *Comparative vertebrate morphology*. New York: Academic Press.

Welty, J. C., and L. F. Baptista. 1988. *The life of birds*. 4th ed. New York: W. B. Saunders.

Northern Gannet
(*Morus bassanus*)

Beneath the Northern Gannet's skin is a system of air sacs that act as shock absorbers, reducing the impact of plunging headfirst into the sea from high above. Note the gannet's streamlined head and exceptionally smooth body contours, which enable the bird to enter the water cleanly. The contour feathers of the gannet's body and head, moreover, are extremely dense, forming a thick outer cushion of air and feathers. Further adaptations to high diving are the gannet's loss of external nares (the nostrils on the outside of the bill) and its development of a strong brow ridge. A flock of gannets feeding is a spectacular sight. Folding their three-foot-long wings behind their bodies, the gannets plunge one hundred feet or more into shoals of fish or squid. Once you have seen them feed, you can easily see how evolution (and a lot of hard water) has molded these birds to fit their unique feeding niche.

The Respiratory System

9

The avian respiratory system is the most efficient in the animal kingdom, and in both its large and small details it is surprisingly unlike that of most other land vertebrates. In spite of their importance, lungs comprise just 2 percent of a birds' body volume (Welty and Baptista 1988). Unlike mammals, birds do not have a muscular diaphragm to power inspiration and expiration; they rely instead on the musculature of the intercostal muscles (Fedde 1975). When a bird *inhales,* air actually leaves its lungs; fresh oxygenated air enters the lungs as a bird *exhales.* Almost every major part of a bird's body is in direct communication with its respiratory system of air sacs, a complex anatomic feature unique among modern vertebrates (Schmidt-Nielsen 1971, 1983). The anatomy and air pathways of the avian respiratory system are described by Powell and Scheid (1989) and Scheid and Piiper (1989).

The primary function of the respiratory system is to supply oxygen to the body tissues and to carry away the carbon dioxide produced by metabolic activity. In birds this process of gas exchange is crucial, because the demand for oxygen produced by active flight is enormous. The flight muscles must receive a large and constant supply of oxygen to maintain flight, and such metabolic wastes as carbon dioxide must be removed quickly. In birds both heat and muscular energy are produced at the cellular level through the oxidation of foods stored primarily as fat within the body tissues. This cellular level of respiration, called **internal respiration,** actually produces metabolic heat and muscular energy. The gross structures of the respiratory system (pharynx, trachea, lungs, and air sacs) comprise the external respiratory system. The external respiratory system brings air into the body, exchanges oxygen from the air with carbon dioxide from the blood, and expels waste-laden air from the body (Lasiewski 1972; Salt and Zeuthen 1960).

Reptiles and mammals move air through their lungs in a tidal-flow pattern in which the lungs are blind sacs and air moves in and out of the lungs through the same pathway. Birds, in contrast, have evolved a complex respiratory system of lungs and auxiliary air sacs that allow a continuous stream of air to pass through the lungs in an efficient one-way flow. At rest, both birds and small mammals of equal body size need about the same amount of oxygen to sustain body temperature and moderate activity. But small mammals cannot match birds in producing the huge extra energy demands needed to sustain flight for long periods. The relative efficiency of the avian respiratory system has been demonstrated by exposing sparrows and mice to a simulated altitude of about nineteen thousand feet. At this height mice become comatose, while sparrows are still able to fly and are apparently unaffected by the low levels of oxygen (Schmidt-Nielsen 1971).

Most flying birds are relatively small, weighing well under a pound. This is a great advantage in flight, but it imposes a difficult physiologic dilemma: small animals have a much higher ratio of body surface area to volume than do larger animals, so they radiate proportionately more heat and must expend relatively more effort to maintain homoiothermy. Most small flying birds face a double respiratory burden—to supply great quantities of oxygen to the flight muscles when in active flight and to feed the metabolic "furnace" that keeps their body temperature from falling dangerously low when they are inactive in colder environments.

The avian respiratory system also cools the body. Birds lack sweat glands and are covered with an insulating blanket of feathers that effectively prevents them from radiating excess heat produced by muscular activity or from shedding heat absorbed from the environment. In hot environments the air sacs of the respiratory system are thought to aid birds in shedding excess heat as they breathe. Cooling apparently takes place as air passes over the walls of the air sacs and absorbs heat from the body, and this lowers the bird's internal temperature. A few of the more ardent advocates of homoiothermy in dinosaurs have theorized that the extensive system of pneumatized bones and body cavities found in birds probably evolved first in dinosaurs. Many theropods, including such well-known genera as *Deinonychus* and *Tyrannosaurus*, had extensively pneumatized skulls and the same type of rigid rib cage found in birds (Bakker 1986; Paul 1988). This suggests (but by no means proves) that some dinosaurs may have had an elaborate birdlike system of air sacs and pneumatized bones, but there is little direct fossil evidence for such a claim (Ostrom 1987).

1 DEEP ORGANS OF THE THORAX AND ABDOMEN

With the heart, liver, stomach, and intestinal tract removed, the more dorsal structures of the thoracic and abdominal cavities are visible. Note the relation of the **trachea**, **syrinx**, and primary **bronchi** that branch off to each **lung** posterior to the syrinx. The **esophagus** and **aortic arch** both pass dorsal to the trachea and bronchi. Before disturbing the lungs try to find and follow the connections between the lung and the more **anterior air sacs**, particularly if you are dissecting a fresh specimen where these delicate structures are more likely to be visible and undamaged. Note the surprisingly small size of the lungs and how they are tucked up in the gutters between the spine and thoracic ribs on each side.

The organs of the urogenital system are studied in chapter 10, but note here the general relations of the major organs of the dorsal abdomen. On each side of the **abdominal aorta** lies a three-lobed **kidney**, with a small **ureter** emerging from the ventral surface and paralleling the aorta down toward the **cloaca**. The condition of your specimen's **sexual organs** will vary depending on when during the year it was obtained as well as what species it is (chicken, pigeon, or something else). The specimen illustrated here is a male pigeon. Note the location of the cloaca and how the **large intestine**, sexual organs (oviduct or vas deferens), and ureters enter it.

Deep Organs of the Thorax and Abdomen
Male Rock Dove (*Columba livia*)

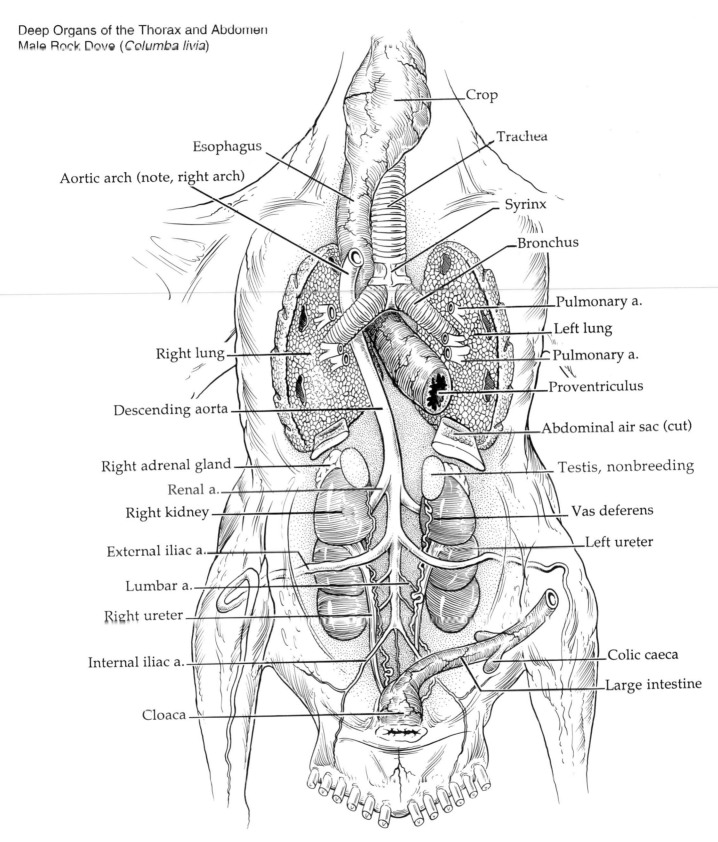

Crop

Trachea

Esophagus

Syrinx

Aortic arch (note, right arch)

Bronchus

Pulmonary a.

Left lung

Right lung

Pulmonary a.

Proventriculus

Descending aorta

Abdominal air sac (cut)

Right adrenal gland

Testis, nonbreeding

Renal a.

Right kidney

Vas deferens

External iliac a.

Left ureter

Lumbar a.

Right ureter

Internal iliac a.

Colic caeca

Large intestine

Cloaca

References: King and Custance 1982; Petrak 1969;
Pettingill 1985; Van Tyne and Berger 1976.

2 OVERVIEW OF THE RESPIRATORY SYSTEM

The illustration opposite is a diagrammatic overview of the major organs of the external respiratory system. The external respiratory system begins in the rear of the bird's mouth, or **oropharynx**. From there the air passes through the **glottis** valve, which prevents swallowed food from entering the respiratory system. The air enters the **larynx** just below the glottis. In most mammals the larynx is structurally complex and contains the vocal cords. In birds, however, the larynx is not involved in sound production and contains no vocal cords.

Most birds have long necks and therefore long, prominent tracheas. The **trachea** is extensively modified in some groups of birds, particularly in ducks, geese, and swans. In male ducks (Family Anatidae) the posterior trachea, just anterior to the syrinx, is often enlarged into a sounding chamber, or **tracheal bulla**. The tracheal bulla apparently modifies or amplifies the sounds produced by the syrinx (Johnsgard 1961). At the posterior end of the trachea lies the main organ of sound production in birds, the **syrinx**. The structure of the syrinx is detailed later in the chapter. At the syrinx the trachea bifurcates into a pair of primary **bronchi,** each leading to a lung. As each bronchus enters the lung it is referred to as the **mesobronchus**.

The mesobronchus continues straight through the lung and delivers air directly to the paired **abdominal and posterior thoracic air sacs**. In the diagram opposite the shape and size of the air sacs have been exaggerated to indicate better the layout of the lung–air sac system. From the posterior air sacs the air then flows forward again through the lung and emerges into the **anterior thoracic, interclavicular, and other anterior air sacs**. From the anterior sacs air is again expelled into the trachea and exhaled from the body.

Observe the following structures

Trachea — The tubular structure connecting the oropharynx to the syrinx.

Syrinx — The organ of sound production in birds, usually located at the base of the trachea, but in some species the syrinx may be located on the bronchi or higher on the trachea. Locate the bifurcation of the trachea into right and left bronchi.

Bronchi — These paired tubes (singular, **bronchus;** also called **primary bronchi**) connect the syrinx to each lung.

Mesobronchi — The tubes continue the bronchi through the lung tissue and posteriorly to the abdominal air sacs.

Interclavicular air sac — A single unpaired sac lying ventral to the syrinx and trachea in the notch between the paired clavicles (the paired clavicles are called the **furcula,** or wishbone).

Anterior and posterior thoracic air sacs — Paired thin-walled sacs within the thorax, lateral to the heart and ventral to the lungs.

Abdominal air sacs — Paired thin-walled structures lying lateral and ventral to the abdominal viscera, generally between the intestines and the abdominal wall.

Diagrammatic View of the Air Sac System
Rock Dove (*Columba livia*)

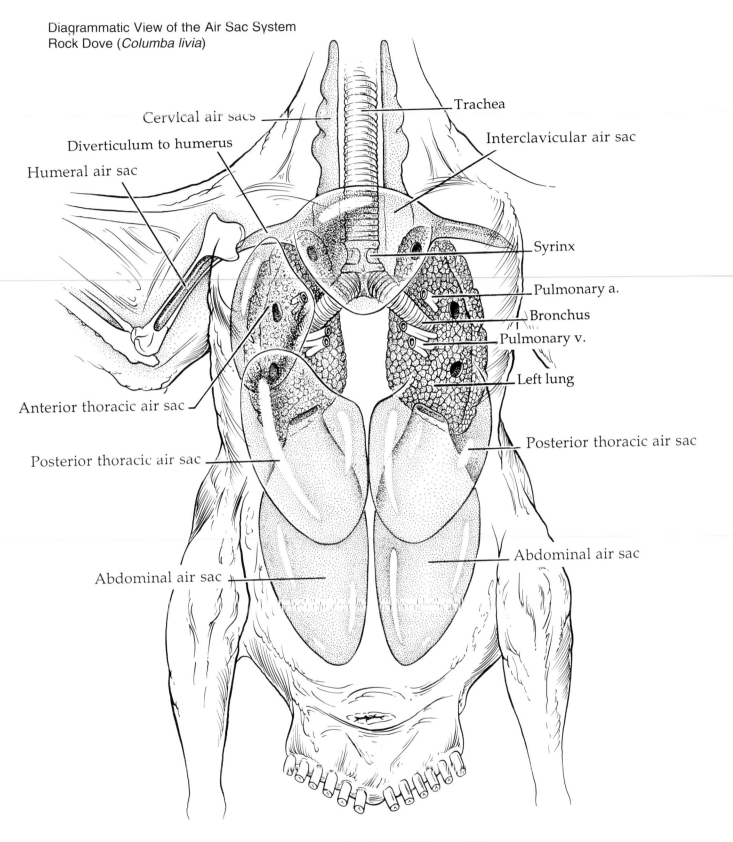

Cervical air sacs

Diverticulum to humerus

Humeral air sac

Trachea

Interclavicular air sac

Syrinx

Pulmonary a.

Bronchus

Pulmonary v.

Left lung

Anterior thoracic air sac

Posterior thoracic air sac

Posterior thoracic air sac

Abdominal air sac

Abdominal air sac

References: Lasiewski 1972; Romer 1962; Schmidt-Nielsen 1971; Salt and Zeuthen 1960; Thomson 1923.

In most species of birds the air sacs occupy about 15 percent of the volume of the thorax and abdomen (Welty and Baptista 1988). In the lateral view of the air sac system depicted below, the air sacs have been inflated to show their positions in the body. Note the smaller air sacs surrounding the pectoral girdle and extending up both sides of the cervical spine. In most of the larger flying species, virtually every major bone is pneumatic to some degree, either perforated with connections to the air sacs or filled with hollow cavities. The remarkable complexity and extent of the air sac system strongly suggests that the sacs have a much greater function beyond external respiration. The air sacs are particularly well placed to aid the bird in shedding excess heat from its body as it breathes, and this thermoregulatory function explains why so much of the bird's body is penetrated with diverticulae from the air sac system. As air flows over the moist surfaces of the air sacs it causes evaporation. Evaporation absorbs heat and cools the tissues that are in contact with the walls of the air sacs.

Lateral View of the Air Sac System
Rock Dove (*Columba livia*)

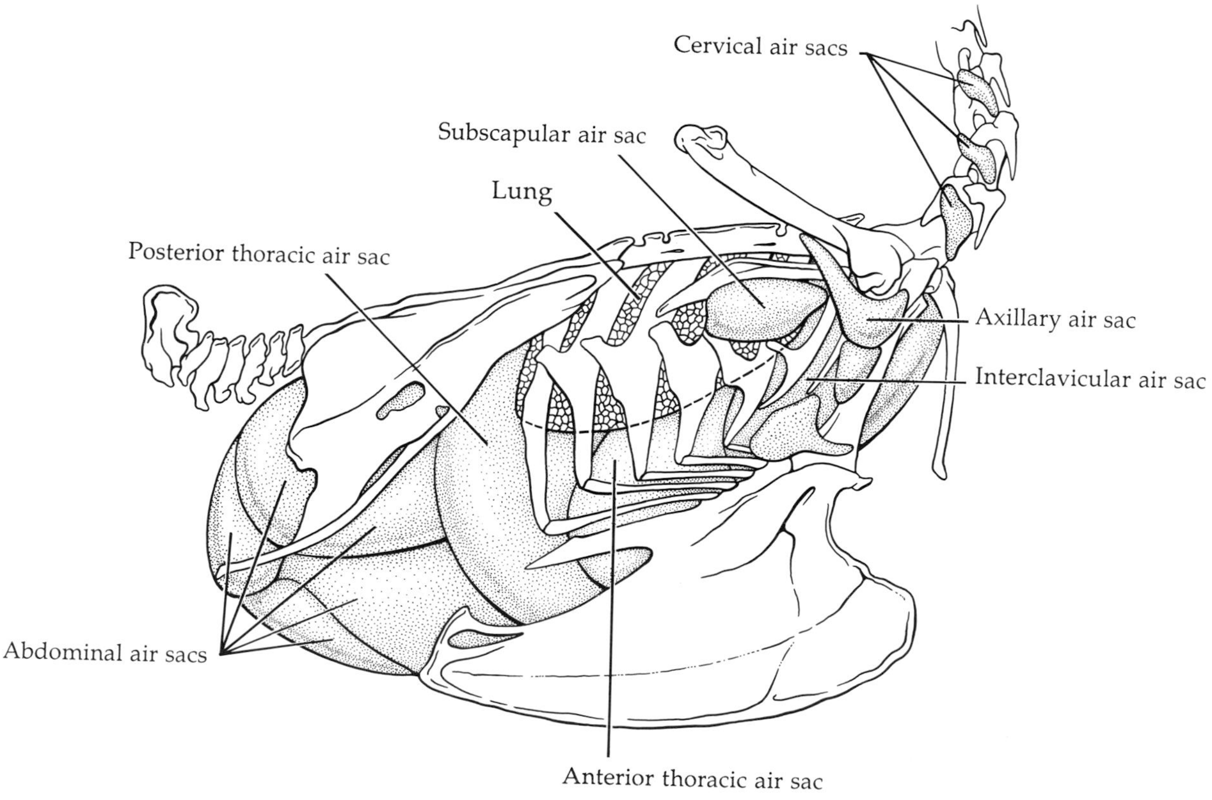

Reference: After Müller, in Neal and Rand 1939.

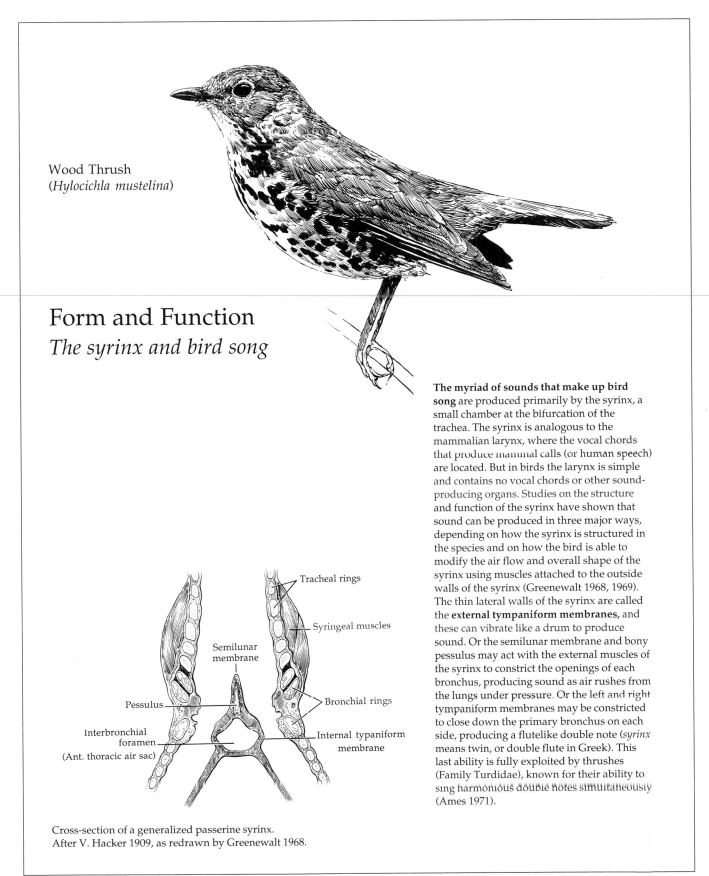

Wood Thrush
(*Hylocichla mustelina*)

Form and Function
The syrinx and bird song

The myriad of sounds that make up bird song are produced primarily by the syrinx, a small chamber at the bifurcation of the trachea. The syrinx is analogous to the mammalian larynx, where the vocal chords that produce mammal calls (or human speech) are located. But in birds the larynx is simple and contains no vocal chords or other sound-producing organs. Studies on the structure and function of the syrinx have shown that sound can be produced in three major ways, depending on how the syrinx is structured in the species and on how the bird is able to modify the air flow and overall shape of the syrinx using muscles attached to the outside walls of the syrinx (Greenewalt 1968, 1969). The thin lateral walls of the syrinx are called the **external tympaniform membranes,** and these can vibrate like a drum to produce sound. Or the semilunar membrane and bony pessulus may act with the external muscles of the syrinx to constrict the openings of each bronchus, producing sound as air rushes from the lungs under pressure. Or the left and right tympaniform membranes may be constricted to close down the primary bronchus on each side, producing a flutelike double note (*syrinx* means twin, or double flute in Greek). This last ability is fully exploited by thrushes (Family Turdidae), known for their ability to sing harmonious double notes simultaneously (Ames 1971).

Tracheal rings
Syringeal muscles
Semilunar membrane
Pessulus
Bronchial rings
Interbronchial foramen
(Ant. thoracic air sac)
Internal typaniform membrane

Cross-section of a generalized passerine syrinx.
After V. Hacker 1909, as redrawn by Greenewalt 1968.

3 FUNCTIONAL VIEW OF THE RESPIRATORY SYSTEM

A diagrammatic view of the avian respiratory system opposite illustrates the path of a single breath of air as it flows through the air sacs and lung. Note that the exchange of carbon dioxide from the blood with oxygen from the air takes place only within the lung. The air sacs are poorly supplied with blood vessels and do not directly aid in gaseous exchange. The total respiratory cycle in birds actually takes two cycles of inhaling and exhaling to complete. The path of a breath through the system can be summarized in four steps:

1. First inhalation (black arrows)

As a breath of air flows down the **trachea** it passes through the **syrinx** and into the left or right **bronchus**. The bronchus brings the air to the **lung**. Once inside the lung the bronchial tube is called the **mesobronchus**, and it passes most of the air completely through the lung and into the posterior air sacs in the abdomen. The **abdominal air sacs** expand on inhalation as the abdomen expands, and this pulls the inhaled breath into the abdominal sacs.

2. First exhalation (black arrows)

As the bird exhales, the abdomen contracts, and this forces the air in the abdominal sacs into the lungs. Within the lungs, the air passes through a progressively finer sieve of tiny air passages, called the **parabronchi**. The smallest air passages within the lung are called the **air capillaries**, and it is along the walls of these blind-ended air capillaries that the exchange of oxygen for carbon dioxide takes place. The blood capillaries channel blood in through the lung tissue along the walls of the air capillaries, and this **countercurrent flow** of blood produces the maximum efficiency in gaseous exchange.

3. Second inhalation (white arrows)

As the bird inhales again the air in the lungs is driven out, and the stale air full of carbon dioxide now passes into the anterior air sacs, principally the **interclavicular and anterior thoracic sacs**.

4. Second exhalation (white arrows)

The anterior sacs contract, and this drives the air out into the trachea, where it passes up and out of the nostrils, completing the respiratory cycle.

References: Goodrich 1930; Lasiewski 1972; Powell and Scheid 1989; Scheid and Piiper 1989; Schmidt-Nielsen 1971, 1983; Salt 1964; Salt and Zeuthen 1960.

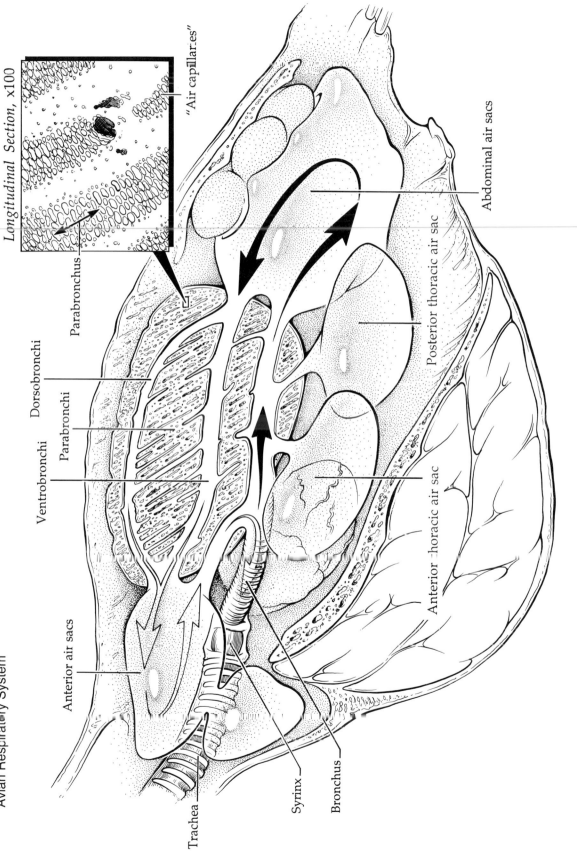

Diagrammatic View of the
Avian Respiratory System

Longitudinal Section, x100

"Air capillaries"

Parabronchus

Dorsobronchi

Parabronchi

Ventrobronchi

Anterior air sacs

Trachea

Syrinx

Bronchus

Anterior thoracic air sac

Posterior thoracic air sac

Abdominal air sacs

For the purposes of this diagram the air sacs are drawn much smaller than they appear in a live bird so that the major sacs will be easier to identify individually. In life the air sacs almost completely surround the abdominal viscera, and the sacs overlap each other extensively. If your dissection specimen was preserved in formalin, it may be difficult to appreciate the complexity and extent of the air sac system. If a fresh specimen is available, carefully open the abdominal wall and examine the air sacs that lie between the viscera and the wall. They will appear as thin, transparent sheets of tissue adhering to both the abdominal contents and the abdominal wall.

4 THE FURCULA AND RESPIRATION

Ornithologists have long speculated on the relation between the avian respiratory cycle and the cycle of downstroke and upstroke in flapping flight. The flight muscles of the breast are so large that they compress the thorax as they contract, and this obviously affects breathing during flight. By studying birds in flight with X-ray cinematography, scientists have learned that the compression and expansion of the thorax in flight complements the respiration cycle of inhaling and exhaling and that birds use some of the power of their flight muscles to help them breath while flying. A surprising by-product of this research showed that the furcula plays an important role in breathing (Jenkins et al. 1988).

It was once assumed that the furcula was simply a static brace to the shoulder joint, helping the coracoid bone stabilize the shoulder against the compression of the flight muscles. But X-ray films of starlings in flight show that the furcula acts like a leaf spring in the pectoral girdle, expanding as the wings are pulled downward and snapping back as they are raised again. This springlike movement of the furcula stores some of the energy of the breast muscles as they contract in downstroke, expanding the shoulders laterally. In upstroke the furcula releases this stored energy as it snaps back into normal position, drawing the shoulders in toward the body midline.

Dynamics of Respiration in Flight

As the furcula expands and contracts during flight it causes the expansion and contraction of the interclavicular air sac, which lies in the notch between the two halves of the furcula. This expansion and contraction is thought to act like a bellows, pumping air through the respiratory system while the bird is flying actively. As the interclavicular sac expands it draws air from the lungs, and as the sac contracts air is forced out of the anterior sacs and back into the trachea to be exhaled.

Reference: Jenkins et al. 1988.

Respiratory System *Chapter Worksheet*

1. Explain briefly the chief ways in which the avian respiratory system differs from that of mammals.

2. What are the two major functions of the respiratory system?

3. Draw a simple diagram of the syrinx, and explain briefly how it operates.

Chapter Worksheet, continued

4. Describe the pathway of air through the avian respiratory system.

5. Define these terms:

a. Parabronchi

b. Pessulus

c. Air sac diverticulae

d. Tracheal bulla

e. Interclavicular air sac

References

Ames, P. 1971. The morphology of the syrinx in passerine birds. *Bull. Peabody Mus. Nat. Hist.* 37:1–194.

Carroll, R. 1988. *Vertebrate paleontology and evolution.* New York: W. H. Freeman.

Fedde, M. R. 1976. Respiration. In P. D. Sturkie, ed., *Avian physiology.* 3d ed. New York: Springer-Verlag.

Greenewalt, C. H. 1968. *Bird song: acoustics and physiology.* Washington, D.C.: Smithsonian Institution Press.

Greenewalt, C. H. 1969. How birds sing. *Sci. Amer.* 221(5):126–39.

Jenkins, F., K. Dial, and G. Goslow, Jr. 1988. A cineradiographic analysis of bird flight: the wishbone in starlings is a spring. *Science* 241:1495–98.

Lasiewski, R. C. 1972. Respiratory function in birds. In D. S. Farner, J. R. King, and K. C. Parkes, eds., *Avian biology.* 8 vols. New York: Academic Press, 2:287–342.

Powell, F. L., and P. Scheid. 1989. Physiology of gas exchange in the avian respiratory system. In A. S. King and J. McLelland, eds., *Form and function in birds.* 4 vols. New York: Academic Press, 4:393–437.

Salt, G. W., and E. Zeuthen. 1960. The respiratory system. In A. J. Marshall, ed., *Biology and comparative physiology of birds.* 2 vols. New York: Academic Press, 1:363–409.

Scheid, P., and J. Piiper. 1971. Direct measurement of the pathway of respired air in duck lungs. *Resp. Physiol.* 11:308–14.

Scheid, P., and J. Piiper. 1989. Respiratory mechanics and air flow in birds. In A. S. King and J. McLelland, eds., *Form and function in birds.* 4 vols. New York: Academic Press, 4:369–91.

Schmidt-Nielsen, K. 1971. How birds breathe. *Sci. Amer.* 225(6):72–79.

Schmidt-Nielsen, K. 1983. *Animal physiology: adaptation and environment.* 3d ed. Cambridge: Cambridge University Press.

Tucker, V. 1969. The energetics of bird flight. *Sci. Amer.* 220(5):70–78.

The Urogenital and Endocrine Systems

In all vertebrates the excretory and reproductive systems are intimately related, though they perform quite different functions. Indeed, the two systems are so intertwined that it is almost impossible to describe one without referring to the other. This odd juxtaposition appears to be a coincidence, for both kidneys and the gonads happen to arise in the same mesodermal layer in the walls of the dorsal coelomic (abdominal) cavity of the embryo. The excretory system removes the nitrogenous and other wastes of metabolism and maintains the critical osmotic fluid balance of the body. The reproductive system produces the male sperm cells and female egg cells that unite in sexual reproduction, and in birds forms the eggs that nourish and protect the developing embryo until hatching (Carey 1983; Romanoff and Romanoff 1949).

As in other avian organ systems, the excretory system has been reduced to a minimum of mass and size, and adapted from the basic reptilian plan to support an active homoiothermic metabolism. Although the excretory systems of birds and mammals share many similarities, there are a few crucial differences in structure and physiology. Unlike mammals, which excrete urea, birds and reptiles excrete the waste products of nitrogen metabolism as urates. Unlike the extremely toxic urea, which must be flushed from the mammal body with great amounts of water, the largely insoluble urates manufactured in bird kidneys require little water to excrete. This is clearly an adaptation for water conservation in dry environments, but there is another perhaps more crucial reason why birds and reptiles excrete urates rather than urea. Both lay eggs, and the embryo within a sealed egg can release only gaseous waste products before it hatches. All other metabolic wastes, including nitrogenous wastes, remain within the egg until hatching. There would be insufficient room within the egg for the large amounts of water needed to excrete urea, whereas the insoluble urates are stored harmlessly within the egg until hatching (Gill 1990). Anyone who has ever washed a car has probably had occasion to contemplate the white-colored urates paste excreted by birds along with darker wastes from the digestive system. No urinary bladder is needed to produce this concentrated form of urates, and except for the small bladders found in Ostriches and rheas, birds have largely dispensed with this organ.

Birds also retain vestiges of the reptilian renal portal system. In reptiles and the lower vertebrates most of the blood that flows toward the heart from the legs and rear of the body is channeled through the capillary network of renal veins before it enters the inferior vena cava. In birds this diversion of blood into a renal portal system is largely bypassed, and blood now flows directly into the inferior vena cava, as it does in mammals. Only a small remnant of the renal portal system remains.

Brown (or Common) Kiwi
(*Apteryx australis*)

Kiwis lay the largest eggs in proportion to their body size of any bird. A fully mature kiwi egg weighs about one pound, or about one quarter of the body weight of an adult female Kiwi. The Kiwi egg also has the highest yolk content of any bird egg, making it a rich source of food for the developing chicks. Kiwi chicks gain so much food from the egg that they usually do not feed for days after they hatch.

Although the Kiwi's enormous eggs are clearly adaptive, it seems likely that the large size of Kiwi eggs is more an accident of their evolutionary history than a specific adaptation to their ground-dwelling ecological niche. Kiwis and their other flightless relatives the giant moas evolved in isolation on the islands of New Zealand. Millions of years ago Kiwis were probably much larger birds than they are now, and laid proportionately large eggs. As the Kiwi evolved into a smaller bird body, the egg size gradually increased relative to the size of the parent bird. Kiwi eggs didn't get big, Kiwi parents just got smaller. The Kiwi's evolution is a reminder that not all physical or behavioral traits of an animal are simply the product of straightforward "Darwinian" adaptation to the current environment. Many adaptive traits, like the Kiwi's giant eggs, may simply be a fortunate accident of the evolutionary history of the species (see Gould 1991c).

The reproductive system in birds shares many similarities with that of reptiles, with significant reductions and adaptations to reduce body mass for flight. For much of the year the avian reproductive organs shrink into dormancy to reduce body mass. But as breeding season approaches, the reproductive organs swell; the male's testes may increase several hundred-fold in size. Birds have retained the oviparous breeding system of reptiles but have greatly increased the quantity of yolk within the egg to serve the needs of a warm-blooded embryo. Unlike the reptile egg, which has a soft, leathery shell, the bird embryo is protected by a complex porous eggshell that allows efficient exchange of gases during the embryo's development (Carey 1983; Rahn et al. 1979).

1 OVERVIEW OF THE UROGENITAL SYSTEM

The avian kidney, like the kidneys of reptiles and mammals, is metanephric and represents one of the most significant adaptations vertebrates made toward life on dry land. In birds, the kidneys excrete dilute urate wastes, which are then dehydrated and concentrated in the cloaca. This saves a great deal of water that might otherwise be lost flushing the body of nitrogenous wastes.

Review the general features of the dorsal abdominal cavity as studied in the previous sections. Here the areas of principal interest are the kidneys and reproductive organs of your specimen. The pigeon depicted here is a male. If your specimen is a female, you should refer to the drawing of the female reproductive organs later in this chapter. To observe the inner structure of the cloaca, carefully cut along one side of the cloaca with a scalpel, then peel open the cloaca to examine the three chambers within it.

Observe the following structures

Note the brown three-lobed **kidneys** lying at the dorsal wall of the abdominal cavity, tucked into concave spaces formed by the ilium and synsacrum of the pelvis. The **ureter** emerges from between the anterior and middle lobes of each kidney and carries uric acid from the kidney down to the **urodeum** (middle) chamber of the **cloaca**. Immediately anterior and dorsal to the most anterior lobe of the kidneys locate the **adrenal glands**. These are small endocrine glands, and in fresh specimens they may be colored yellow or slightly orange. In male specimens locate the bean-shaped **testes**, which are usually just medial to the most anterior lobe of the kidney. The convoluted tubes of the **vas deferens** emerge from the testes and pass posteriorly along the ureters to enter the urodeum chamber of the cloaca. If your specimen is male and was collected outside the breeding season, the testes and vas deferens may be so small as to be impossible to locate. Very prominent testes and swollen vas deferens ducts indicate a male in breeding condition. In females locate the **ovary** just medial to the most anterior lobe of the left kidney and follow the **oviduct** posteriorly to the cloaca. Note the asymmetrical development of the female reproductive system—most species have no right ovary or oviduct. The male and female reproductive systems are detailed on the following pages.

Deep Abdominal Viscera
and Urogenital System
Male Rock Dove (*Columba livia*)

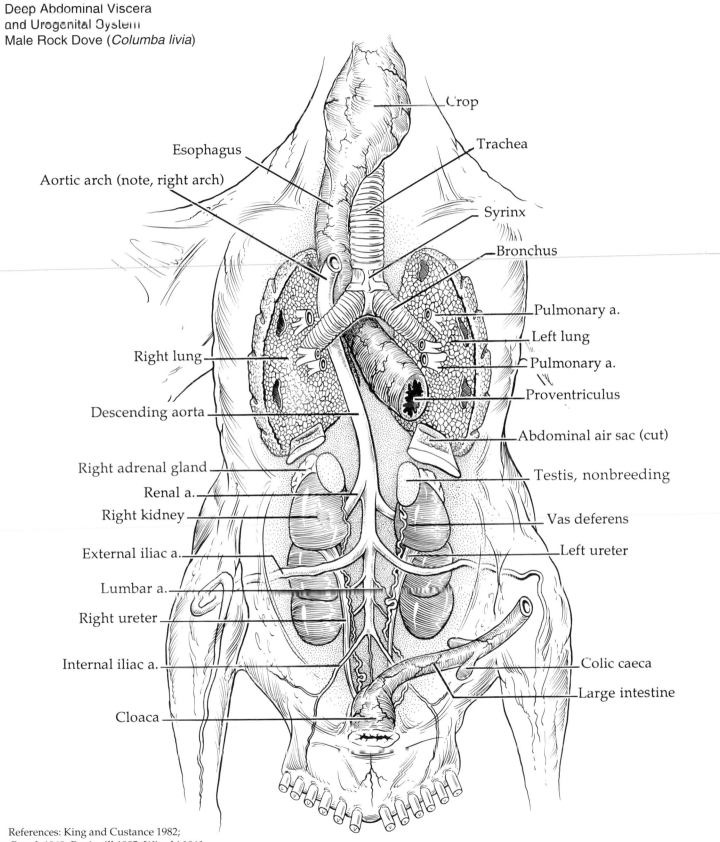

Crop

Trachea

Esophagus

Aortic arch (note, right arch)

Syrinx

Bronchus

Pulmonary a.

Left lung

Right lung

Pulmonary a.

Proventriculus

Descending aorta

Abdominal air sac (cut)

Right adrenal gland

Testis, nonbreeding

Renal a.

Right kidney

Vas deferens

External iliac a.

Left ureter

Lumbar a.

Right ureter

Internal iliac a.

Colic caeca

Large intestine

Cloaca

References: King and Custance 1982;
 Petrak 1969; Pettingill 1985; Witschi 1961.

2 THE MALE UROGENITAL SYSTEM

Male birds have paired testes within the abdominal cavity just anterior and ventral to the lobes of the kidneys. During much of the year the testes may be difficult or impossible to find because of their small size, but during the breeding season the testes may grow to several hundred times their nonbreeding size, resembling two bean-shaped organs lying next to the kidneys on the dorsal abdominal wall (Campbell and Lack 1985). As in mammals, avian sperm cells cannot develop fully at the high temperatures found within the body cavity, where the male bird's testes are located. Some birds experience a nightly drop in body temperature that allows sperm development in males. In other species the male develops a cloacal protuberance, a swelling of the terminal end of the vas deferens. This functions like a mammalian scrotum, holding the developing sperm away from the high temperatures within the abdomen. In most species the male bird lacks a penis or other copulatory structure; to transfer sperm to the female he simply approximates his cloaca to hers. Ostriches and rheas, storks, flamingos, ducks, and a few other families, however, do have an erectile grooved penis on the ventral wall of the cloaca.

Observe the following structures

In the diagram opposite, the male urogenital system is shown isolated from the structures that surround it in the abdominal cavity. Begin by locating the **kidneys**, which lie in recesses under the synsacrum and iliac of the pelvis. Note the three-lobed structure of each kidney and the tube of the **ureter** emerging from between the anterior and middle lobes. The ureter carries urine from the kidney posterior to the **cloaca**. Just anterior and dorsal to the kidneys lie the **adrenal glands**. These bilaterally paired endocrine glands are often bright yellow, orange, or even reddish or pink, but in preserved specimens the natural color may be dulled or lost because of the preservative formalin.

In the male bird the testes and other reproductive structures are bilaterally paired, whereas in the female bird only the left-side structures usually develop and become functional. Locate the paired **testes**, two whitish bean-shaped organs that are found near the anterior end of the kidney, near the midline of the body. Do not confuse the testes with the adrenal glands, which are usually much more colorful and tend to lie more lateral and dorsal than the testes. The testes of male pigeons are always fairly prominent, even out of breeding season, but be aware that in some species (or in immature males of any species) the testes virtually disappear when not breeding and may be difficult or even impossible to locate with certainty. The tubule of the **vas deferens** emerges from the testes and carries the sperm posteriorly to the cloaca. As the vas deferens approaches the cloaca it widens to form the **seminal vesicle**, a storage area for sperm cells. Again, the vas deferentia and seminal vesicles are particularly prominent in breeding birds but may be quite reduced in size otherwise. The male cloaca is slightly smaller than the female's, with a more prominent and protruding rim, but it is otherwise identical in structure.

Urogenital System
Male Rock Dove (*Columba livia*)

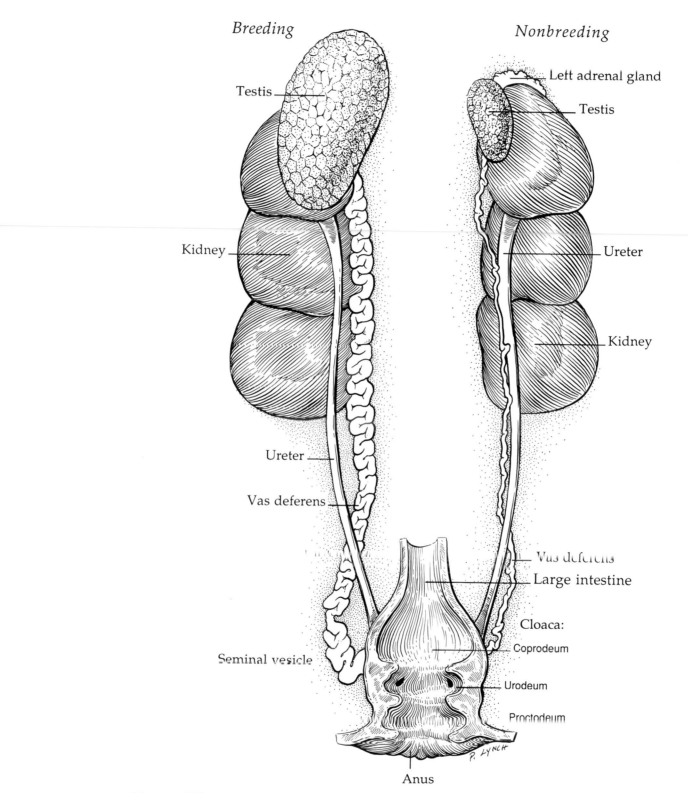

Breeding   *Nonbreeding*

Testis

Left adrenal gland

Testis

Kidney

Ureter

Kidney

Ureter

Vas deferens

Vas deferens

Large intestine

Cloaca:

Coprodeum

Seminal vesicle

Urodeum

Proctodeum

P. LYNCH

Anus

References: King and Custance 1982;
Lofts and Murton 1973; Petrak 1969; Witschi 1961.

3   THE FEMALE UROGENITAL SYSTEM

The female reproductive system in birds is reduced in most species to a left ovary and oviduct (see Blackburn and Evans 1986 on the lack of vivipary in birds). There are exceptions to this unilateral development, notably in birds of prey, in which some females develop both right and left ovaries and oviducts, though it is not known if the right ovary is functional in all cases (Kinsky 1971). This unilateral reduction of the female reproductive system is thought to have two major benefits: it reduces the female's body and it prevents the potential problem of simultaneously carrying two large, fragile eggs within the abdominal cavity. The ovary enlarges tremendously as breeding season approaches, growing from ten to as much as fifty times its nonbreeding weight (Gill 1990; Welty and Baptista 1988).

Observe the following structures

As in most other bird species female pigeons develop only the left side of the reproductive system. The right-sided structures are vestigial in all but a few groups of birds. Locate the kidneys and other structures of the excretory system, as described for males on the previous pages.

Just anterior and medial to the left kidney lies the white-colored **ovary**, with its many round **follicles** resembling a tiny bunch of grapes. The large tubular structure running posteriorly from ovary to cloaca is the **oviduct**, which begins next to the ovary in a funnel-shaped structure called the **ostium** (or, sometimes, the **infundibulum**). The large, mature follicles of the ovary each contain a potential ovum, although only a few follicles ever mature to be laid as eggs. The ovum is released from the ovary and enters the ostium of the oviduct. From there it travels down through the various segments of the oviduct—the magnum, isthmus, uterus, and vagina. Each segment has a distinct histological structure and function, but the exact divisions between sections are not visible in the gross specimen.

The **magnum** area of the oviduct is where the egg (ovum) receives its surrounding jacket of albumen, the "white" of the egg. Glandular cells of the magnum walls manufacture the albumen from water and proteins absorbed from the bloodstream. The egg then passes to the **isthmus,** where it receives its coating of keratin shell membranes. From the isthmus it passes to the **uterus,** where the hard, limy shell layers are added onto the egg before hatching. The **pigment glands** that color the eggshell are also located in the uterus, and it is the motion of the egg within the uterus and how long the egg remains within the uterus that determine what colors and patterns (streaks, spots, patches, and so on) the shell will show when laid. The **vagina** holds the egg until it is laid and has no function in forming the egg. The entire process of ovulation through egg laying takes about forty-one hours in the Rock Dove but just twenty-four hours in the domestic chicken.

Urogenital System
Female Rock Dove (*Columba livia*)

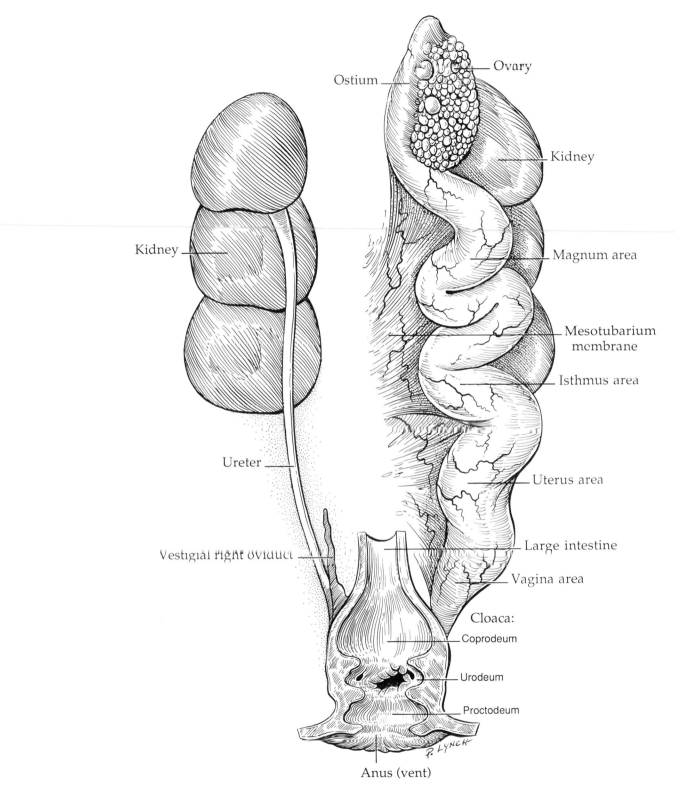

References: King and Custance 1982;
Lofts and Murton 1973; Petrak 1969; Witschi 1961.

4 THE OVIDUCT AND THE FORMATION OF EGGS

Ostium, or infundibulum—Thirty minutes

The ovum is released from the ovary and immediately enters the funnel of the ostium (or infundibulum), at the anterior end of the oviduct. Over the next half hour the egg is moved by ciliary action into the magnum region of the oviduct.

Magnum region—Three hours

Within the magnum the ovum receives a coating of albumen, typically over a three-hour period.

Isthmus region—One hour

Next the ovum and its albumen coating pass into the isthmus region, where the shell membranes are deposited onto the egg. After about an hour the ovum passes posteriorly through the oviduct into the uterus region at the posterior end of the oviduct.

Uterus region—Twenty hours

In the uterus the ovum receives its outer shell and, in most species of birds, its pigmentation. As the egg moves through the uterus region it may twist and turn (see the arrow in the diagram). Pigment-secreting areas of the uterus mark the eggshell as it passes, making the egg patterns a visual record of egg movement within the uterus. Rapid movement results in streaks or elongated tracks on the egg surface. Slow periods are marked with round spots, or even bands around the egg. It takes about twenty hours for the final deposition of the hard outer eggshell. The egg then passes through the vagina area of the oviduct and into the cloaca and is immediately laid.

Reference: Sturkie 1986.

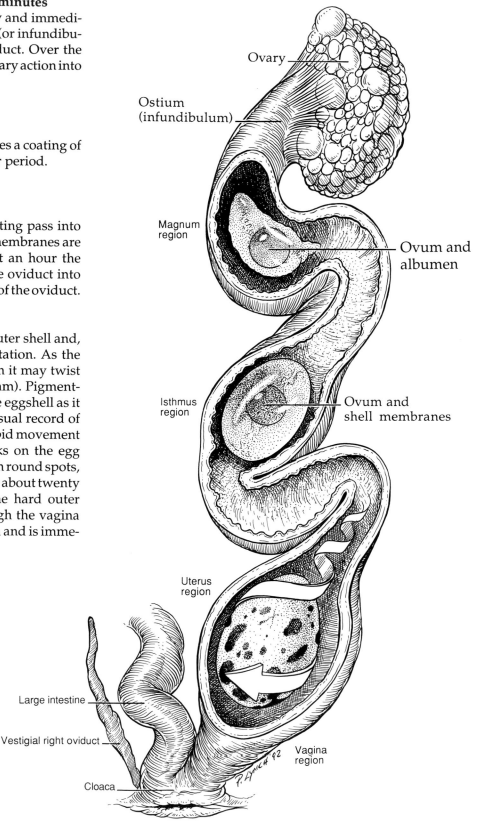

Form and Function
Salt glands in birds

Salt (supraorbital) gland

Ring-billed Gull
(*Larus delawarensis*)

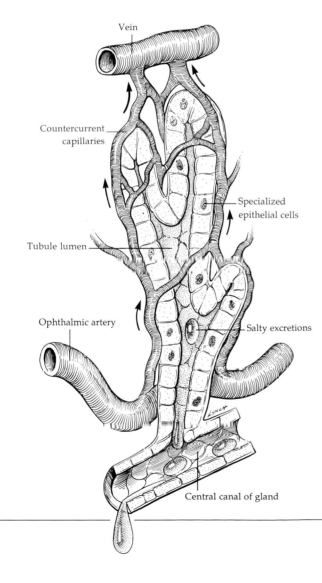

Vein

Countercurrent
capillaries

Specialized
epithelial cells

Tubule lumen

Ophthalmic artery

Salty excretions

Central canal of gland

What do seabirds drink? A child's question, but one that long puzzled ornithologists as well. "Water, water everywhere, nor any drop to drink," lamented Coleridge's sailor in *The Ancient Mariner*, and anyone else who has drunk the sea's salty water, for drinking saltwater actually makes you thirstier. The reason has to do with how the kidney (avian or mammalian) controls the body's osmotic balance and handles excess salts accumulated in the bloodstream. To flush the body of excess salt the kidney must use great quantities of water, more water than is gained by drinking seawater in the first place. It takes about one and a half quarts of fresh water to flush the body of the salt accumulated from drinking one quart of seawater. Clearly, drinking seawater is a quick route to physiologic disaster without a highly effective method for secreting salt from the body (Schmidt–Nielsen 1959).

Seabirds such as petrels, albatrosses, gulls, terns, and sea ducks have developed special salt-secreting glands located above the eye, over the bony orbit, and they secrete collected salt into the nasal passages just anterior to the eye. From there the salty liquid drips from the beak or, as in petrels, is forcibly "sneezed" from the nostrils. These salt glands (sometimes called **nasal glands** or **supraorbital glands**) have a microstructure similar to the kidney and use a system of countercurrent blood flow to remove and concentrate salt ions from the bloodstream. Tiny capillaries carry blood along the secretory tubules of the gland. The tubules have walls just one cell thick and form a simple barrier between the salty fluid within the tubules and the bloodstream. The gland removes salt efficiently: a gull given one-tenth its body weight in seawater can secrete about 90 percent of the salt in three hours. The secreted fluid is concentrated by physiological standards, containing about 5 percent salt. This is five times the normal concentration of salts in body fluid and thus allows seabirds to drink the water they live in.

Diagrammatic view of the microstructure
of an avian salt gland.

5 THE ENDOCRINE SYSTEM

The endocrine system is composed of a number of glands scattered throughout the body. These glands are called **endocrine** because they have no ducts and secrete their hormones directly into the bloodstream (Assenmacher 1973; Epple and Stetson 1980; Hodges 1979). This contrasts with the **exocrine** glands like the liver and the pancreas, which deposit their secretions into the digestive tract through major ducts. But the liver and pancreas are also endocrine glands because they also secrete hormones and other compounds (such as insulin from the pancreas) directly into the bloodstream. Collectively, the endocrine glands function in many ways as an adjunct to the nervous system. The brain and nervous system control individual parts of the body quickly and directly; the endocrine system acts through the bloodstream more slowly to make global changes simultaneously in many parts of the body. The endocrine glands are thus responsible for the body's reactions to stress, reproduction and breeding, mineral metabolism, and general body growth patterns. Note that all major functions of the endocrine system involve many parts of the body and many organ tissues simultaneously and that the best method for coordinating such global changes is to distribute those "messages" through the bloodstream.

The names, locations, and functions of the major endocrine glands are listed below and opposite. Except for the adrenal glands and gonads, the endocrine glands are small and hard to locate, particularly in small birds or preserved specimens.

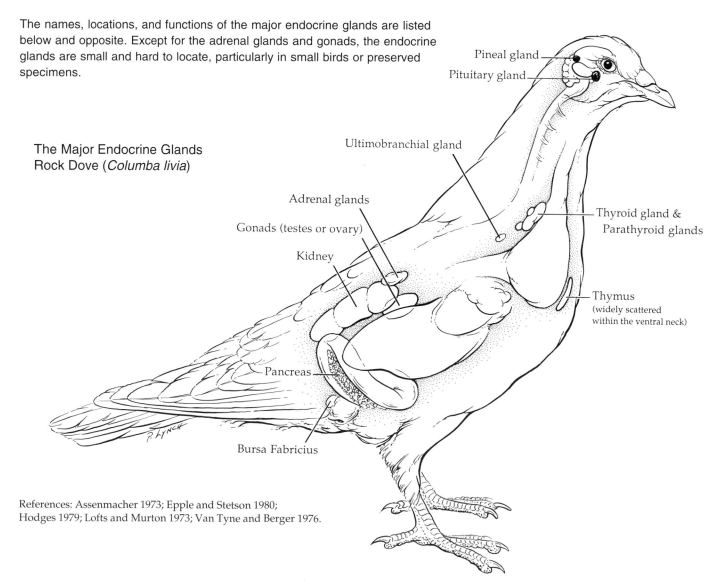

The Major Endocrine Glands
Rock Dove (*Columba livia*)

Pineal gland

Pituitary gland

Ultimobranchial gland

Adrenal glands

Gonads (testes or ovary)

Kidney

Thyroid gland &
Parathyroid glands

Thymus
(widely scattered
within the ventral neck)

Pancreas

Bursa Fabricius

References: Assenmacher 1973; Epple and Stetson 1980; Hodges 1979; Lofts and Murton 1973; Van Tyne and Berger 1976.

The Major Endocrine Glands

Gland or organ	Location, hormones	Function of gland [hormones secreted]
Pituitary	Ventral surface of the brain TSH, ACTH, FSH, LH, prolactin, antidiuretic hormone, and oxytocin	The pituitary has two lobes. The anterior lobe secretes the "master" control hormones of the body, regulating the actions of the thyroid [TSH], adrenals [ACTH], and gonads [FSH, LH, prolactin]. The posterior lobe produces hormones that act on the kidney to conserve water [antidiuretic hormone] and a hormone that stimulates uterine contractions during egg laying [oxytocin].
Thyroid	Ventral base of neck Thyroxin	The thyroid gland produces the hormone [thyroxin] that controls most aspects of growth and development of the body, feathers, and pigmentation patterns. Thyroxin also helps regulate the molting of feathers and in some species apparently stimulates the urge to migrate.
Parathyroid	Ventral base of neck Parathormone	The parathyroid gland consists of three to four tiny lobes just dorsal to the body of the thyroid gland. It produces a hormone [parathormone] that helps regulate the levels of phosphorus and calcium in the body. This is crucial in the female, where the parathyroid and the ovary regulate calcium for eggshell production.
Adrenal	Just anterior to kidneys Cortex: mineral corticoids, cortisone, androgens, estrogen, progesterone, corticosterone Medulla: epinephrine and norepinephrine	The paired adrenal glands consist of two functionally unrelated tissues that produce different sets of hormones. The adrenal cortex secretes hormones that control mineral concentrations in the body [mineral corticoids] and regulate blood sugar levels and liver sugar metabolism [cortisone], as well as sexual hormones in both sexes [male, androgens; female, estrogens and progesterones] and a hormone related to stress [corticosterone]. The medullary tissue of the adrenals is derived from the same tissue that makes up the sympathetic neurons of the autonomic nervous system, and it secretes hormones that help the body react to stress [epinephrine (adrenaline) and norepinephrine], which can increase the basal metabolic rate and blood sugar levels during stressful periods or in sudden emergencies.
Gonads	Adjacent to kidneys Testes: testosterone Ovary: estradiol, progesterone, testosterone	The gonads consist of the paired testes in males and the single left ovary in females of most bird species. The testes produce the male sex hormone [testosterone]; the ovary secretes estrogens [estradiol, progesterone] and some testosterone. Both male and female hormones regulate the secondary sexual characteristics of the bird and control behavioral responses to the opposite sex.
Thymus **Bursa of Fabricius**	Ventral area of neck Pouch on wall of cloaca, found only in very young birds, atrophies in adults	These widely separated glands are both involved with forming and stimulating the cells of the immune system, particularly in young birds. Although no secretions have been isolated, both glands are assumed to secrete hormones that stimulate the production of antibodies to infections and the production of lymphocytes.

6 EGG STRUCTURE AND DEVELOPMENT

The avian egg is a miracle of natural engineering. Light and strong, it provides everything a developing bird embryo needs from just after the ovum is fertilized until the chick hatches. Bird eggs share many similarities with reptile eggs; a few crucial adaptive changes over the reptile egg, however, have allowed birds to exploit a wider range of habitats than do reptiles. In the reptile egg the principal nutrients are proteins, whereas in the bird embryo the principal stored nutrients are fats within the yolk sac. When metabolized the fats in a bird egg yield more water and a higher relative amount of energy than do the proteins of reptile eggs. Birds thus can nest in a wider, drier range of habitats than reptiles. The shell of the bird egg is also much more complex than that of reptiles, because birds are homoiothermic and so have a much higher metabolic rate, even in the egg. The bird egg must allow an efficient exchange of gases to support the growing chick, allowing waste gases out and oxygen in. The avian eggshell is not simply a coating of calcium carbonate but a complex laminate of mineral crystals embedded within a weblike matrix of material similar to collagen fibers. Reptiles, birds, and mammals are called **amniotes** because the eggs of all three groups develop within an amniotic membrane. The amniotic membrane forms an enclosed, watery environment that protects the developing fetus and allows it to exchange gases (and in mammals, nutrients) with the surrounding world (Carey 1980, 1983; Romanoff and Romanoff 1949).

Carefully crack a hen's egg and gently pour the contents onto a clean, smooth surface to examine some of the structures within the egg. Save the eggshell parts. Remember that hens' eggs from the supermarket are not fertilized and will not show more than rudimentary development of the germinal spot on the yolk.

Observe the following structures

The difference between the runny **thin albumen** and the **thick albumen** should be readily apparent. The thick albumen forms a loose cloud around the **yolk sac**, and the thin albumen fills out the rest of the space within the immature egg. Note the **chalaza**, a twisted, stringlike, whitish structure attached to both sides of the yolk sac. The chalaza suspends the yolk and allows it to rotate freely, so that the embryo developing on the surface of the yolk sac is always kept above the yolk, not below it. Look carefully at the yolk sac, and find the small white circle marking the **germinal spot**, or blastula. In the unfertilized hen's egg the blastula is featureless and will not develop further. Examine the eggshell, and notice that the more rounded end of the shell contains within its cavity a small airspace between the inner **keratin shell membranes** and the shell itself. Note the smooth, clear membrane that coats the inner shell. These are layers of keratin laid down by the isthmus of the oviduct; they form the first layer of the shell. Outside the keratin layers lie the layers of **calcium carbonate** that make up the hard eggshell. A network of protein fibers runs through the calcium layers, and the columns of calcium are separated by minute pores that permit gaseous exchange. The structure of the eggshell is discussed on the following pages.

The color of the yolk in bird eggs is heavily influenced by the bird's diet. The common yellow yolk color is from yellow-red carotene compounds picked up in vegetable foods and seeds. The chicken yolk is actually composed of concentric alternating layers of white and yellow yolk. This layering is analogous to tree rings—as the hen eats during the day yolk layers are colored by the yellow carotene compounds in the food, and at night the yolk layers are clear.

Schematic of the Internal Anatomy of a
Domestic Hen's Egg (*Gallus* sp.)

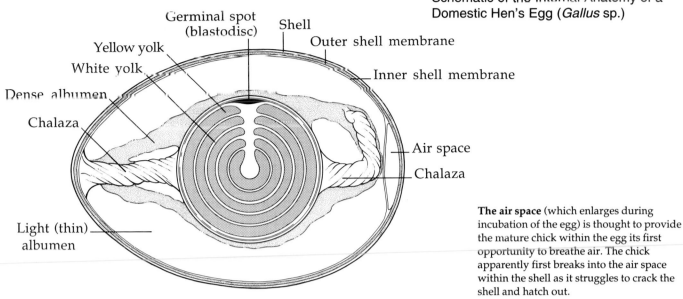

Germinal spot
(blastodisc)
Yellow yolk
White yolk
Dense albumen
Chalaza
Light (thin)
albumen
Shell
Outer shell membrane
Inner shell membrane
Air space
Chalaza

The air space (which enlarges during incubation of the egg) is thought to provide the mature chick within the egg its first opportunity to breathe air. The chick apparently first breaks into the air space within the shell as it struggles to crack the shell and hatch out.

References: Walter and Sayles 1949; Witschi 1956.

Form and Function
The shape of bird eggs

Golden Plover

Common Loon

Tawny Owl

Common Murre

Red-necked Grebe

Bird eggs vary widely in shape, from the spherical form of owls' eggs to the long, narrow eggs of cormorants and swallows. Egg shape is determined by such factors as the size and natural history of the hen, as well as by environmental forces that have shaped the evolution of the species. The muscles of the magnum and isthmus areas of the oviduct form the egg as the eggshell is laid down. The shape and size of the pelvis also affect egg shape. Species with deep pelvis bones, such as owls, tend to lay round eggs, and birds with long, shallow pelvis bones, such as cormorants, lay long, elliptical eggs.

Cavity nesters like owls and kingfishers tend to lay round, white eggs, as they have no need to conceal the eggs and there is little danger of the eggs rolling out of the nest. The classic oval shape of the chicken egg is typical of most bird species, although many groups, such as ducks and pigeons, lay oval eggs that are equally rounded at either end. Both types fit well in cup-shaped nests and do not roll as readily as round eggs would. Cliff-nesting species like murres and guillemots lay long, pointed eggs that roll in tight circles when disturbed. This prevents the eggs from rolling off the narrow ledges on which these species breed. Long, narrow eggs may also be a compromise between a narrow pelvis and the need to maximize the size of the egg to ensure the embryo's survival. The narrow shape allows a much larger egg to be laid through a narrow pelvis opening. Many shorebirds and other ground-nesters lay "pyriform," or top-shaped, eggs that have a very sharp point on one end. Three to four pyriform eggs fit tightly into a nest with pointed ends in, helping the female to cover that maximum number of eggs efficiently.

References: Palmer 1976; Preston 1969; Van Tyne and Berger 1976.

7 THE EGGSHELL AND EMBRYONIC MEMBRANES

The avian egg is a self-contained womb in which the embryo is protected and provided with the food and nutrients it needs to develop and hatch. But the egg is not a completely closed system, shutting off all outside contact. Birds, even as they are developing in the egg, are warm-blooded animals, with all of the metabolic activity that homoiothermy demands. The chick within the egg must be able to respire—to exchange gases and water vapor with the outside world—or it will suffocate.

The hard mineral shell of the egg is deposited in the uterus area of the oviduct as the egg descends toward the cloaca. A basal layer of mammillary crystals (pure calcium carbonate, $CaCO_3$) grow up from the mammillary knobs and are the primary components of the shell. They are arranged in vertical palisades and interwoven with a light network of collagen fibers. On the outer surface of the egg a fine layer of crystalline calcite is overlaid with a protein cuticle that lends most eggs a smooth, slightly lustrous appearance. The shell thus has both rigid elements (the mineral layers) and a flexible component (the collagen fibers) that bind the brittle calcite crystals. Flaws between the calcite crystals form tiny pores through the calcite layer, and it is through these pores that the embryo "breathes" passively while developing (Rahn et al. 1979).

The calcium metabolism of the breeding female is obviously crucial in egg production. Hens entering the breeding season may transfer as much as 12 percent of their bone calcium to the production of eggshells while laying. The major bones of females are particularly dense with calcium deposits as they enter the breeding season. A special tissue called **medullary bone** develops within the marrow cavities of the major bones. This medullary bone tissue (see figure at right) serves as a substrate for crystal spicules of calcium carbonate. The mineral is stored within the bones until eggshell production begins and is then rapidly reabsorbed into the bloodstream and transferred to the uterus.

One of the world's best-documented environmental disasters occurred because of the sensitivity of the avian calcium metabolism. In the 1950s and 1960s the pesticide DDT devastated many populations of birds worldwide, including such well-known species as the Peregrine Falcon (*Falco peregrinus*) and the Osprey (*Pandion haliaetus*). Aside from the outright poisoning of many birds, DDT and its metabolite derivative DDE block the formation of a crucial carbonic anhydrase enzyme that controls the deposition of calcium carbonate within the uterus. Proper shell thickness is crucial to the success of the egg—a loss in shell thickness of 20 percent is usually fatal, because the parent birds eventually crush eggs with such thin shells (Peakall 1970). DDT can also cause flaws in the calcium structure of the shell, resulting in many fewer shell pores for gas exchange and a high rate of suffocation within the egg (Fox 1976). Predatory birds at the top of the food chain were especially hard hit by DDT-related eggshell thinning. Partly through human assistance, Osprey and Peregrine populations are now starting to recover from this disaster, almost twenty years after DDT was banned.

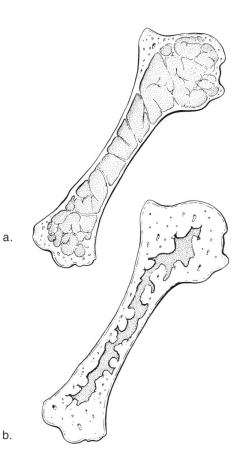

a.

b.

Domestic chicken hens (*Gallus* sp.) store calcium by depositing it within the marrow cavities of their bones (*a.*). The calcium is held within an open, spongy medullary bone tissue that lines the walls of the long bones. To store calcium, specialized marrow cells called **osteoblasts** accumulate and deposit calcium (*b.*). When calcium demand is heavy (as in egg laying), **osteoblast** cells reabsorb the calcium crystals and secrete it back into the bloodstream for use in the ovaries (Taylor 1970).

The diagrammatic view of an eggshell, opposite, shows the structure of the outer calcium layers, the inner shell membranes, and the chorioallantois membrane of the developing embryo. The pores of the eggshell permit gaseous exchange between the embryo and the outside environment. (After Rahn et al. 1979)

References: *above*, redrawn after Taylor 1970; *right*, redrawn after Rahn et al. 1979.

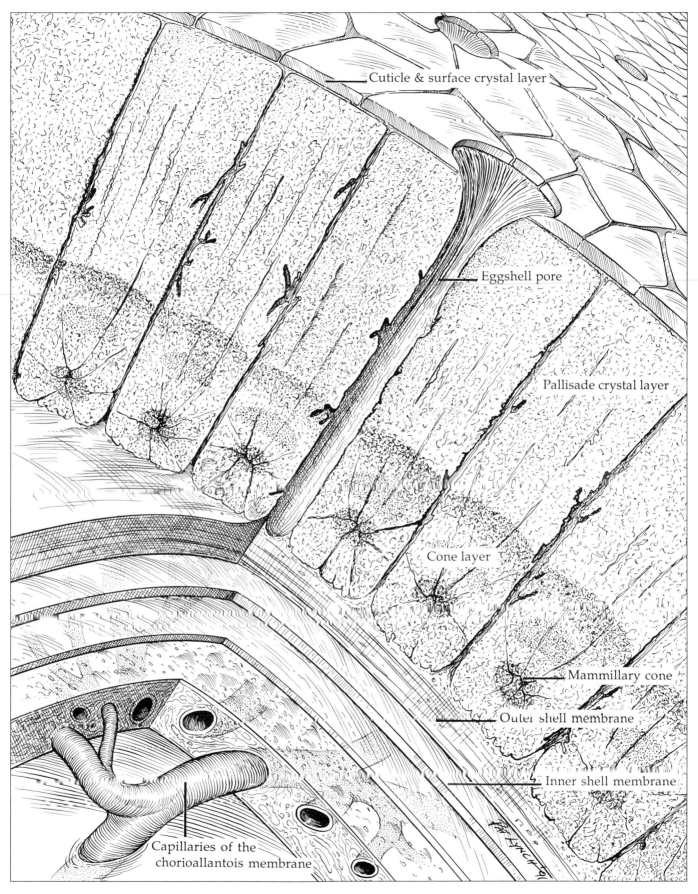

Cuticle & surface crystal layer

Eggshell pore

Pallisade crystal layer

Cone layer

Mammillary cone

Outer shell membrane

Inner shell membrane

Capillaries of the
chorioallantois membrane

8   THE DEVELOPING EGG

The egg of a bird consists of the **ovum,** a series of layers of **albumen,** and the **shell** with its internal membranes (Portmann and Stingelin 1964). Like the fetuses of reptiles and mammals, the avian embryo is surrounded by amniotic membranes. The **ovum** is the large single-celled structure commonly called the **yolk** and is produced by the ovary. The embryo is formed from a fertilized **blastoderm,** or germinal spot on the ovum. Three layers of tissue, the **amnion, chorion,** and **allantois,** protect and segregate the developing embryo. The extraembryonic membranes consist of two layers—the ectoderm and mesoderm, or endoderm and mesoderm, depending on the interface of the membrane.

As the blastoderm matures, the **amnion** grows out around the embryo to form the amniotic fluid. The embryo develops suspended in amniotic fluid, which is in turn surrounded by the amniotic membrane. The second concentric membrane, the **chorion,** expands to line the inner shell wall. The third, the **allantois,** develops from the hind gut of the embryo and also lines the inner shell. The combined membranes are called the **chorioallantois,** and this highly vascularized surface acts as both respiratory and excretory systems for the embryo. The blood vessels of the chorioallantois carry oxygen from the shell lining to the embryo and bring carbon dioxide back to the shell surface. This inner shell lining then acts like a large, passive lung. Metabolic wastes are deposited as urate salts crystals within the allantois. The insoluable urates are the most biologically inert form of nitrogenous waste. If birds, like mammals, excreted nitrogenous wastes as ammonia (urea), they could not reproduce by laying eggs: the toxic ammonia within the egg would quickly poison the embryo.

The yolk of the egg supplies the embryo with nutrients, storing food until the chick hatches. The yolk sac surrounding the yolk is another embryonic membrane, sometimes called the **vitelline membrane.** The yolk is approximately 30 percent lipids (fats), 15 percent protein, and 55 percent water. Fat, not protein, is the primary food in bird eggs. This gives bird eggs a significant advantage over reptile eggs, in which the primary nutrients are stored as protein, because when broken down, fats yield more metabolic energy and water per unit than do proteins. They can thus survive in much drier environments than reptile eggs. The yolk color is influenced by the hen's diet. Yolks are generally pale to bright yellow, but orange and even bright red can occur when the hen eats foods rich in red-orange carotenes (Carey 1983; Lofts and Murton 1973; O'Connor 1984).

Once the ovum bursts from the ovary it begins its journey through the oviduct. The first significant pause is at the magnum, where it receives the albumen. This can take three hours in the domestic hen. The albumen, or white, is named for the main protein component in the clear liquid surrounding the yolk, but the thick albumen also contains a significant portion of the protein mucin, which gives it its characteristic viscosity. The albumen is the egg's primary water supply and functions as a shock absorber for the yolk sac and embryo. The mass of the albumen also stabilizes the temperature of the embryo, retaining the hen's body heat when she leaves the nest (Van Tyne and Berger 1976). The egg next pauses at the isthmus of the oviduct, where it receives the shell with its internal membranes. In the final stage, the fully formed egg moves on to the uterus, where it is held for a time before it is passed through the muscular vagina and laid.

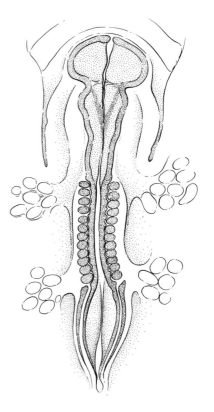

Dorsal view of a twelve-somite chicken (*Gallus* sp.) embryo. (After Portmann and Stingelin 1964)

Embryonic Development
Common Tern (*Sterna hirundo*)

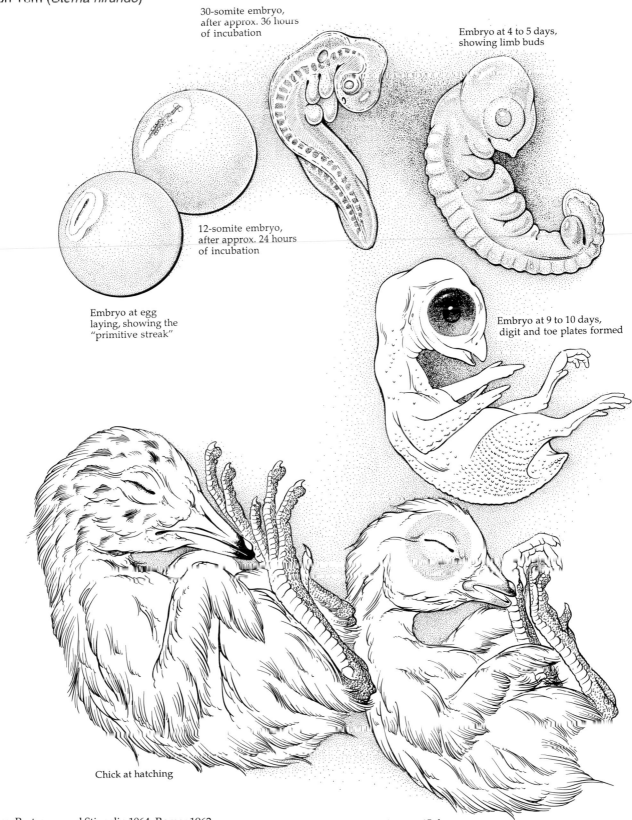

30-somite embryo, after approx. 36 hours of incubation

Embryo at 4 to 5 days, showing limb buds

12-somite embryo, after approx. 24 hours of incubation

Embryo at egg laying, showing the "primitive streak"

Embryo at 9 to 10 days, digit and toe plates formed

Chick at hatching

Embryo at 15 days

References: Portmann and Stingelin 1964; Romer 1962.

9 THE CHICK AND HATCHING

At the end of its development within the egg the incipient hatchling begins one of the most strenuous events of a bird's life—hatching from the egg. Throughout the development of the embryo the egg steadily loses water by transpiration through the chorioallantois membrane. Because of this water loss and the loss of yolk fats metabolized during development, the egg is much lighter at hatching than when it was laid. The eggshell, too, is thinner than when it was laid down because the chick has absorbed much of the calcium from the inner shell lining, but the shell still represents a substantial barrier to the hatchling. The chick also faces several major physiologic hurdles before hatching. Most critically, it must shift from respiring via the chorioallantois to breathing with its lungs. The chick takes its first breaths of air from the air space within the shell as it struggles to break free. It also absorbs the remaining yolk sac into the abdominal cavity and begins to swallow any remaining amniotic fluid before hatching.

Two specialized structures found only in hatchlings aid the chick in its struggle to break open the shell. A small, sharp egg tooth develops on the tip of the upper beak; the pre-hatchling rasps the inner shell wall to weaken the surface, repeatedly extending its head to drive the egg tooth against the wall. A substantial enlargement of the complexus muscle (the "hatching muscle") at the nape of the chick's neck helps brace and cushion the head as the chick forces the egg tooth through the shell (Bock and Hikida 1969). After hatching most of the fluid within the complexus muscle is reabsorbed, and the muscle continues to function as a head extender in most adult birds. The egg tooth is lost in the weeks after hatching.

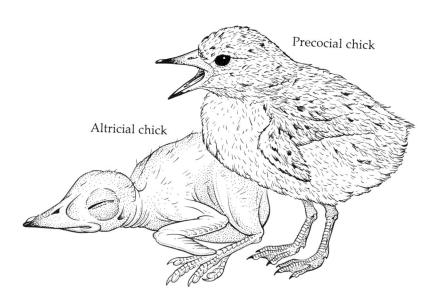

Precocial chick

Altricial chick

Precocial versus Altricial Chicks

Precocial chicks hatch from the egg with open eyes and thick coats of natal down; they are strong enough to leave the nest within one or two days. Most shorebirds, gamebirds (including the domestic chicken), and ducks have precocial chicks that can walk away from the ground nest within hours of hatching. Altricial chicks, by contrast, are hatched blind and virtually naked; they require long periods of feeding before they are strong enough to leave the nest on their own. Most tree-nesting and cavity-nesting birds have helpless altricial young. Not surprisingly, the eggs of precocial species are much larger and richer in nutrients than those of altricial species, and this is the best clue as to why such different egg-laying strategies have evolved. In ground-nesting precocial species, the hen must gather abundant resources before egg laying, "preloading" her investment in the chicks before they hatch. Precocial chicks are able to scatter soon after hatching, reducing the chances that a predator will find and destroy the hen's brood in one meal. Because the nests of altricial species are more sheltered from predators, they can afford a longer period of nesting care after the chicks hatch and their eggs can be correspondingly smaller.

Urogenital System   *Chapter Worksheet*

1. Why do birds and reptiles excrete uric acid from their kidneys instead of urea?

2. What is the major function of the adrenal glands? Where are they located?

3. How can seabirds drink saltwater, and how do they rid their bodies of excess salt?

Chapter Worksheet, continued

4. What is the function of the medullary bone in the female bird?

5. Define these terms:

   a. Endocrine gland

   b. Exocrine gland

   c. Chorioallantois membrane

   d. Vas deferens

   e. Thyroxin

References

Blackburn, D. G., and H. E. Evans. 1986. Why are there no viviparous birds? *Amer. Nat.* 128:165–90.

Carey, C. 1983. Structure and function of avian eggs. In R. F. Johnson, ed., *Current ornithology.* Vol. 1. New York: Plenum Press, 69–103.

Collias, N., and E. Collias. 1984. *Nest building and bird behavior.* Princeton, N.J.: Princeton University Press.

Drent, R. H. 1975. Incubation. In D. S. Farner, J. R. King, and K. C. Parkes, eds., *Avian biology.* 8 vols. New York: Academic Press, 5:333–420.

Epple, A., and M. H. Stetson, eds. 1980. *Avian endocrinology.* New York: Academic Press.

Fox, G. 1976. Eggshell quality: its ecological and physiological significance in a DDE-contaminated Common Tern population. *Wilson Bull.* 88:459–77.

Lofts, B., and R. Murton. 1973. Reproduction in birds. In D. S. Farner, J. R. King, and K. C. Parkes, eds., *Avian biology.* 8 vols. New York: Academic Press, 3:1–107.

O'Connor, R. 1984. *The growth and development of birds.* New York: John Wiley and Sons.

Portmann, A., and W. H. Stingelin. 1964. Development, embryonic. In A. L. Thomson, ed., *A new dictionary of birds.* New York: McGraw-Hill.

Rahn, H., A. Ar, and C. V. Paganelli. 1979. How eggs breathe. *Sci. Amer.* 240(2):46–55.

Romanoff, A., and A. Romanoff. 1949. *The avian egg.* New York: John Wiley.

Romer, A. S. 1962. *The vertebrate body.* 3d ed. Philadelphia: W. B. Saunders.

Schmidt-Nielsen, K. 1959. Salt glands. *Sci. Amer.* 200(1):109–16.

Skutch, A. F. 1976. *Parent birds and their young.* Austin: University of Texas Press.

Van Tyne, J., and A. J. Berger. 1976. *Fundamentals of ornithology.* 2d ed. New York: John Wiley and Sons.

Webster, D., and M. Webster. 1974. *Comparative vertebrate morphology.* New York: Academic Press.

Witschi, E. 1961. Sex and secondary sexual characters. In A. J. Marshall, ed., *Biology and comparative physiology of birds.* 2 vols. New York: Academic Press, 2:115–68.

Great Gray Owl
(*Strix nebulosa*)

The Great Gray Owl's most distinguishing characteristics are the huge facial disks around its eyes, which act like funnels to focus sounds into its ears when hunting. The Great Gray Owl hunts lemmings and other small rodents, using its keen ears to listen for the sounds the lemmings make as they tunnel under the snow cover. Although it appears to be huge, the Great Gray is actually smaller and lighter than the Great Horned Owl of temperate North America. It looks bigger because of its thick gray coat of feathers, evolved to shield the owl against the cold of the northern boreal forests of North America and Eurasia.

The Sensory and Nervous Systems

11

If an animal's central nervous system exists to receive, analyze, and coordinate the body to react effectively to external stimuli, then birds should surely rank among the most intelligent of animals. Yet as a group birds have traditionally been regarded as much less intelligent than mammals, and their mental abilities have been long slighted in such popular insults as "bird brain" and "dodo." Many of these misconceptions arose in the study of biology during the nineteenth century and before, when natural historians were keen to compose a comprehensive ranking of everything in the natural world, from the most humble algae and microorganisms at the bottom to humans (of course!) at the top of the evolutionary scale. Birds fit into the grand scheme somewhere between reptiles and mammals, so they were also ranked between reptiles and mammals in intelligence. It is not that the earlier scientists were dead wrong on the matter; the most intelligent nonhuman mammals do seem to be more intelligent than the smartest of birds. But a simple ranking of a few representative species from either group slights the true complexity of the natural world. Many types of birds, particularly hunters and other active foragers, exhibit complex intelligence and learning abilities. Simple comparisons of avian versus mammalian intelligence often fail to credit the extraordinary abilities of the avian brain in physical coordination, long-distance navigation, and vision (Hoage and Goldman 1986).

The nervous system of a bird consists of two major parts: the central nervous system of the brain and spinal cord, and the peripheral nervous system of cranial and spinal nerves (Breazile and Hartwig 1989; Pearson 1972; Portmann and Stingelin 1961). The fundamental unit of the nervous system is the **neuron,** a specialized cell able to conduct electrical impulses. The tissues of the nervous system are composed of groups of neurons and other cells that physically support and nourish the neurons. The central nervous system receives stimuli from within the body and from the outside world through the sense organs and peripheral nervous system. After processing the information, the central nervous system sends motor signals through the spinal cord and peripheral nervous system to control the bird's movements and reactions to the world around it.

The avian brain shows many structural similarities to the brains of reptiles and is now known to be quite unlike the mammalian brain in physical and functional organization (Goldby 1964). Both birds and mammals have relatively large brains in comparison to body size, and their cerebella and cerebral hemispheres are proportionately much larger than those found in reptiles. But neurologists have discovered that, unlike mammals, birds do not depend on a complex cerebral cortex for learning and memory. Instead, the avian cerebrum is composed primarily of the **corpus striatum,** a complex structure underlying a thin cerebral cortex and occupying most of the anterior volume of the avian brain. This dependence on the corpus striatum is thought to explain why much of bird behavior is rather stereotypical and based primarily on inborn instincts and fixed behavior patterns rather than on learning and memory. Crows, parrots, and other more intelligent groups of birds show greater development in the cerebral hemispheres of the brain than groups with less capacity to learn quickly. The evidence is that most birds are capable of learning new tasks and are reasonably flexible and adaptable to changing circumstances; many species compare well with mammals in overall learning ability.

1 | THE BRAIN AND CRANIAL NERVES

Thoroughly remove any skin or muscle tissue from the back of the skull of your dissection specimen, and carefully cut away the thin occipital, parietal, and frontal bones covering the brain. Try to preserve the thin meningeal tissues that lie between the brain and the inner surface of the skull. Be particularly careful in dissecting fresh specimens; an unpreserved brain is often so soft and fragile that it cannot support its own weight and will break apart if handled casually.

Observe the following structures

Note the nearly **vertical orientation** of the pigeon's brain within its skull. This odd arrangement is caused by the need to house the enormously **enlarged optic lobes** of the brain, which could not fit into the thin interspace between the eyes. The whole brain has thus been shifted back and angled upward within the skull. The orientation of the brain varies considerably in birds, from nearly horizontal to angles beyond the vertical—almost "upside down" within the skull. The thin layers of tissue covering the brain are the **meninges** and are composed of a tough outer layer, the **dura mater,** and a fine inner layer, the **pia mater**. Remove these meningeal layers before proceeding. Then carefully remove the brain from the skull, taking care to preserve as much of the posterior brain as you can manage. Note the following general features of the avian brain:

The tiny **olfactory lobes** form the most anterior and dorsal part of the brain. The large, smooth **cerebral hemispheres** are situated just posterior to the olfactory lobes. Just lateral and posterior to the cerebral hemispheres are the large **optic lobes** of the brain. Note the thick fiber tracts of the **optic chiasma** on the ventral surface of the brain.

The large **cerebellum** sits at the posterior midline of the brain. The cerebellum controls muscular coordination and, not surprisingly, is quite large in birds. The **medulla oblongata** lies at the base of the brain just below the cerebellum and extends posteriorly into the spinal cord. If possible, note and identify the tiny **cranial nerves** exiting the medulla area.

Slice one of the cerebral hemispheres and the cerebellum along the sagittal fissure at the midline (as in the diagram to the right) and note the **lateral ventricle** within the tissue of the cerebrum. Locate the **third ventricle** at the posterior of the cerebrum. Just anterior and medial to the lateral ventricle lies the mass of the **corpus striatum**, the center of most instinctive behavior patterns in birds. A very thin **cerebral cortex** covers the dorso-medial surface of the cerebrum. Between the medulla oblongata and the cerebellum note the **Aqueduct of Sylvius,** or **fourth ventricle**.

References: Goldby 1964; Portmann and Stingelin 1961; Webster and Webster 1974.

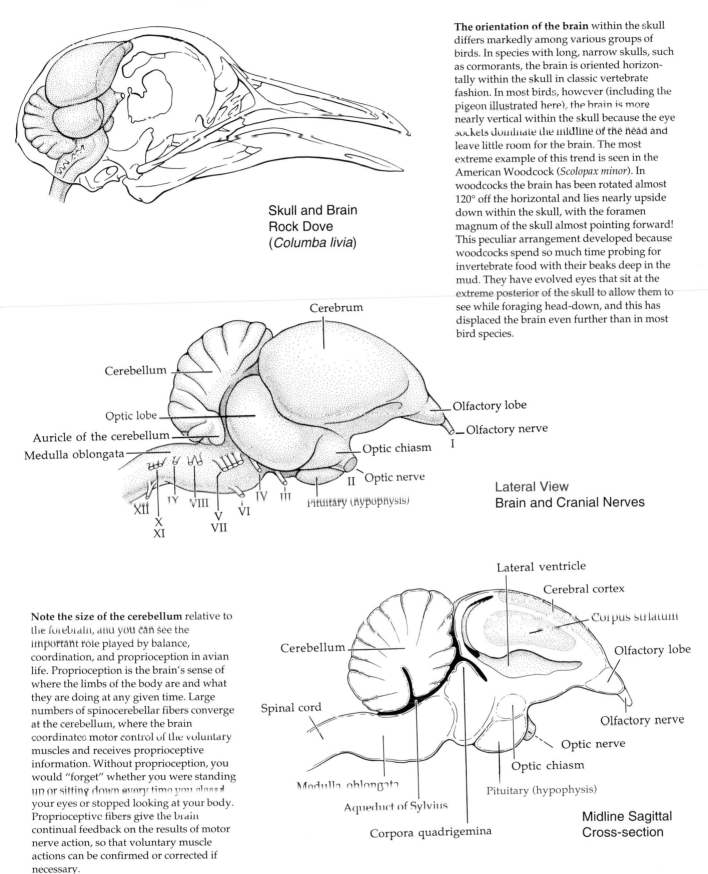

Skull and Brain
Rock Dove
(*Columba livia*)

The orientation of the brain within the skull differs markedly among various groups of birds. In species with long, narrow skulls, such as cormorants, the brain is oriented horizontally within the skull in classic vertebrate fashion. In most birds, however (including the pigeon illustrated here), the brain is more nearly vertical within the skull because the eye sockets dominate the midline of the head and leave little room for the brain. The most extreme example of this trend is seen in the American Woodcock (*Scolopax minor*). In woodcocks the brain has been rotated almost 120° off the horizontal and lies nearly upside down within the skull, with the foramen magnum of the skull almost pointing forward! This peculiar arrangement developed because woodcocks spend so much time probing for invertebrate food with their beaks deep in the mud. They have evolved eyes that sit at the extreme posterior of the skull to allow them to see while foraging head-down, and this has displaced the brain even further than in most bird species.

Cerebrum

Cerebellum

Optic lobe

Auricle of the cerebellum

Medulla oblongata

XII

X   IY   VIII   IV   III
XI   V   VI
         VII

Olfactory lobe

Olfactory nerve

I

Optic chiasm

Optic nerve

II

Pituitary (hypophysis)

Lateral View
Brain and Cranial Nerves

Note the size of the cerebellum relative to the forebrain, and you can see the important role played by balance, coordination, and proprioception in avian life. Proprioception is the brain's sense of where the limbs of the body are and what they are doing at any given time. Large numbers of spinocerebellar fibers converge at the cerebellum, where the brain coordinates motor control of the voluntary muscles and receives proprioceptive information. Without proprioception, you would "forget" whether you were standing up or sitting down every time you closed your eyes or stopped looking at your body. Proprioceptive fibers give the brain continual feedback on the results of motor nerve action, so that voluntary muscle actions can be confirmed or corrected if necessary.

Lateral ventricle

Cerebral cortex

Corpus striatum

Cerebellum

Olfactory lobe

Spinal cord

Olfactory nerve

Optic nerve

Optic chiasm

Pituitary (hypophysis)

Modulla oblongata

Aqueduct of Sylvius

Corpora quadrigemina

Midline Sagittal
Cross-section

1 THE BRAIN AND CRANIAL NERVES, *continued*

Viewed from above, the smooth, unfolded surfaces of the avian cerebral hemispheres show the similarity of the bird's brain to those of crocodiles and other reptiles. Not all mammals have heavily folded cerebral hemispheres, but most larger mammals do. As mentioned earlier, this is a reflection of the profound structural and functional differences between avian and mammalian brains. In birds the cerebral cortex (the upper and outer surfaces of the cerebral hemispheres) is much less important than it is in mammals, and it is therefore less complex in structure; birds rely instead on the corpus striatum. The large optic lobes are very complex in structure and may play an important associative and coordinating function for visual information, much as the mammalian cerebral cortex functions to integrate and coordinate visual stimuli in mammals.

Observe the following structures

View the posterior-dorsal surface of the brain as shown below. Note in this view the smooth surfaces of the **cerebral hemispheres**, divided along the midline of the body by the **mid-sagittal groove**. The **olfactory lobes** at the anterior tip of each cerebral hemisphere are tiny, reflecting the poor development of the sense of smell in most birds. At the posterior end of the mid-sagittal groove lies the **epiphysis**, between the cerebral hemispheres and the cerebellum. Just behind and lateral to the cerebral hemispheres are the two **optic lobes**. These huge lobes are the most distinctive feature of the avian brain. The **cerebellum** is composed of a large lobe on the midline of the brain, the **cerebellar vermis**, and smaller lateral **flocculi**, or **auricles**. Posterior to and below the cerebellum lies the **medulla oblongata**, where many of the basic metabolic functions of the body are regulated and where most of the **cranial nerves** exit the brain.

The medulla oblongata forms the rear of the brain and merges into the spinal cord. In birds (as in humans) the medulla is responsible for many basic "housekeeping" tasks— monitoring and maintaining a constant body temperature and adequate respiration rates. It also monitors other fundamental physiologic conditions crucial to maintaining life in a warm-blooded animal.

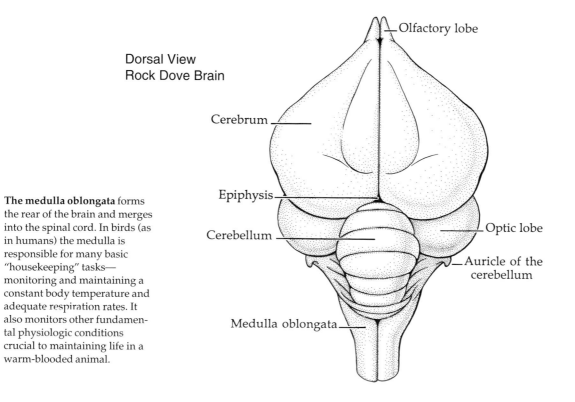

Dorsal View
Rock Dove Brain

Olfactory lobe

Cerebrum

Epiphysis

Cerebellum

Optic lobe

Auricle of the cerebellum

Medulla oblongata

The Cranial Nerves

Number	Name	Type	Function
I	**Olfactory**	Sensory	Fibers of the olfactory carry impulses from the nose to the olfactory lobe of the brain. Small in birds.
II	**Optic**	Sensory	Fibers of the optic carry impulses from the retina of the eye to the optic lobes of the brain. The optic "nerve" is actually a fiber tract of the brain, not a true nerve.
III	**Oculomotor**	Motor	Fibers of the oculomotor raise and lower the eyelids, open and close the iris of the eye, and focus the lens of the eye. The oculomotor also supplies the eye muscles of the orbit.
IV	**Trochlear**	Motor	Fibers of the trochlear supply the muscles of the eye, primarily the superior oblique muscle.
V	**Trigeminal**	Mixed	
	Ophthalmic division		Sensory fibers from the eye region, eyelids, forehead, and crown region.
	Maxillary division		Sensory fibers from the bill, upper jaw, mouth lining, and facial and lores region of the anterior head.
	Mandibular division		Sensory fibers from the lower jaw and bill. Motor nerves to the anterior head, bill, and mouth regions.
VI	**Abducens**	Motor	Motor fibers of the abducens control muscles of the eye and eyelid, particularly the nictitating membrane.
VII	**Facial**	Mixed	Sensory fibers of the facial collect impulses from the mouth lining and anterior tongue. Its motor fibers control the salivary glands and muscles of the anterior face and head.
VIII	**Acoustic,** or **Vestibulocochlear**	Sensory	Sensory fibers of the acoustic control muscles of the eye and eyelid, particularly the nictitating membrane.
	Vestibular branch		Carries impulses from the semicircular canals of the inner ear to support equilibrium and balance.
	Cochlear branch		Carries impulses from the cochlea of the inner ear to provide hearing.
IX	**Glossopharyngeal**	Mixed	Sensory fibers of the glossopharyngeal collect impulses from the pharynx, posterior tongue, and taste buds of the tongue. Its motor fibers control the pharynx and salivary glands.
X	**Vagus**	Mixed	Sensory fibers of the vagus collect impulses from the pharynx, esophagus, neck, thorax, and abdomen. Its motor nerves control the same regions.
XI	**Accessory**	Motor	
	Cranial branch		Motor control of pharynx and posterior mouth area.
	Spinal branch		Motor control of some neck and back musculature
XII	**Hypoglossal**	Motor	Motor branches of the hypoglossal control the tongue and syrinx.

References: Goldby 1964; Pearson 1972; Portmann and Stingelin 1961; Stettner and Matyniak 1968; Webster and Webster 1974.

2 SPINAL CORD & PERIPHERAL NERVOUS SYSTEM

The brain and spinal cord form the central nervous system. The brain sends signals out to the body and receives information from the body and the surrounding environment through the peripheral nervous system, which joins the central nervous system at the spinal cord. The spinal cord runs the length of the vertebral column in birds and is protected from mechanical damage by the surrounding bony **lamina** and **spinous processes** of the vertebrae. Within the **vertebral canal** the cord is protected by two **meningeal membranes**: the **dura mater** and **pia mater**. The meningeal layers, fat pads, and other connective tissue within the vertebral canal protect the delicate spinal cord from damage. In a feature unique to birds, a **rhomboid sinus** filled with a gelatinous, fatty tissue runs the length of the spinal cord, dividing the cord's dorsal horns. A small **central canal** also runs the length of the spinal cord.

The **peripheral nervous system** consists of the **spinal nerves**, which exit in pairs along the entire length of the spinal cord. At the cord each spinal nerve root consists of two branches that join to form the spinal nerve: a **dorsal sensory nerve root** and a **ventral motor nerve root**. Sensory roots tend to be smaller in birds than in other vertebrate groups. This is because birds have few sensory cells in their skin layers. The skin is covered by feathers, and does not need the elaborate networks of sensory cells found, for example, in human skin. This relative paucity of sensory cells throughout the body places fewer demands on the sensory tracts of the nervous system.

Birds typically have **thirty-eight pairs of spinal nerves:** twelve cervical pairs, eight thoracic pairs, twelve sacral pairs, and six caudal pairs serving the tail musculature. In the brachial and pelvic regions, large nets of interjoined spinal nerves control and help coordinate the wings and legs. At the base of the cervical spine, on each side of the spine, the **brachial plexus** is formed by the last two cervical nerves and first three thoracic spinal nerves. Fibers from the brachial plexus enter the wing as the **superior and inferior brachial nerves** and continue into the forearm area as the **radial, median, and ulnar nerves** of the wing. The **lumbar**, or **lumbo-sacral plexus,** forms from the first seven lumbo-sacral spinal nerves exiting the **synsacrum** (fused pelvic and sacral vertebrae). These fibers enter the thigh as the **sciatic nerve** and continue distally through the leg as the **tibial and fibular nerves**. A smaller **pudental plexus** of sacral nerves serves the lower abdomen and vent area. The spinal cord enlarges considerably within the synsacrum, forming a **lumbo-sacral enlargement**.

The meninges are a double layer of tissues that protect the brain and spinal cord from injury and provide nourishment and support to the spinal cord. The outer **dura mater** is a tough, white outer meningeal layer surrounding the brain and spinal cord, and is largely composed of collagen and other fibrous tissues. It shields the brain and cord from mechanical damage and helps support the soft tissues of the central nervous system. The dura is rich with blood vessels and actually forms part of the venous drainage system of the brain (the dural sinuses) along the sagittal midline of the brain, between the two cerebral hemispheres. The **pia mater** is the deepest meningeal layer, lying under the dura mater at the surface of the brain and spinal cord. It is heavily supplied with nerves and blood vessels and helps nourish the underlying brain and spinal cord tissues. Clear cerebrospinal fluid circulates in the space between the dura mater and pia mater.

References: Goldby 1964; Pearson 1972; Portmann and Stingelin 1961; Stettner and Matyniak 1968; Webster and Webster 1974.

Spinal Cord, Spinal Nerve Roots,
and Meningeal Layers of the Spinal Cord

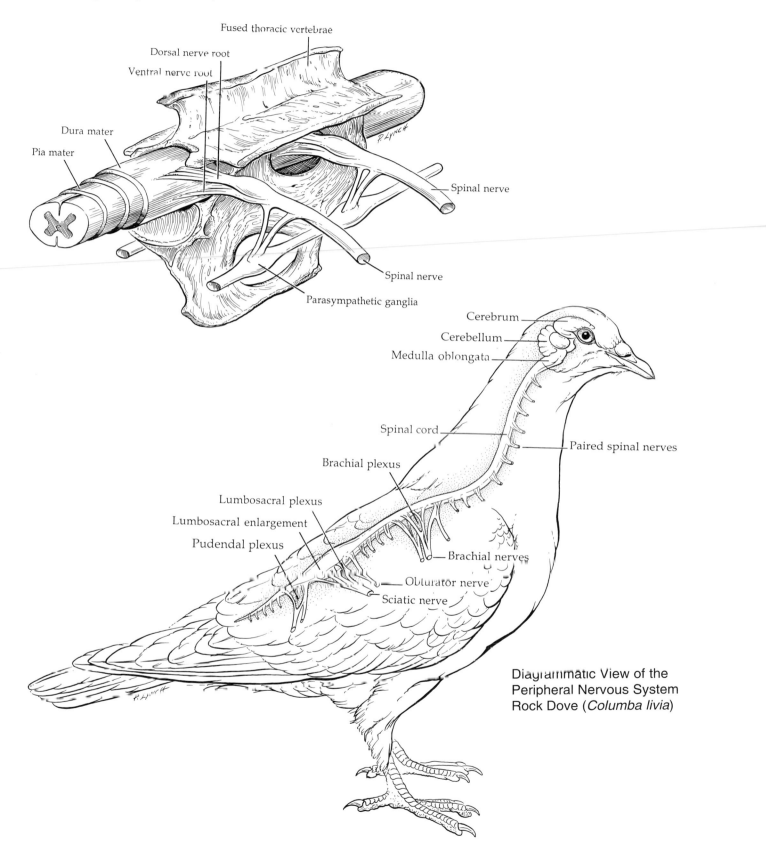

Fused thoracic vertebrae

Dorsal nerve root

Ventral nerve root

Dura mater

Pia mater

Spinal nerve

Spinal nerve

Parasympathetic ganglia

Cerebrum

Cerebellum

Medulla oblongata

Spinal cord

Paired spinal nerves

Brachial plexus

Lumbosacral plexus

Lumbosacral enlargement

Pudendal plexus

Brachial nerves

Obturator nerve

Sciatic nerve

Diagrammatic View of the
Peripheral Nervous System
Rock Dove (*Columba livia*)

3 | THE AUTONOMIC NERVOUS SYSTEM

As in reptiles and mammals, the avian autonomic nervous system controls and maintains the basic reflexes of the circulatory system, respiratory system, and abdominal viscera (Akester 1979; Bennett 1974). The medulla oblongata and other higher brain centers act through the autonomic system to excite emotional and fright reactions and to control and regulate the body's basic metabolic functions. The various parts of the autonomic nervous system are connected to the spinal cord and brain through fiber networks from the spinal and cranial nerves. The autonomic system consists of two parts—the sympathetic and parasympathetic systems— with opposite physiologic functions in regulating metabolism:

The **sympathetic system** supports the "fight or flight" reflexes of the bird, heightening respiration and heart rate and constricting blood flow to the intestines when the bird faces predatory challenges or other sudden dangers. It works closely with the adrenal glands of the endocrine system to stimulate the production of epinephrine (adrenaline) and norepinephrine, hormones (or neurohumors) that create the heightened physiologic responses seen in emergencies.

The **parasympathetic system** consists of scattered groups of ganglia found close to the organs they regulate. It acts to stimulate digestion and food processing in the intestines and also slows the heartbeat and respiration.

Observe the following structures

The **sympathetic system** is most visible in the thoracic region between the lungs and the rib cage as two chains of ganglia, called the **sympathetic trunks,** which run along the ventral (inner) surface of the ribs, one on each side of the spinal cord. Substantial plexus formations of sympathetic nerves may be seen near the aorta and dorsal to the intestines, but these may be difficult to distinguish readily from surrounding tissues, particularly in preserved specimens.

Cranial nerves are also involved in autonomic activity, particularly the **vagus nerve**, the tenth cranial nerve. Try to locate the vagus nerve as it passes near the jugular veins in the neck and continues posteriorly to send branches to the heart, aorta, esophagus, crop, proventriculus, gizzard, and duodenum.

Diagrammatic View of the Autonomic Ganglia along the Spinal Cord

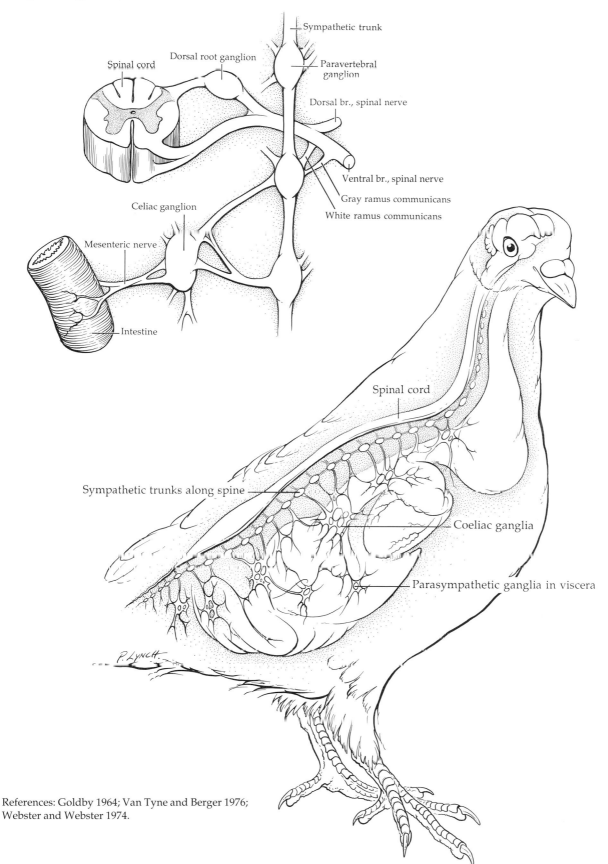

Sympathetic trunk

Paravertebral ganglion

Dorsal root ganglion

Spinal cord

Dorsal br., spinal nerve

Ventral br., spinal nerve

Gray ramus communicans

White ramus communicans

Celiac ganglion

Mesenteric nerve

Intestine

Spinal cord

Sympathetic trunks along spine

Coeliac ganglia

Parasympathetic ganglia in viscera

P. LYNCH.

References: Goldby 1964; Van Tyne and Berger 1976; Webster and Webster 1974.

4   THE SENSORY SYSTEM: ANATOMY OF THE EYE

The gross anatomy of the avian eye is quite similar to the eyes of mammals but retains a few distinctly reptilian characteristics, such as the pecten structure found within the globe of the eye and the sclerotic ring of bony plates that support the shape of the globe (Martin 1985). Relative to their body size birds have enormous eyes. The Ostrich has the one of the largest eyes of any land animal, and many birds of prey have eyes as large or larger than an adult human's. The particularly important role vision plays in the life of birds is reflected in every facet of the eye's anatomy and physiology. Birds have by far the keenest vision of all vertebrates. The Golden Eagle (*Aquila chrysaetos*) can reportedly see the movements of a rabbit from more than a mile away, and many birds have retinas with two to three times the visual acuity of the human eye (Brown 1977).

To dissect the eye of your specimen, first remove the skin around the eye to expose the bony rim of the orbit. Note the **lacrimal (tear) gland** at the ventral rim of the orbit. The eye is held in the orbit by a series of six eye muscles, which are homologous to those that move the human eye within its eye sockets. Free the globe of the eye from its muscular attachments carefully. Try to avoid puncturing the globe as you work back around it toward the **optic nerve** deep within the orbit. Cut the optic nerve and remove the eye from the eye socket, then transect the eye as shown in the diagram at the right.

Observe the following structures

Cornea — The transparent tissue covering the exposed area of the avian eye. Light passes through the cornea and into the **pupil** behind it.

Anterior chamber — The space formed by the curve of the cornea, between the cornea and the **iris** of the eye. Filled with a clear, thin liquid, the **aqueous humor**.

Iris — A thin sheet of striated muscle fibers and connective tissue that form a diaphragm in front of the lens, controlling the amount of light entering the **posterior chamber** of the eye. Contains pigment cells that give the eye its color. The circular open space at the center of the iris is the **pupil**.

Ciliary body — The structure at the base of the iris, containing muscle fibers that contract to alter the shape of the lens, changing the focal point of the eye. The fibers between the ciliary body and the lens are the **zonular fibers**, and the ring around the inside surface formed by the ciliary muscle groups is sometimes called the **ora serrata**.

Lens — A transparent optical lens composed of a tough outer capsule and a more flexible series of layers within. The lens is suspended at the center of the anterior eye by the **zonular fibers** from the **ciliary muscles**. The ciliary muscles expand and contract to change the shape of the lens to focus on near or far objects. This focusing process is called **accommodation** and is more readily done in birds with flexible lens adapted for a variety of viewing distances. Predatory birds such as owls have long, tubular eyes shaped to act like telephoto lenses and cannot readily accommodate their eyes to focus on very near prey. Owls are thus sometimes forced to back away from wounded or dead

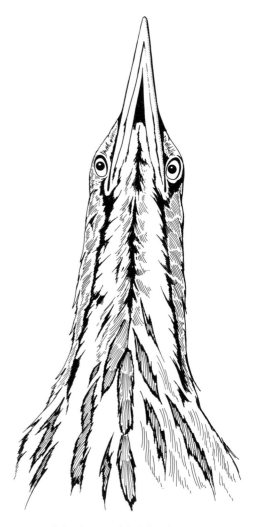

Because of the shapes of their heads, most birds have a more limited range of binocular vision than do humans. The American Bittern (*Botaurus lentiginosus*) has a most unusual (if somewhat comical) way of gaining additional binocular vision. When frightened, it freezes in an upright posture to blend in with the vertical stems of reeds in the marshes it inhabits. In this camouflaging posture the bittern can still see easily with both eyes, and this ventral binocular vision is also helpful in finding and striking small aquatic prey.

prey to better see the animal before attacking or eating it.

Vitreous body — The clear, jellylike substance that fills the posterior chamber of the eye. The viscosity of the **vitreous humor** supports and maintains the shape of the eye.

The globe of the eye is enclosed in three distinct layers, or tunics: the **outer tunic** of the cornea and the sclera, the **middle tunic,** or choroid layer, and the **inner tunic**, or retina.

Sclera — A white layer composed of tough collagen fibers, forming the "white" of the eyeball. The sclera supports the shape of the eye and serves as the attachment point for the eye muscles.

Choroid — The dark middle layer of the globe, richly supplied with blood vessels. The choroid carries nutrients to support both the sclera and especially the retina. It is heavily pigmented to absorb light that passes through from the retina.

Retina — The inner surface of the eye, sensitive to light. It contains photoreceptive **rod and cone cells** and is continuous with the **optic nerve** at the posterior surface of the globe. The microstructure of the retina is complex, composed of a pigmented epithelium layer, a layer of photoreceptor cells, a

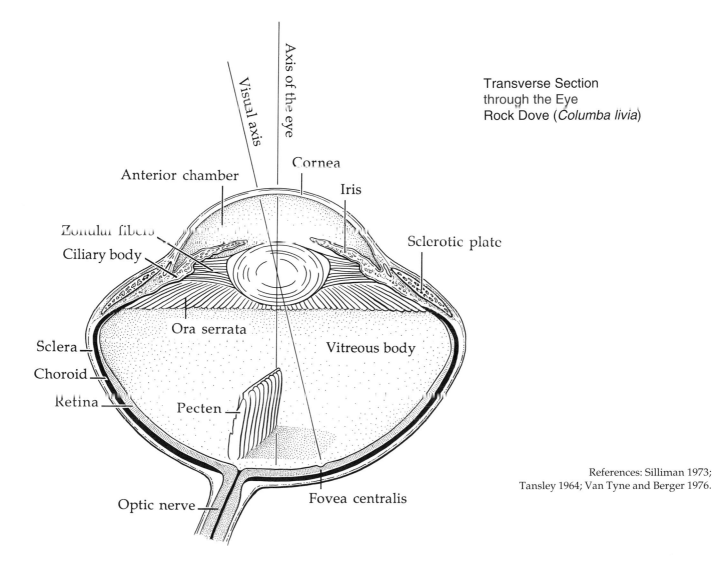

Transverse Section
through the Eye
Rock Dove (*Columba livia*)

Visual axis

Axis of the eye

Cornea

Anterior chamber

Iris

Zonular fibers

Ciliary body

Sclerotic plate

Ora serrata

Sclera

Vitreous body

Choroid

Retina

Pecten

Optic nerve

Fovea centralis

References: Silliman 1973;
Tansley 1964; Van Tyne and Berger 1976.

layer of bipolar neurons, and a deep layer of ganglion cells and nerve fibers that exit the retina through the optic nerve. The surface of the retina is not equally sensitive to light, and the areas of sharpest visual acuity are centered in a **macular disk** where the visual axis of the eye meets the retina. At the center of the macular disk a tiny pit, the **fovea centralis**, marks the point of sharpest visual acuity. Some bird groups, notably hawks, swallows, and kingfishers, have two fovea on their retinas; a **lateral fovea** for vision to the side of the head and a **temporal fovea** that gives these birds better forward binocular vision.

Pecten — A highly vascular, elaborately folded structure projecting from the avian retina into the vitreous humor of the posterior chamber. The pecten is richly supplied with blood vessels and is thought to aid the choroid layer in supplying the retina with nutrients and in transporting oxygen and carbon dioxide (Wingstand and Munk 1965).

Optic nerve — Carries nerve fibers from each retinal photoreceptor to the optic tracts and optic lobes of the brain. A small blind spot, the **optic disk**, occurs where the optic nerve exits the retina.

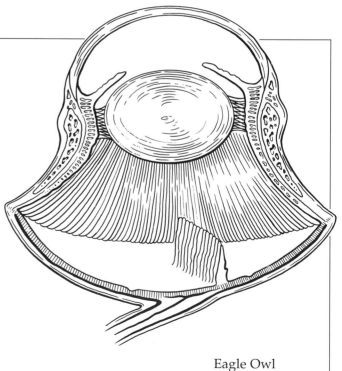

Owl Eyes

One explanation for the extraordinary eyesight of owls is that their eyes are shaped very differently from human eyes. Pictured here is a cross-section of an Eagle Owl's eye. Notice that the eye is very long, not spherical like the human eye. This "tubular" eye gives the bird greater telescopic vision, helping it pick out small shapes (like prey) from far away. Note the near-circular lens and relatively large corneal area of the owl's eye. These adaptations let maximum light into the eye, but the large cornea also requires a longer "focal length." The circular lens compensates for this, shortening the focal length and keeping the retinal image as bright as possible. The effect of the elongated eye with a large front lens element is very much like putting a "fast" telephoto lens on a camera—the bird sees details of faraway objects in dim light with great acuity.

Eagle Owl
(*Bubo bubo*)

References: Burton 1973; Tansley 1964; Welty and Baptista 1988.

Form and Function
Vision in vultures

White-backed Vulture
(*Gyps africanus*)

The central area of the vulture's field of vision is slightly magnified, allowing the vulture to see long distances over the African plains.

To locate kills the vultures of East Africa circle high on mid-day thermal currents rising above the savanna that often take them to heights of five thousand feet or more. From this vantage point circling vultures can survey vast stretches of territory, scanning the ground below for signs of the previous night's kills. Several adaptations in the eyes of vultures and other birds of prey allow them to see at heights of a mile or more and give these birds eyesight that may be more than twice as keen as the sharpest human eyes.

In both vultures and humans the retina contains a special area near the optic center of the eye called the fovea centralis, where the rod and cone photoreceptor cells are especially dense and where the eye's highest visual acuity is obtained. The foveal area of the human retina contains roughly 200,000 rod and cone photoreceptors. The same foveal area in the Common Buzzard (*Buteo buteo*) has been measured as containing over one million cones alone, theoretically giving the buzzard visual acuity six to eight times that of humans (Walls 1942). Like a fine-grained photographic film that captures the smallest details, the avian eye simply has more photoreceptors with which to build an image than our own coarse retinas allow.

Vultures, hawks, and eagles also take advantage of an optical characteristic of the retinal walls to magnify the image projected onto the retina by the lens. The fovea centralis is a shallow pit in the retina. As rays of light strike the walls of this pit they are refracted at the retinal surface by the difference in density between the retina and the clear vitreous humor within the globe. This causes the rays to bend outward enough to magnify the image by about 30 percent as it hits the cone cells. Because the fovea is a pit with sloping sides, it also has a slightly greater surface area than areas of flat retina, and this distributes the projected image within the fovea over more photoreceptors, enhancing the magnification. The effect must be much like having a modest set of binoculars built into the eye, magnifying the central images in the field of view by 30 to 50 percent more than the rest of the retina.

In the Field   BIRDING FOR OWLS

Owling, or going on "owl prowls," has become one of the most popular aspects of birding. The excitement of the unknown as one goes afield in darkness is accentuated by the variety of sounds to be heard in the night forest. A mouse foraging in the leaf litter sounds much larger than it really is. A creaking tree limb or squeaks from a flying squirrel seem much more mysterious than they would in daylight. Perhaps the best-known sounds of the night are those produced by owls. In particular, the soft wail of the Eastern Screech-Owl (*Otus asio*) has accented innumerable Hollywood movies to depict the eeriness of the night woods. If we can hear such wails so clearly, imagine how much better the owls themselves must hear each other's calls!

Owls are well adapted to their roles as night hunters and have the keenest night vision and hearing in the bird world. Large ear openings frame the facial disks, acting as parabolic reflectors to focus sound waves into the inner ear. Asymmetrical ear openings further aid most owls in locating their prey in three-dimensional space. A simple twist of the head, causing each ear to receive and process a slightly different sound pattern, allows the owl to triangulate the source of the noise and home in with deadly precision on even the slightest rustle of leaves. The primary feathers of the wing have soft, downy leading to dampen the sound of air rushing over them. Most obvious of all, owls have huge, searching eyes, with retinas densely packed with rod cells allowing vision at extremely low levels of light.

While afield at night, try the following exercises to gain some idea of what sensory adaptations the owls have made to nocturnal hunting. Hold a cupped hand up to your ear as you listen for owls, and note how much more you can hear when your hand acts like the owl's facial disks, focusing waves of sound into your ear. The muscles around an owl's ears allow the bird to change the shape of the ear opening at will, enhancing the differences between sounds entering each ear and aiding in triangulating the source of the sound (Schwartzkopff 1973). Wear a noisy nylon jacket on an owl prowl and it will be painfully obvious why owls have such soft, sound-dampening feathers. On an owl prowl it is best to wear wool or other soft, sound-dampening clothing and avoid nylon or other fabrics that tend to make a loud "whisking" sound as you move. All one needs to do is stand beside someone with a nylon jacket on a still night to appreciate the importance of sound dampening to owls or people. The nylon cloth produces sound at every move (even breathing!), but the soft surface of wool damps out almost all sound, even when walking quickly. The incredible night vision of owls is our greatest handicap in this sensory comparison, for owls have so many more rod cells in their retinas that they can see almost as well in a moonlit forest as humans see during the day. Our retinas quickly tire under the strain of low-light vision. Try picking out an object just barely visible in the dark. Stare intently at it and in a few seconds the image will fade from your retina and the object will seem to disappear. Shift your eyes slightly (moving the projected image to a fresh area of the retina) and the object pops back into view again. It is hard to imagine flying at full speed through a dense forest at night, pursuing prey and avoiding branches and tree trunks, yet owls do it routinely with barely a noise. Fortunately for us, owls do make one noise (their calls) that makes them possible to find, and we can even imitate owl calls to draw them in close enough to be seen.

With either tape-recorded calls or live imitations, even a novice birder can get owls to call back, and sometimes calls will even bring owls close enough to be viewed with a flashlight. Some calls, such as the deep, booming five hoot call of the Great Horned Owl (*Bubo virginianus*) is effective throughout North America. But it can be difficult to get such a deep bass call to carry through the forest. Try creating squeaking sounds by loudly kissing the back of your hand, which imitates the sound of a rabbit in distress. This tactic will draw in several of the big owls to investigate the noise. In southern swamps and bottomland areas throughout eastern North America, the Barred Owl (*Strix varia*) can often be heard giving its characteristic nine-note hooting, "Who cooks for you, who cooks for you-all!" The Barred Owl is probably the easiest owl to imitate. Any reasonable facsimile of all or part of its call will often bring in this common species. In western states Northern Pygmy-Owls (*Glaucidium gnoma*) give a repetitive single-note call at twilight, and Screech-Owl can also be heard just after sunset. Note that Eastern and Western Screech-Owls have distinctly different calls: be sure you have the correct "dialect" before launching into a Screech-Owl imitation. In southwestern canyons you can chirp, purr, hoot, and wail to more than a dozen species of owls for an exciting night of investigation.

Your first "contact" with an owl calling back is an unforgettable moment, like contacting a world unknown even to most daytime birders. Yet owling is easy, and it is certainly fun. Generate a list of owl species in your area, and make a cassette tape of their calls from one of the bird song albums, such as Roger Tory Peterson's *Field Guide to Bird Songs*. Even without a tape

Eastern
Screech-Owl
(*Otus asio*)

The Screech-Owl is the most common small owl in temperate North America. It is nocturnal (not all owls are) and eats small birds, rodents, insects, and even fish. Even experienced birders are often surprised at how common Screech-Owls are. They survive quite well in suburban areas close to humans. There may well be an owl hunting regularly in your backyard or local park.

In the Field   BIRDING FOR OWLS, *continued*

recorder you can probably imitate the calls of several species without too much practice. Pick a driving route through habitats with a good mixture of open field, edge habitat, bottomland forest, and large wooded upland tracts. Overgrown orchards, weedy fields near water, and wooded windbreaks between fallow farm fields are all excellent spots for owl calling. Remember to respect private property, and stay away from populated areas where people are likely to hear and be disturbed by the calls. Never "overcall" a location by bombarding it with calls for long periods of time. Limit your calling time to a minute or two at any location, and keep the volume on your tape recorder down to about the natural volume of the real owl. Remember that owls respond to calls because they are defending their home territories. A few soft calls are just a gentle challenge to the local birds to investigate the boundaries of their territories. But constant, loud tape-recorded calls can drive a bird to distraction, and even chase it from its home territory. When calling for a number of owl species, start by calling the smallest birds first, and progress up to the largest species. Remember that large owls *eat* small owls, so you won't have much luck finding Screech-Owls after playing a Great Horned Owl tape!

In cold weather, dress more warmly than you do for daytime birding, because owling involves a lot of standing still and won't keep you warm the way hiking will. Bring a good flashlight, but don't use it unless absolutely necessary to light your way or when an owl is close enough to see with the flashlight. If you rely too much on the flashlight your eyes will never fully adjust to the darkness, and you will see much less of your surroundings if you limit yourself to the tiny area of light illuminated by the flashlight. Keep quiet! No whispering, no joking, no clod-hopping down the trail breaking every branch under foot. Every noise is a siren, every whisper a yell to an animal that can hear a hundred times better than a human can, so keep the noise to a minimum or the rest of your efforts will be in vain.

Form and Function
The sense of smell in vultures

King Vulture
(*Sarcoramphus papa*)

The large eyes and colorful plumage of birds point to a way of life that depends on keen eyesight. With such groups as owls, hearing is the principal sense used for hunting. The question of how well most birds can "smell" has been debated for years (Duncan 1964). The size of the olfactory lobes are one of the few purely anatomic clues to how well a particular species can smell. Kiwis, for example, with their long bills have the nares placed uniquely at the tip of this bill and "sniff" for worms and grubs in the soil (Sturkie 1954). Murphy (1936) noted that such tubenose species as shearwaters and petrels quickly congregate over areas where feeding schools of predatory fish attack anchovies, leaving slicks of oily prey remains in the water. Experiments have shown that petrels are unable to relocate their nesting burrows once their nostrils are plugged (Bang 1960).

The sense of smell in vultures in particular has been hotly debated. Stager (1964) has given the greatest insight into the vulture's ability to locate food by smell. After twenty-five years of careful research, Stager concluded that the Turkey Vulture (*Cathartes aura*) primarily uses its sense of smell to locate food. Not surprisingly, the Turkey Vulture also has the largest olfactory lobes among the New World vultures (Bang and Cobb 1968, Bang and Wenzel 1985, Wenzel 1970). In contrast, the Black Vulture (*Coragyps atratus*) depends much more heavily on sight to locate prey. Black Vultures will also follow Turkey Vultures to find carcasses (Wilber and Jackson 1983). In Old World vultures, such as the complex of vulture species found on the Serengeti Plains, sight is apparently the sole method of locating carcasses. But that is just what you might expect in open grasslands, where most carcasses can be located by sight alone. Stager's investigations showed that the second largest olfactory system among the New World vultures is found in the King Vulture (*Sarcoramphus papa*) of Mexico and Central and South America. The King Vulture is the only vulture that consistently hunts over the dense tropical forest canopy, where it is able to locate prey hidden from sight below the treetops. With its keen sense of smell, the King Vulture has exploited a scavenging niche unavailable to vultures that must hunt by sight alone.

5 | THE SENSORY SYSTEM: ANATOMY OF THE EAR

In most ways birds are better equipped than humans to sense the world around them, but birds and people share the distinction of being the most vocal vertebrates, and most birds have ears that match the sensitivity range of the human ear (Kühne and Lewis 1985). A few species actually have greatly extended ranges of hearing, both in frequency range and in the ability to make fine distinctions in the timing of sounds arriving at the ears. The Barn Owl (*Tyto alba*) has ears so finely tuned that it can locate prey in total darkness (Payne 1971). Several birds, notably the Oilbirds (*Steatornis caripensis*) of South America and the Cave Swiftlets (*Callocalia linchi*) of Asia, have gone even further and are known to use echolocation to navigate in the dark caves in which they nest and roost. Both Oilbirds and Cave Swiftlets use clicking sounds within the range of human hearing. Although these sounds are well below the high-frequency squeaks bats use to hunt insects, they apparently keep the birds from hitting cave walls.

Observe the following structures

Just below and slightly posterior to the eye is the opening of the **external ear**, beneath a ruff of **auricular feathers** below the eye. Trim the auricular feathers on your specimen to expose the opening of the outer ear, called the **external auditory meatus**. Carefully remove the skin and any muscular tissue posterior to the external auditory meatus to expose the structures of the **middle ear**. A tiny semi-transparent membrane, the **tympanum** (eardrum), marks the division between the outer and middle ear chambers. On the inner surface of the tympanum locate the slender **columella**, a single **middle ear ossicle** that transmits the vibrations of the tympanum to the cochlea of the inner ear. The inner end of the columella ends at the **oval window** (fenestra vestibuli) of the **cochlea**, within the inner ear chamber. The cochlea converts the vibrations of the tympanum and columella into nerve impulses that are sent to the brain via the **cochlear nerve**, a branch of the **eighth cranial nerve** (auditory, VIII). The inner ear also houses the **semicircular canals**, the organs of balance found in all higher vertebrates.

Contrary to popular belief, the "ear" tufts of owls have nothing to do with owls' sense of hearing. They probably have some function in display behavior, but no one knows why some owls have tufts and others do not. Other feathers on an owl's head do have something to do with hearing. Just as cupping your hand near your ear will amplify faint sounds, the concentric rings of feathers around the owl's eyes focus sound into the ear openings below the eyes.

Diagrammatic View of the Ear
Rock Dove (*Columba livia*)

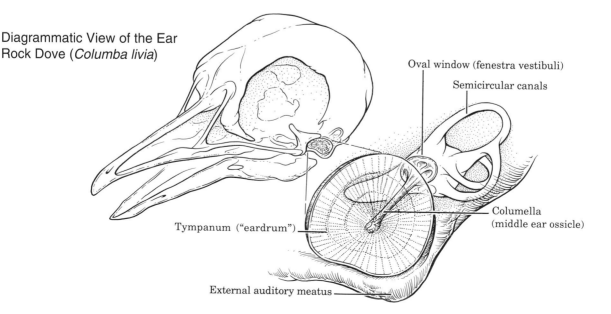

Oval window (fenestra vestibuli)

Semicircular canals

Columella (middle ear ossicle)

Tympanum ("eardrum")

External auditory meatus

Nervous System   *Chapter Worksheet*

1. Compare and contrast the avian corpus striatum with the mammalian cerebral cortex.

2. Name three major plexus groups of spinal nerves and the body areas they control.

3. Explain the functions and importance of the avian cerebellum.

Chapter Worksheet, continued

4. Name two ways the fovea centralis of vultures helps them to locate food.

5. Define these terms:

   a. Arachnoid layer

   b. Medulla oblongata

   c. Cone cells of the retina

   d. Cochlea

   e. Abducens

References

Breazile, J. E., and H.-G. Hartwig. 1989. Central nervous system. In A. S. King and J. McLelland, eds., *Form and function in birds.* 4 vols. New York: Academic Press, 4:485–566.

Carroll, R. 1988. *Vertebrate paleontology and evolution.* New York: W. H. Freeman.

Goldby, F. 1964. Nervous system. In A. L. Thomson, ed., *A new dictionary of birds.* New York: McGraw-Hill.

Hoage, R., and L. Goldman. 1986. *Animal intelligence: insights into the animal mind.* Washington, D.C.: Smithsonian Institution Press.

Kreithen, G., and G. Caeler. 1979. Infrasound detection by the homing pigeon. *J. Comp. Phys.* 129:1–4.

Kühne, R., and B. Lewis. 1985. External and middle ears. In A. S. King and J. McLelland, eds., *Form and function in birds.* 4 vols. New York: Academic Press, 3:227–71.

Martin, G. R. 1985. Eye. In A. S. King and J. McLelland, eds., *Form and function in birds.* 4 vols. New York: Academic Press, 3:311–73.

O'Connor, R. 1984. *The growth and development of birds.* New York: John Wiley and Sons.

Pearson, R. 1972. *The avian brain.* New York: Academic Press.

Portmann, A., and W. Stingelin. 1961. The central nervous system. In A. J. Marshall, ed., *Biology and comparative physiology of birds.* 2 vols. New York: Academic Press, 2:1–36.

Sillman, A. J. 1973. Avian vision. In D. S. Farner, J. R. King, and K. C. Parkes, eds., *Avian biology.* 8 vols. New York: Academic Press, 3:349–87.

Stettner, L., and K. A. Matyniak. 1968. The brain of birds. *Sci. Amer.* 218:64–76.

Van Tyne, J., and A. J. Berger. 1976. *Fundamentals of ornithology.* 2d ed. New York: John Wiley and Sons.

Walls, G. 1942. *The vertebrate eye and its adaptive radiation.* Bloomfield Hills, Mich.: Cranbrook Institute of Science.

Webster, D., and M. Webster. 1974. *Comparative vertebrate morphology.* New York: Academic Press.

Field Techniques

12

Perhaps the most exciting time for ornithology students comes when they first watch birds in the field or assist ornithologists or bird banders in a field project. Learning to identify and count wild birds is itself no small feat, but that is just the first step. Field biology encompasses many skills beyond those learned in the laboratory and lecture hall. Field ornithologists must be experts in how weather, geography, local ecology, and other environmental factors affect wild bird populations. And skill in capturing and handling wild birds is needed—ornithologists must be able to gather information quickly and accurately without harming their subjects.

The most basic skill is the ability to identify bird species in the field, and it comes only with practice and time. You cannot become an expert at anything quickly or without training, and your first birding trip may convince you that it is impossible to identify all the bird species you see. But after a few weeks you will become accustomed to the common birds of your area, and this will allow you to begin to recognize and concentrate on new or unusual species. The key is not to become too frustrated in the early stages of birding. Novice birders can find it discouraging to go afield with experts and listen to them confidently reel off species after species, even though many of the sightings were mere flashes of color in the underbrush.

Expert birders are not performing magic tricks or making wild guesses. Although more than seven hundred bird species occur in North America, only a tiny fraction of those species are likely to show up in a given area at a particular time. Through years of experience most birders know what kinds of birds are likely to occur at each time of year in their area and which species favor certain habitats. They have also learned the crucial field marks that separate one species from another. The job of identifying a given bird is thus not a matter of picking a single species from a list of seven hundred possibilities—rarely are there more than half a dozen likely choices if you know the area well and have at least a little information on where the bird was, how it was behaving, and some indication of what it looked like. Identifying birds is much more than memorizing color patterns from a field guide. Most birders go as much by the *context* of the sighting (time of year, habitat type, current weather) and the *behavior* of the bird as by what color patterns they are able to see through binoculars. Here are a few ways you can shorten the process of becoming an expert birder:

Know the Birds of Your Area

Most states and many counties have compiled checklists of the common and unusual birds that occur in the area. Check with your ornithology professor, local bird clubs, your state fish and wildlife agency, or wildlife biology professionals for information on local checklists. Check whether anyone has written books on the birds of your state or region. These books

U.S. Fish and Wildlife Service biologist Jeff Spendelow bands a Common Tern (*Sterna hirundo*) on Falkner Island, Connecticut. The colony of Common and Roseate terns (*Sterna dougalii*) on Falkner Island has been studied extensively for more than ten years, and banding data from the study have yielded a great deal of new information on the breeding biology, reproductive success, and longevity of these threatened seabirds.

can help you work out the timing of migrations in your area, so you will know what species to expect at a given time of year. Such information can greatly simply the task of identifying birds in the field by reducing the list of possible species to a manageable level.

Train Your Ears

In woodland areas experienced birders do most of their "watching" by ear, not by eye. Wild birds are often secretive, and small animals are always difficult to see in heavy foliage. But most birds actually *advertise* their presence in an area through territorial songs or warning calls and thereby tell you exactly which species are present even if you can't see them. The human brain is much better at memorizing visual patterns than aural ones, so learning the songs of the two dozen or so most common forest birds of your area will take a bit more time than learning what the birds look like. But once your ears are trained you will enjoy an enormous advantage over birders who rely on sightings alone. If you can bird by ear you will always know what common birds are around you, and you will have a head start in looking for and identifying singing birds hidden in woodland foliage, because you will already know which birds to look for.

Know the Seasonal Patterns in Your Area

The change of seasons affects every aspect of birdwatching. Learn which birds are found in your area at a given time of year. Fewer species will usually be resident in winter, shortening your list of birds. Spring and fall are often the most interesting seasons for birdwatching: this is when many species migrate through your area. Knowing *when* to look for a bird can make the difference between a successful sighting and years of frustration chasing down a species that visits your area only a few times a year. For example, in most of the United States large falcons are rarely seen in spring, summer, and winter, but they are often seen during fall migration, especially along the east and west coasts.

Know Your Local Habitats

Know the species that occur commonly in the habitat in which you bird; this is one of the best ways to reduce the number of species you must memorize.

Red-eyed Vireo
(*Vireo olivaceus*)

Although the Red-eyed Vireo is one of the most common birds in the eastern deciduous forest (Robbins et al. 1983), it is not often seen by inexperienced birders because singing male vireos tend to stay high up in the foliage of the forest canopy. If you do not know the songs of this common bird, you might never realize it existed in such numbers. Although secretive, the Red-eyed Vireo is a tireless singer; one (indefatigable) ornithologist recorded 22,197 songs from a single male bird (Lawrence 1953).

Wet bottomland forests echo with bird song in late spring. Red-eyed Vireos and Scarlet Tanagers are both quite common in northeastern forests, but unless you know the calls of these birds you will rarely locate them high in the forest canopy.

Along the shoreline, for example, most birds you encounter will be gulls, shorebirds, and ducks. Prepare yourself in advance for the likely species in those groups and it will be much easier to identify what you see. Similarly, before birding in forest areas, consult a list of local woodland birds in your area to narrow down the likely possibilities. Birds also have preferences *within* a given habitat. The Scarlet Tanager (*Piranga olivacea*) is quite common in eastern forests but is not often seen by novice birders because it favors the high canopy and rarely drops below fifteen feet above the ground. Knowing how a species uses its habitat can often be crucial in locating the bird.

Censusing, mist netting and banding, field collecting, nesting studies, and behavioral studies are all fascinating aspects of bird study that you will experience if you remain active in field ornithology. *Mastery of these skills comes only with time spent in the field.* Studying laboratory manuals and memorizing the paintings in field guides will not in themselves make you a competent field biologist. The time you spend actually finding, identifying, and studying wild birds in their habitats will be the most important skill you can develop in field ornithology.

1 BIRDS AND THE ATMOSPHERE

Just as the lives of pelagic fish are shaped and bounded by the movements of the great ocean currents, so the lives of birds are profoundly affected by events in the earth's atmosphere. The annual changing of the seasons and more local outbursts of violent weather affect all wildlife, but as the most aerial group of warm-blooded vertebrates, birds are uniquely sensitive to severe weather. Naturalists and ornithologists have used weather forecasts to predict bird movements since the earliest days of modern meteorology, but a weather map is no precise predictor of migration patterns. Many complex, interrelated factors govern the movements of birds through the atmosphere.

Wind strength and direction are the most important influences on the timing and direction of bird migration (Bellrose 1967; Bruderer 1975; Eastwood 1967). Birds tend to migrate in great numbers only when moderate tailwinds blow them in the general direction they want to go, in North America generally north in the spring and south in the fall. But some bird migration occurs, regardless of wind direction, in all but the most severe weather conditions. Once a bird has begun its migration only prolonged severe weather will deter it from continuing. Birds will rarely migrate into a strong headwind and will sometimes reverse migration if the prevailing winds threaten to blow them out over large bodies of water. After the winds shift direction the birds reverse course once again and fly in the usual direction.

Bird migration is largely an invisible phenomenon. Most songbirds migrate at night, passing overhead without a trace of their presence to daytime birdwatchers. Only when inclement weather or unfavorable winds force the birds to delay their migrations do birders see most flocks of migrating passerines. If you are lucky enough to live within a major migration flyway you can hear these massive movements of passerines late on spring or fall nights, after the normal sounds of human activity have ceased. On nights when the cloud ceiling is low, flocks can be heard clearly, sometimes just above the treetops. The invention of radar in World War II provided a means for regular monitoring of these huge nocturnal migrations (Bellrose 1967).

Remember as you learn to check weather maps before going out to bird for migrants that weather patterns can only *suggest* future migration patterns over a *wide geographic region*. Other influences can override the effects of local weather conditions, so the local movement of birds in your immediate area may not always accurately reflect the extent of migration in the wider region. On a clear fall morning following the passage of a cold front you might expect the sky to be full of migrating raptors, and it may well be— a hundred miles away. If local winds push the hawks down the coastline, what should have been a spectacular show could turn out to be a complete bust.

What is good for birdwatching, ironically, is often terrible for the birds themselves. Birders love storms and other major shifts of weather patterns, because turbulence in the atmosphere often brings in birds that might

Fall Cold Fronts

These are the weather patterns that sweep birds down from the north in the fall and effectively halt most migration from the south in the spring. In either season a strong cold front on the weather map is a fairly reliable predictor of birdwatching conditions in the days ahead. A fall cold front followed by moderate easterly or northeasterly winds is a sign of good birding, particularly for hawks and other large diurnal migrants moving along the shores of the Great Lakes and Eastern Seaboard (Dunne et al. 1988; Mueller and Berger 1961). Fall cold fronts usually originate with high pressure systems in Canada (the so-called Canadian highs), which push south from the northwestern interior of the continent, driving millions of migrants before them as they move southeastward across North America (Baird and Nisbet 1959). But the same cold fronts that sweep birds south in the fall can form considerable barriers to migrants returning north in the spring. Spring migrants are often blocked along the length of an advancing cold front by northerly or north-westerly winds, concentrating in great north-south bands along the edge of the advancing weather system. Although this barrier sometimes results in springtime "windfalls" of migrants into your area, birding is usually poor until the wind shifts to the south (Bagg et al. 1950; Dennis 1954).

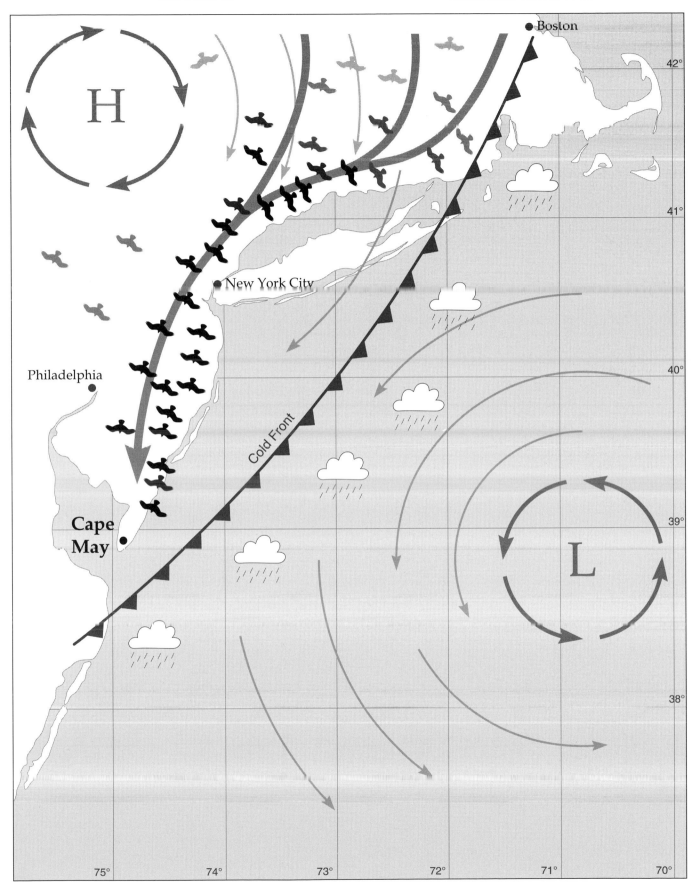

1 BIRDS AND THE ATMOSPHERE, *continued*

otherwise never be seen by ground observers. Bad weather can drive birds hundreds of miles beyond their normal ranges, creating the sort of oddball sightings and "rare bird alerts" on which dedicated birders thrive. For birdwatchers a "great day in the field" often occurs because the migrating birds were forced down into the trees by bad weather, adverse winds, or hunger.

Knowledge of weather patterns can enhance your time afield by allowing you to anticipate which bird groups will be favored or hampered by the coming day's weather. For the field ornithologist, weather patterns often dictate where banding, censusing, or other studies will be made on a given day and where the day's best birding will be, based on local geography, wind direction, and wind speed. Winter ocean storms can force pelagic birds ashore in seabird "wrecks," in which severe onshore winds sweep the birds down onto beaches and headlands, sometimes killing thousands of small seabirds. Spring storm fronts may bring in waves of migrants or stop all movement completely for days. Heavy summer rains may have a devastating effect on the nesting season, and fall cold fronts bring with them massive bird movements composed mainly of young birds hatched that year. Hurricanes roaring north from equatorial oceans can lead to spectacular coastal sightings of such normally pelagic birds as albatrosses, skuas, frigatebirds, and petrels. Each season brings unique weather patterns, and experienced field biologists and birdwatchers learn to read these signs in the sky for help in predicting both where particular birds will soon occur and the extent and direction of migratory movements.

Spring Warm Fronts

A warm front moving up from the south in spring tends to bring many migrants with it. Overcast conditions with occasional sprinkles may occur in the initial weather formation, but if the southerly wind is moderate, warm, and steady, conditions are right for a large movement of songbirds (Gauthreaux 1971). Gentle rains and drizzle do not deter most migrating passerines, though rain does force them below the cloud ceiling and into binocular range. Moderate wet weather can thus produce some of the best spring birding conditions. These foggy, drizzly mornings are ideal times to get out and immerse yourself in the "warbler waves" from the south. When the sun burns off the dawn fog and cloud cover and strikes the treetops, the songbirds really begin to move. At times the trees seem filled with birds as they pass in distinct pulses or waves. These are the "warbler mornings" birders eagerly anticipate each year all across North America.

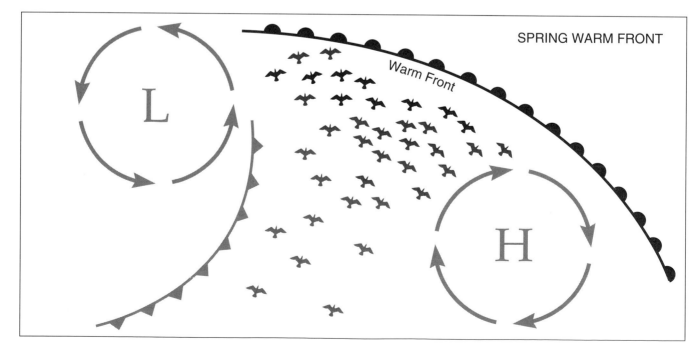

SPRING WARM FRONT

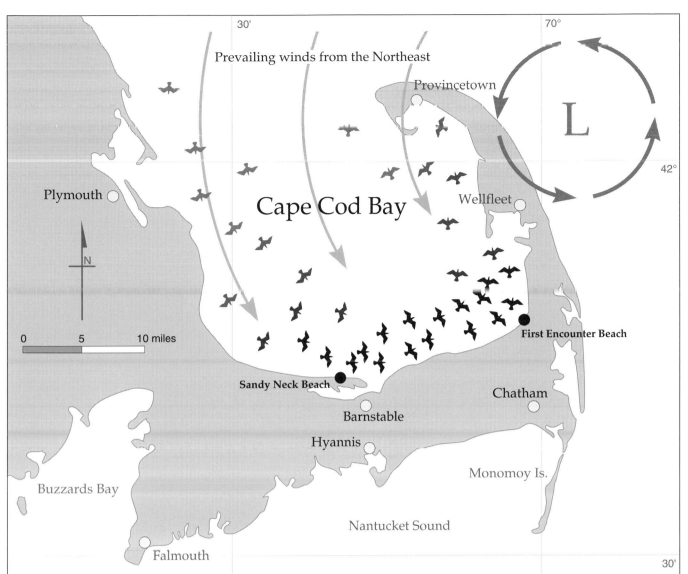

Northeasters

Each fall and winter, groups of such seabirds as alcids (auks, puffins, dovekies), sea ducks, and phalaropes (pelagic members of the shorebird family) wander the open ocean. These relatively weak fliers are often at the mercy of wind and weather. During New England winters, storms called **northeasters** (or Nor'easters) may sweep in from the northeast Atlantic, causing spectacular seabird sightings along the coast from Maine to Long Island. Severe northeasters can devastate pelagic bird populations, often pushing birds ashore or even inland in what are known as seabird **wrecks**. A wreck can extract a heavy death toll as well as displace tremendous numbers of birds. Dovekies turn up in coastal ponds, and phalaropes land in backyard pools. Birds that normally never come within sight of shore become commonplace along the coast for days after a northeaster has passed.

Seabird wrecks usually occur at capes or other prominent headlands and along the shores of enclosed bays. Cape Cod Bay is one such location, and northeasters often create excellent birding conditions along its shores. Seabirds swept from the ocean by strong winds are forced into the mouth of the bay. Once trapped there, they parallel the coast in a counterclockwise pattern. Birdwatchers at sites like Sandy Neck Beach or First Encounter Beach have ringside seats as normally inaccessible pelagic birds pass in review. It is not uncommon to observe Northern Gannets (*Morus bassanus*) diving for fish mere yards beyond the surf or to see thousands of shearwaters, alcids, sea ducks, and phalaropes flying within binocular range. When the storm abates, these birds slip out of the mouth of Cape Cod Bay and disappear back into the open ocean. Similar spectacular avian displays can be seen along both coastlines of North America; study the shapes of your local ocean or lake coastal headlands and bays to identify the besting birding sites in your area.

2 GEOGRAPHY AND BIRD MIGRATION

We know that avian migratory patterns are influenced by regular seasonal weather changes, the broad geography of the continents, and local weather events and landforms. But our understanding of the evolutionary origins of bird migration remains incomplete. Probably because of the way the continents are shaped, migrations occur primarily in birds that breed in the northern hemisphere. North America and Asia are broad continents offering bird species a wealth of space and temperate habitats, whereas South America and Africa each form narrow capes as they enter the moderate climate latitudes of the southern hemisphere. The southern hemisphere offers much less temperate habitat than does the northern hemisphere, and northern birds probably began migrating to Africa and South America to escape the food shortages and severe winter weather of the cold months in their northern hemisphere breeding grounds (Dorst 1962; Griffin 1964; Mead 1976).

Continental Patterns of Migration

Most North American landforms reinforce and channel the major north-south migration routes (Bellrose 1967). The Rocky Mountains, Appalachians, and Sierra Nevadas all run roughly north-south, as does the Mississippi River valley and both the east and west coasts of the continent. Texas, Central America, and the Isthmus of Panama act as a giant funnel for many migrating North American birds, concentrating flocks in both spring and fall migrations.

The broad patterns of bird migration in North America have been described as four major routes, or flyways, as shown in the illustrations opposite (Lincoln 1935, 1950). The concept of flyways is useful, but when taken too strictly it is a limiting and simplistic description of the migratory patterns of most birds. The conventional concept of flyways largely reflects observations of waterfowl and shorebird movements and tends to focus on the fall migration, which is much more visible than the more geographically diffuse spring migration. Human intervention also plays a part in the flyway model of migration. If you viewed an overlay of the wildlife refuge system you would see that refuges have been developed along these main waterfowl flyways to provide shelter and food for the migrating waterfowl (and many other bird species that favor wet habitats), reinforcing the flyway model. The misleading impression flyway maps give is that *all* birds travel along these well-defined routes and that between these flyways lie large gaps of "nonmigratory" ground. In fact, most birds do not follow these flyways, particularly the passerines, which migrate primarily at night. Radar studies of bird migration in both spring and fall show that movements occur across broad regional fronts and are not channeled into a few relatively narrow flyway areas (Richardson 1974, 1976; Bingham et al. 1982). In particular, nocturnal migrants tend to fly at altitudes where they are much less affected by local landforms; they concentrate along natural barriers like coastlines only when inclement weather forces them to migrate during the day (Richardson 1974, 1976).

The variety of bird species in spring and fall migrations will vary noticeably in your area. In fall migration birds tend to concentrate along the continental coastlines and major river valleys, where the large bodies of water

The North American Flyways

The four major North American migration flyways, a model proposed by U.S. Fish and Wildlife biologist Frederick Lincoln (1935, 1939, 1950). Since its inception, the flyway model of migration has been influential in federal waterfowl conservation efforts and has been reinforced over the years with the siting of many federal and state wildlife refuges along the four routes. Although the flyway concept has served the needs of waterfowl and shorebird conservation, it is an overly simplistic description of the behavior of most other migratory birds, particularly the passerines. More recent studies (Bellrose 1968; Richardson 1974, 1976) suggest that most species migrate over broad geographic fronts, particularly in spring, and do not follow narrow migratory channels or corridors, as suggested in the classic flyway model.

Flyway graphics drawn by Bob Hines (Lincoln 1950).

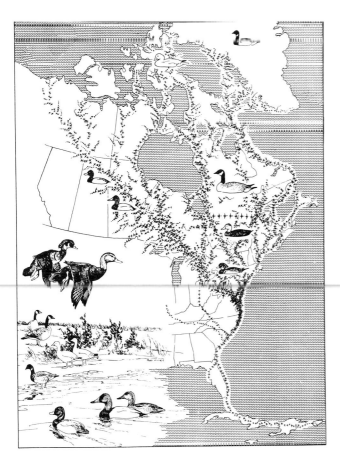
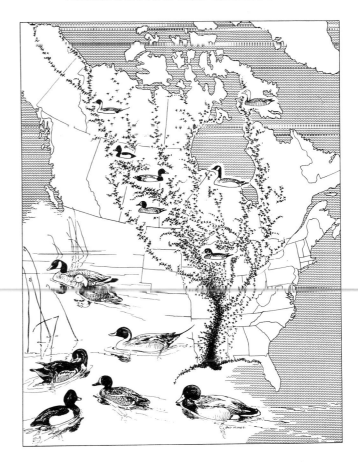
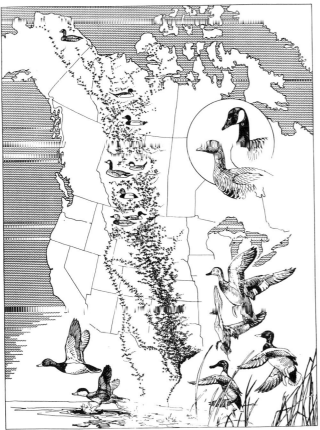
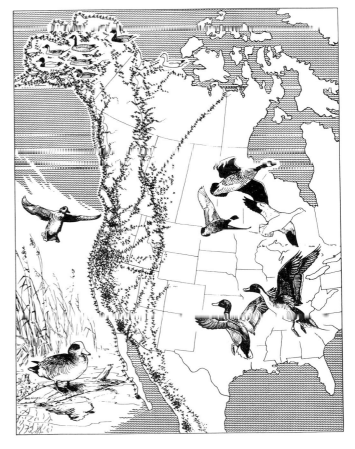

moderate the drop in temperature and food and shelter are often abundant well into late fall and early winter. But the northern coastlines of North America tend to have cold, foggy spring seasons and offer fewer food sources to birds returning to northern forests and tundras to breed. In spring, bird migrations occur across a broad front favoring the interior of North America, because spring warming comes sooner to inland areas and food is therefore more abundant inland in early spring. Spring migration is less dramatic than fall migration, not only because the routes of travel are more diffuse but also because fewer than half the young birds that begin the migration south in the fall will survive to return to their breeding grounds the following spring (Gill 1990).

Local Geography

Prominent geographic features often attract migrating birds, but the number and types of birds to be found in an area depends on the season and the direction the birds are traveling, the migratory preferences of particular

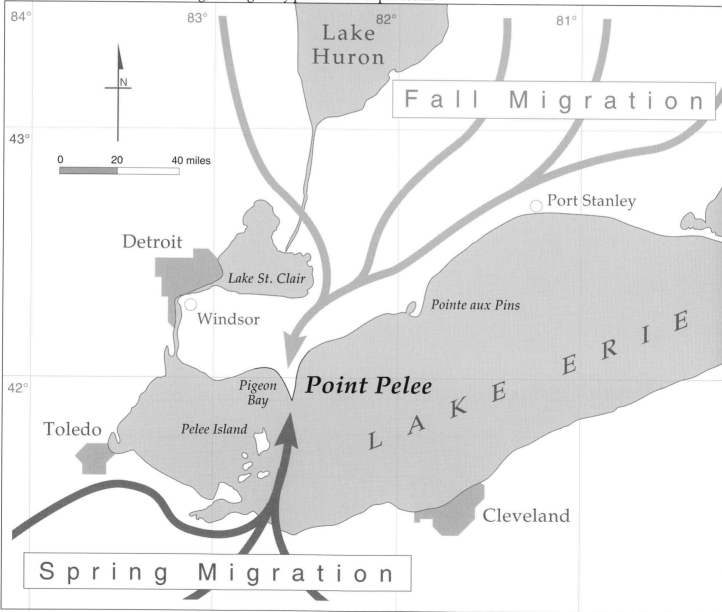

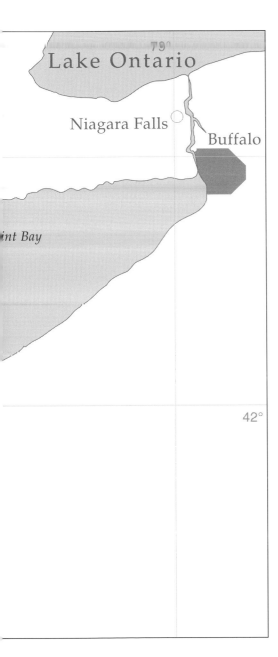

Lake Ontario

Niagara Falls

Buffalo

int Bay

42°

79°

species, and on whether suitable roosting and feeding habitats are available in or near the area.

Coastal peninsulas, such as Point Pelee in Ontario, Cape May in New Jersey, the Delmarva Peninsula along the Chesapeake Bay, and Point Reyes in California, are some of the most prominent and productive birding areas in North America. Many birds are reluctant to cross large bodies of water, so peninsulas become natural funnels to species migrating along the coast For birds that do fly out over water, peninsulas and coastal headlands are often the first landfall the migrating birds see, and flocks of exhausted migrants will often literally drop out of the sky into woodland areas on peninsulas. The north-south peninsula of Point Pelee attracts a wide variety of birds in both spring and fall migrations. During the fall birds move along the coast and south down Point Pelee. Some then strike out over Lake Erie when they reach the southern end of the peninsula, but many birds (most hawks, for instance) retreat northward back up the peninsula to avoid flying over the lake, giving birders on the peninsula a double chance to see migrating birds of prey. In spring, tired migrants cross Lake Erie during the night and at dawn are attracted to Point Pelee as the first landfall they encounter after their long flight over the lake. The reluctance of most hawks to cross open water also makes New Jersey's Cape May Point one of the premier hawk-watching spots along the east coast. In the fall, accipiters, buteos, and falcons follow the New Jersey coast down to Cape May but must then double back north and west to avoid having to fly south across Delaware Bay, where they might be blown out to sea (Dunne et al. 1988; Stone 1965). In general, south-pointing peninsulas are likely to be more productive in fall migration, while north-pointing peninsulas will attract and funnel more birds during the northward spring migration.

Other geographic features such as mountain ridges, river valleys, and such large bodies of water as bays and large lakes can create "corridor" effects, channeling flocks of migrants along coasts or ridgelines. High ridgelines or mountain slopes are attractive to migrating hawks, which use thermal updrafts along the highlands to soar from ridge to ridge. Ridges and other prominent landforms also affect nocturnal passerine migrants. The song-birds drop down low at dawn in search of suitable feeding and resting habitat, and in both spring and fall migrations songbirds often concentrate in forests and wetlands along the windward side of ridges early in the morning. River valleys, especially north-south valleys like the Hudson and Mississippi, create natural migration corridors. Islands also act as way stations for birds traveling over water. Coastal barrier islands and smaller islands in lakes and along rivers are often excellent areas for birdwatching, offering natural refuges to migrating birds. Offshore islands along the east coast of North America and the island chains of the Bahamas and West Indies are major way stations for flocks of passerines and shorebirds that fly out over the Atlantic and Caribbean in both migration seasons (Richardson 1974, 1976; Williams et al. 1977).

3 FIELD MARKS AND FIELD GUIDES

Many factors influence the identification decisions of an experienced birdwatcher: the time of year, the range of species known to be in the area, and the habitat in which the bird is seen all contribute to the identification of a bird in the field. Most important are the **field marks**—the unique combinations of color, pattern, and shape that mark and separate each species. Field marks are not always visual; behavior plays an important role in most identifications and is often more important than visual details in separating one species from another. For example, on a field trip you might see a tiny bird feeding in the top of a tree. No marks can be seen on the bird's body, but the bird is constantly twitching or flicking its wings. That behavior clinches the identification: the bird is a Ruby-crowned Kinglet (*Regulus calendula*), a tiny bird that favors treetops and the high canopy of woodlands. Ruby-crowned Kinglets are the only North American bird of similar size with that habit of constantly twitching their wings. Or consider the following scenario: You are scanning with your scope and you see a bird sitting upright on a breakwater a mile from shore. Although you can see no features of the bird, you know instantly it is a cormorant, because cormorants are the only coastal birds that habitually sit bolt upright. No body markings were used in these examples; both identifications depended on behavioral cues that set the target bird apart from other similar species.

Jizz

As you gain experience in the field, your identifications of many birds will depend as much on subtle differences in shape and behavior as on any visible color patterns. This is especially true when you are birding along seashores, marshes, or in other situations where birds are far away and plumage details are difficult to see clearly. The British use the term **jizz** to summarize the unique combination of silhouette and behavioral idiosyncrasies that mark each species in the field. The word **jizz** supposedly originated with British aircraft spotters during World War II, many of whom were also experienced birdwatchers. Instead of collecting volumes of detailed descriptions of planes, the aircraft spotters were asked to give a concise "General Impression of Shape and Size" (G.I.S.S., or "jizz") to identify each type of aircraft they saw. In birding, jizz describes the birdwatcher's gut reaction to the total visual impression a bird makes on the viewer and is not concerned with minute descriptions of plumage. For example, in North America there are two small falcons of roughly the same size, shape, and overall plumage; the American Kestrel (*Falco sparverius*) and the Merlin (*Falco columbarius*). If you rely solely on a color plate in a field guide, you might think the two birds are hard to separate in the field, yet the jizz of each species is so different that the two are not often confused. In flight Kestrels give a light, bouncy, windblown impression, whereas Merlins fly like miniature jet aircraft, cruising stiff-winged in a straight line, seemingly impervious to the wind. The distinctive behavior patterns make plumage details almost irrelevant in distinguishing these two small falcons in the field.

Not that plumage details are unimportant in all birdwatching; in forest or thicket birding, tiny variations of color or pattern may be your only clue to identifying a bird with certainty. This is especially true when birding for small songbirds like sparrows, where several dozen species all have roughly the same size, shape, and coloration. Experienced birdwatchers will even admit that most of those "little brown jobs" look pretty much alike in the field and that nailing down an identification is often a matter of collecting that lone plumage detail that distinguishes one sparrow species from another. Part of becoming a good birder is training your eyes to pick out small details of color and pattern quickly when sighting a bird in the field, then forming a solid mental "snapshot" of your sighting to compare against field guide references. In his original *Field Guide to the Birds*, Roger Tory Peterson's great innovation was to group the birds visually, not in taxonomic order. Peterson placed all the birds that look roughly alike together on a page, then pointed out the most important marks (the field marks) that allow you to separate one species from another. Train your eyes to seek out those field marks—the concentration and discipline will make you both a better birdwatcher and a much more observant person in other areas of your life.

Selecting a Field Guide

Most people are visual in their overall approach to buying field guides and simply look for the book with the most pleasing and detailed paintings. But in the bush those intricate field guide plates may actually make it harder to identify birds by presenting you with irrelevant information. Advanced birders may appreciate elaborate paintings, but novice birdwatchers need simple, patternistic diagrams that are carefully organized in logical visual groups. Fortunately, there are enough field guides to North American birds to suit a wide range of abilities and preferences. Here are some factors to consider when shopping for a field guide:

1. Organization. Most novices do better with a field guide that is organized visually rather than taxonomically. Peterson's Eastern and Western editions of *A Field Guide to the Birds* are organized so that birds that look roughly alike are grouped on pages regardless of whether the species are related. Most other field guides to North American birds are organized by taxonomic order. Many experienced birders prefer the taxonomic field guide arrangement, however, because once you know the taxonomic order of bird families, it is easier to locate particular groups of birds quickly.

2. Geographic area. The guides that cover all the species of North America are valuable to the experienced birdwatcher, but they usually offer too many confusing options to the beginning birder. Beginners rarely pay adequate attention to range maps and lack field experience to help them pinpoint what birds are most likely to occur in the local area. Field guides that concentrate on either the eastern or western United States may be easier and less confusing to use.

3. Location of range maps and descriptive text. Maps illustrating the overall geographic range and the summer and winter range of each species are an essential feature of the better field guides. Ideally, the range map and the text describing each species should be on the same or opposite page as the main illustrations of the bird, so that all of the essential information on a species is presented in a single spread.

4 CHOOSING AND USING BINOCULARS

Of all the equipment used by the ornithologist and birder, binoculars rank as the most important. A poor pair of binoculars can frustrate the beginning birder to the point of giving up. How many times have we seen the change in attitude when someone unhappy with their "beginner's" binoculars has looked through a fine piece of optical equipment. The exclamation, "Wow, I can really see the birds well!" leads to the question, "What type of binoculars should I invest in?" The key word is "invest," which means different things to different people, and can mean the difference between a mediocre pair of binoculars or an outstanding optic that will last a lifetime if treated properly. People usually favor whatever binoculars they are currently using and are willing to defend them against all comers. Rarely will a birder who has invested a considerable amount of money in a "glass" turn around and condemn its quality. Some baseline knowledge will help lead a birder to an educated choice. Here are some features of binoculars to consider before choosing a pair:

Magnifying Power
The first number written on your glass indicates the magnification power. Hence, the designation 7 x 35 indicates that the binoculars render an image that appears seven times larger than it would to the unaided eye. This number often leads beginners astray. "If 7x brings things in 7 times closer, why not use 20x?" Don't get power crazy. The 7x or 8x range will usually be the best overall glass for most birdwatching situations, including all forest birdwatching. With the new techniques used in binocular construction and glass preparation, however, the 10x glass is now of high quality and might be your first choice if most of your birdwatching is done along the shoreline or in other situations where the extra magnifying power might help in identifying distant birds. Avoid binoculars with a power rating greater than 10. The light-gathering power of binoculars drops considerably in the high-power range. High-power binoculars are also difficult to hold steady enough to use the extra magnifying power gained—the image jumps around so much that you will not really see much more than you would with a lower power glass.

Light Transmission
This is the second number on your glass. It gives you the millimeter (mm) diameter of the objectives, or front lenses of the binocular. For example, a 7 x 35 glass is 7x and has an objective lens diameter of 35 mm. The wider the diameter of the objective lens, the more light comes through the binocular. The low-light performance of a glass can be critical in early morning birdwatching, in heavy forest vegetation, or when birding at dusk. A basic rule of thumb states that the objective diameter should be at least five times the magnifying power of the binocular. If you are using a 7x glass the objective lenses should be at least 35 mm in diameter. If you do much birdwatching at dawn or dusk you might choose a glass with more light-gathering power, with an objective lens 50 mm in diameter. But new optical coatings have improved the light-gathering efficiency of binocular optics, making 7 x 35 glasses almost as bright as older 7 x 50 glasses. Cheap binoculars usually have a dark, blue-tinged field of view. Poorly made, uncoated optics are never as efficient at gathering light as quality lenses, so don't go by the numbers alone. Take a long, critical look through any pair

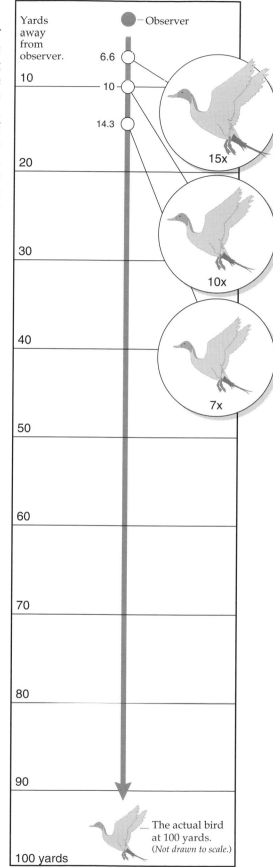

Yards away from observer.

Observer

10 — 6.6
— 10
— 14.3

20

30

40

50

60

70

80

90

100 yards

15x

10x

7x

The actual bird at 100 yards.
(*Not drawn to scale.*)

The diagram at the left explains a curious point of diminishing returns in binocular magnifying power and explains why so few birders bother to use binoculars with a magnifying power greater than 10x. As magnification increases, the bird appears to be closer to the observer. With 7x binoculars the bird seems to be about 14 yards away from the observer (100/7, or 100 yards divided by 7x binoculars). But note how little apparent gain there is when the magnifying power of the binocular is doubled. Although the 15x binoculars do produce a bigger image, the image is not seen as being twice as big but merely a few yards closer. High-powered binoculars are also much heavier and produce a darker image than 7x or 10x binoculars. If you need magnifying power greater than 10x, a spotting telescope mounted on a sturdy tripod would be a better choice than 15x binoculars. (Diagram adapted from Connor 1988.)

of binoculars before buying them. Compare the colors of objects around you both through the glasses and with your unaided eye. A really good pair of binoculars seems to *improve* the amount of light under marginal viewing conditions and never noticeably darkens the view.

The weight of binoculars can be a significant comfort factor to some people. The better brands of binoculars in the 7 x 35 range are lightweight and rarely bother most people once they have used the glass for a while. The larger objective lenses of high-power glasses increase the weight considerably, so if you are sensitive to the weight of binoculars you should stick with the lighter models.

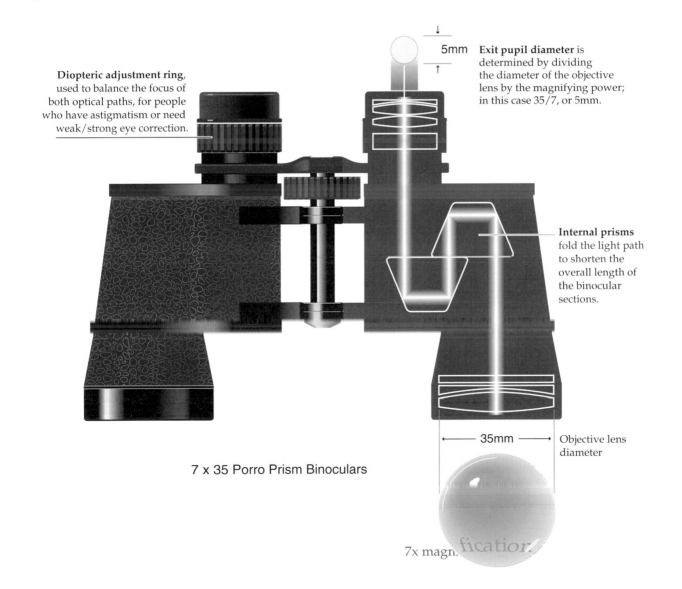

5mm

Exit pupil diameter is determined by dividing the diameter of the objective lens by the magnifying power; in this case 35/7, or 5mm.

Diopteric adjustment ring, used to balance the focus of both optical paths, for people who have astigmatism or need weak/strong eye correction.

Internal prisms fold the light path to shorten the overall length of the binocular sections.

35mm   Objective lens diameter

7 x 35 Porro Prism Binoculars

7x magnification

Focus

Two types are available: individual eyepiece focus and center focus. Individual eyepiece focusing is a cumbersome procedure ill-suited to birding and is now rarely seen in mainstream binocular design. Central focus with a single knob or lever is the modern standard. Quality binoculars also have a diopter adjustment ring on one eyepiece of the binocular to allow the viewer to compensate for differences in eye strength. The close-focusing distance of a glass is important in birdwatching. In the tropics or in thickets birds often appear in a frustrating zone that is too close for binoculars and too far for the naked eye. When considering a glass, see how close it will focus. Some new models focus as near as six or eight feet and can make thicket birding much more enjoyable. With Porro prism binoculars (see Prism Type, below), close-focusing distance is rarely a problem, but check carefully the close-focus distance on cheaper roof prism models, which often do not perform as well close up.

Durability

Testing for water resistance can be impossible when purchasing a glass, but do consult articles that have tested the durability and watertightness of various brands of binoculars (Dunn 1985; Bonney 1988). You do not want to go afield and find your glasses fogged up after being caught in a torrential rain. Too many birders learn this the hard way on extended trips to the tropics, where the daily deluge of afternoon rain leaves some birders with binoculars hopelessly fogged up and useless for hours or even permanently damaged.

Prism Type

The two major types of binoculars widely used by birdwatchers are distinguished by the arrangements of optical prisms within the binoculars. The most common is the Porro prism. Porro designs give a wide, bright field of view and focus quickly; they dominate the mid-range of binocular price and quality. Roof prism designs result in a trimmer, lighter binocular; they dominate the high end of the market. Neither prism design is inherently better than the other, and each have their proponents among the birding community. Beware of roof prism binoculars that have not been designed especially for birding, because many roof prism binoculars have minimum focusing distances too long (twenty-five feet or more) for birdwatching. Binoculars for birdwatching should focus down to at least fifteen feet—closer, if possible.

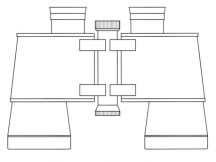

Porro prism binoculars

The prism system of any quality binocular should be solidly locked in, so that the optics will not readily shift out of alignment when the glass is dropped. Lower priced binoculars go out of alignment quickly and are difficult to realign or repair once damaged. Poor alignment forces the eyes to adjust constantly for the optical mismatch in the twin images of the binoculars. Severe eyestrain headaches after a day in the field, as well as missed sightings, can result from poorly aligned optics. A good pair of binoculars will take years of knocks and bumps and still stay in alignment. To test the alignment of a pair of binoculars, look through them at a straight horizontal line with the glass held up to your eyes in the normal viewing position. Then slowly draw the glass away from your eyes, still peering through the openings, and watch to see if the line divides or suddenly jumps out of position. If it does, the glasses are out of alignment and need

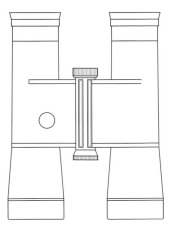

Roof prism binoculars

Common Types of Binoculars for Birding

Power x Objective Diameter (mm)	Exit pupil	Notes
Porro Prism Models		
7 x 35	5.0 mm	Porro prism binoculars of this power and size are the most widely used optics in birding. Most birders find 7 x 35 binoculars the best overall balance of magnifying power, price, and image brightness.
7 x 50	7.1 mm	The wide diameter of the objective lens gives 7 x 50 glasses a bright field of view. Good for birding at twilight or under dense forest foliage.
10 x 50	5.0 mm	These models are best for birding along coastal areas or in any situation where the birds are likely to be quite distant from you. At 10x these binoculars give the maximum magnification that is still easy to hold steady. Above 10x it becomes difficult to hold the glass steady enough to benefit from the extra magnifying power.
Roof Prism Models (always check the close-focus distance of roof prism designs)		
8 x 22, 8 x 24	2.75 mm, 3.0 mm	These mini or compact models are light and easy to carry but tend to give a dark, narrow field of view. Popular for casual birding only.
7 x 35	5.0 mm	Roof prism glasses in this power range have all the advantages of 7 x 35 Porro prism designs but tend to be smaller and more lightweight than equivalent Porro models.
10 x 40, 10 x 42	4.0 mm	Binoculars of this type, made by famous German and Japanese optical firms, are the ultimate status symbols in birding. They combine superb optics with a slim, lightweight design. Advanced lens coatings give these glasses as much or more brightness as cheaper 7 x 35 models. Buy models that focus to at least 15 feet.

References: Bonney and Baringer 1988; Connor 1988; Dunn 1985.

repair and realignment by a professional.

Price

The price range of acceptable to high-quality glasses is wide—from less than US$100 to well over US$1,000. If you plan on birding for a long time, a more expensive glass may be your best investment. Most people start with a medium priced model and move up when they can afford better binoculars. Remember, this is the "tool" you will use most; perhaps more than any other factor it will determine how much enjoyment you derive from birdwatching. With poor optics you will miss much of the visual richness and detail that birdwatching offers.

5 SPOTTING SCOPES AND TRIPODS

Once you begin to spend more time afield you will realize that you need at least one more tool for enjoyable and efficient birding—a spotting telescope. A good telescope and tripod can give you long, steady close-up views of interesting birds and free your hands to consult field guides or take notes on the birds you see. A telescope and tripod also make it easier to share your observations with others, because viewers can take turns looking through a scope that is locked on a stationary bird. As with binoculars, a wide array of spotting scopes is on the market, ranging in price from $200 to more than $4,000. With this range of price and quality, there are several factors you should consider before deciding which telescope best fits your needs and budget.

Catadioptric Telescopes

Catadioptric telescopes (often called "cats" or "fat cats") are famed for their high magnifying power and image resolution, allowing some of the finest views imaginable of distant birds. Under optimal conditions, the views of birds through a cat are often spectacular. Cat designs have a few distinct drawbacks, however. Although their image quality is superb, catadioptric models are the most expensive telescopes on the market, costing several thousand dollars and up. Catadioptric scopes are also heavy, with some models weighing in at more than twelve pounds. They are also bulky, although some cat scopes can be packed for travel in a relatively small metal case. The viewing system of cat scopes reverses the image in the eyepiece, and this takes some time to get used to, especially when following birds in flight. Catadioptric telescopes often have eyepieces angled at either 90° to the visual axis of the scope (you look straight down into the eyepiece), or set at a 45° angle to the axis of the telescope. Angled eyepieces allow the scope to stay lower on the tripod, and this often makes for a steadier field of view on windy days. The high magnifying power of cat scopes is also a mixed blessing in many instances, as high-powered scopes are more vulnerable to wind vibration and optical distortion due to "heat waves" rising from the ground (even in winter). In general, only expert birders get substantial benefits from the detailed views such expensive telescopes provide.

Standard Spotting Telescopes

This is the type of telescope most people think of when spotting scopes are mentioned. Two conventional telescope designs are popular: a straight-through design, which is basically a long metal tube with optics at either end, and the more compact offset design, in which the light path is bent by a small internal prism and the rear eyepiece is offset slightly above the front (objective) lens tube. Neither design has inherent optical advantages over the other. Offset designs tend to be shorter and more compact and are therefore slightly more popular. Standard spotting telescopes of either design usually have an objective lens diameter of 60 mm. Beware of telescopes with an objective lens diameter less than 60 mm, because the field of view will probably be too dark for birdwatching. The rear eyepiece optic determines the magnifying power in conventional scopes, and almost all spotting telescopes allow easy unscrewing of the eyepiece to change the magnifying power. Most birders find the ideal balance of magnifying power and practicality to be a 20x eyepiece with a wide field of view. Most

telescope makers also offer zoom eyepieces that allow you to vary the magnifying power of the eyepiece from about 15x to more than 40x. Examine zoom optics carefully before buying. Most experienced birders avoid zoom optics of any kind, but a few swear by them and find the flexibility and extra magnification of zooms worth the sacrifice in clarity and brightness of view.

One of the best things you can do when in the market for a scope is to ask other birders what they like or dislike about their scopes. And be sure to purchase from a dealer who allows you to make comparisons. Some scopes give greenish tints, others bluish, and still others will seem just right to your eye. Try the scope on several objects and be sure it focuses sharply. Be sure, too, that it gathers enough light to allow an identification under low light conditions. You might try the scope late in the day before making a decision. Check the scope's weight. Although most scopes run in the range of 38–45 ounces, some seem much heavier because of their awkward weight balance. Most of all, you must feel comfortable in using it. Remember, it is a big investment and should provide years of enjoyment.

The Tripod

The view through a spotting telescope is only as good as the tripod the scope is mounted on. If your tripod is flimsy and vibrates with the slightest breeze, the best scope in the world will not improve the accuracy of your birdwatching. Purchasing a tripod for birding is basically a matter of balancing the need for sturdiness against the weight of the tripod. The ideal tripod is strong and stable, has a single control arm on the tilt head, sets up quickly, and doesn't weigh so much that you dread carrying it along. Try extending and collapsing a tripod model several times before purchasing it, to be sure it works conveniently and consistently. Avoid fussy leg–locking mechanisms or latches that will jam up quickly when sand or dirt inevitably gets into them. Tripods with a sturdy central post elevation crank are the most useful, although some excellent field tripods have a simple twist-lock central post that raises and lowers rapidly.

Catadioptric telescopes originally designed for amateur astronomy have become popular among advanced birders. These cat scopes are much bulkier and heavier than conventional models but offer unparalleled sharpness over long distances. They are most popular for shoreline birding, in which the smallest details in plumage may separate one sandpiper or gull from another.

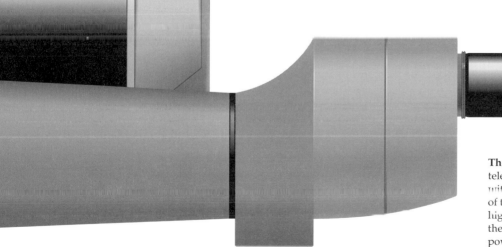

The most popular type of birdwatching telescope uses an optical path that is folded with pentaprisms to reduce the overall length of the scope. The eyepiece is thus typically higher than the objective lens. In this design the rear eyepiece determines the magnifying power. These telescopes typically have removable 15x–25x eyepieces. Higher magnifications are available but tend to be too dark for most birdwatching.

6 PHOTOGRAPHING BIRDS

The experience of watching wild birds is often so beautiful that it is only natural to want to share that pleasure with others through photography. Nature photographers and film makers have added immeasurably to our knowledge of the natural history of wild birds, bringing intimate details of their lives into our homes and libraries. Photography has strengthened our understanding of how to identify birds in the wild, by supplying accurate field references to the artists and naturalists who write field guides. Basic documentary photography is an essential skill in field biology, where photographs are often used to demonstrate bird behavior, to record the details of bird nesting and feeding habitats, and to preserve many other kinds of visual references in natural history and biology. Most published photographs of wild birds are made by professional wildlife photographers who often spend weeks on location in the wilderness to obtain the spectacular photographs that appear in magazines like *Natural History, Audubon,* or *National Wildlife.* The most important skills in nature photography are patience and a thorough understanding and appreciation of the natural history and habits of the animals you wish to photograph.

Cameras

Most wildlife photography is done with 35mm single lens reflex (SLR) cameras, and long focal length telephoto lenses. The SLR style of camera allows you to change lenses easily, but the primary advantage of an SLR camera is that you focus through the same lens that will form the picture on film; therefore, the composition you see in the viewfinder is exactly what you will see after the pictures are developed. Most of today's moderately priced SLR cameras offer automatic exposure control, but most experienced photographers strongly prefer the option to use manual exposure control as well. Most nature photographers use motor-driven SLR camera bodies made by companies that cater to the professional photographic market, such as Canon, Leica, Minolta, Nikon, Olympus, and Pentax. These manufacturers supply a wide range of top-quality lenses and accessories for their cameras, and there are dozens of smaller companies that make lenses and accessories compatible with these major brands (Sacilotto 1989).

The characteristic shape of an SLR camera, with its distinctive pentaprism above an interchangeable lens mount.

Lenses

The most pervasive myth about wildlife photography is that purchasing a long telephoto lens will solve the problem of how to get close to wild animals. Although it is true that most photographs of wild birds are made with telephoto lenses, a telephoto lens alone will not give you the tight, full-frame photographs of birds you see in magazines and books. For example, 400mm focal length lenses are the most popular all-around telephoto in bird photography. But even with a huge 400mm telephoto lens you must still be closer than twenty feet to shoot a full-frame picture of a songbird. Wild birds will often flush at distances of less than fifty feet, so clearly a 400mm lens alone is not the answer to successful close-up pictures of songbirds (Conway 1989).

Blinds

The key to successful wildlife photography is in getting as close as possible to the animals. Most close-up pictures of wild birds are taken from blinds (called "hides" in Great Britain), which are small tents specifically designed

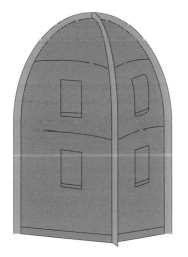

A photographic blind is typically taller and narrower than a camping tent.

for photography or animal observation. Well-designed photographic blinds are taller than tents made for camping, so that the photographer can sit comfortably on a stool or even stand almost upright while shooting. A blind must have a floor area large enough to accommodate a stool, a tripod, and a reasonable amount of photographic gear. The canvas or nylon walls of blinds usually have at least one small window flap on each wall to slip a telephoto lens or spotting scope through. Canvas blinds made for duck hunting usually have elaborate camouflage color schemes, but this is not necessary for photographic work. In nature photography the only advantage of camouflage is that a blind is less likely to be seen by passersby, reducing the likelihood of that you will be disturbed. Avoid blinds with flaps that close with noisy strips or zippers. These closures make loud ripping sounds when opened and will be a constant nuisance while working near wild animals. The best way to enter a blind is to bring along a companion to accompany you to the tent entrance. After you enter the blind your companion should leave the area. The birds (which do not count) will assume that both of you have left and will settle down and ignore the blind much faster.

35mm Lenses Useful in Bird Photography

Focal length (mm)	Notes
Wide-Angle Lenses	
24mm–35mm	Although wide-angle lenses are seldom used to shoot close-ups of wild birds, they are useful for collecting detailed photographs of avian habitats. The broad perspective a wide-angle lens provides makes it easy to combine in one photograph a detailed view of a bird's nest with a good view of the nesting habitat in the background.
Macro Lenses	
50mm–135mm, macro-focusing	Macro lenses are essential tools for the field biologists. They allow close focusing down to distances of just a few inches. Most true macro lenses have a fixed focal length between 50mm and 135mm; macro zoom lenses are optically inferior to true macros.
Telephoto Lenses	
135mm–200mm, 70mm–200mm zoom lenses	Medium-length telephoto lenses are useful for documentary work around nests and for shooting relatively tame birds in enclosures, zoos, or parks. Zoom telephoto lenses can be especially useful in photographing nests from blinds.
300mm–1,000mm, conventional designs	These classic "long" telephoto lenses are the workhorses of nature photography. Most people can hand-hold a 300mm lens reliably, but beyond 300mm it is wiser to use a tripod, monopod, or gunstock mount to steady the camera and lenses.
500mm–1,000mm, catadioptric designs	Catadioptric telephoto lenses ("cat" or "mirror" lenses) fold the light path through the lens with mirrors, creating a short, thick telephoto design. Cat lenses produce characteristic "doughnut" highlights in photographs and so are less popular despite their size. They also transmit less light than do conventional lenses with the same focal length.

7 RECORDING BIRD SONGS IN THE FIELD

Many naturalists find that listening to bird song is one of the most enjoyable aspects of going afield. Inevitably, many want to record the songs and calls they hear on field trips. A tremendous amount of excellent recorded material of common North American species is available, but for many people, recording the sounds of wildlife compares with wildlife photography as a serious hobby or avocation. Some birders never go afield without a tape recorder and have produced high-quality audio material that in some cases surpasses available professional recordings. Sound recordings by amateur naturalists have contributed significantly to the library of bird sounds available to scientists. Many regions of the world (particularly in the tropics) are still largely unexplored by sound recordists, and recordings made by amateur birdwatchers may be the only records we have of bird song from the fast-disappearing rain forests.

Audio Recordings as Documentation

Just as a photograph or specimen can document the occurrence of a species in an area, so can a tape recorder be used to document unusual birds. For example, the Tropical Kingbird (*Tyrannus melancholicus*) and Couch's Kingbird (*T. couchii*) look almost exactly alike and can be separated only by listening for their unique song patterns. With a sound recording the species can be positively identified, and in this case it would be better supporting evidence of the species' occurrence in an area than a documentary photograph.

Recordings as Attractants for Censusing

Birdwatchers sometimes use taped playbacks of bird song as a way to lure species that are usually very hard to see in the open. Taped calls trigger territorial responses in many birds, which seek out the source of the calls and thereby expose themselves to view. Taped calls are frequently used when owling at night and can be the only way to attract shy species of tropical birds within viewing range. They are also an important tool for ornithologists conducting census studies of bird populations. As with any potential disturbance to wild animals, the following caution should be observed: the *overuse* of taped calls, especially during the breeding season, can lead a bird to abandon its nesting area. Tapes should not be used during the peak of breeding season and should never be played within the nesting territories of rare and endangered species. Overuse of this field technique can cause terrible harm to breeding birds.

Audio Equipment

Most naturalists who record bird song use relatively inexpensive audio equipment to make recordings. Good-quality portable audio cassette recorders with noise-reduction circuitry are the most popular choice for most amateur sound recordings. Many excellent cassette recorders are no bigger than personal hand-carried stereo systems and can be taken on field trips without much bother. Choose a recorder that has Dolby noise reduction and is resistant to wow and flutter distortions when jostled while recording. Field recording conditions are seldom ideal. If the recorder will not tolerate a bit of movement during a recording it will be a constant frustration in the field. Better quality "semi-professional" stereo cassette

Small personal stereos with recording capabilities are ideal for experimenting with recording bird songs or other natural sounds. Special microphones like the one shown opposite must be used to record distant singing birds with sufficient fidelity and volume.

recorders have large VU meters for each stereo channel to allow close monitoring of sound levels by eye as well as ear. These larger cassette models will also have many more input and output jacks for connecting the better types of microphones without the use of adapters, and they usually have small built-in loudspeakers that can be used to broadcast songs to attract birds in territorial census studies (Davis 1978).

Microphones are a crucial component of successful field recording, and knowledgeable audio hobbyists often spend much more on a good microphone than they spend on the tape recorder. Two types are widely used in making wildlife recordings: shotgun microphones and parabolic reflector microphones. Shotgun microphones are long, tubular devices that accept sound only from a narrow beam directly in front of the microphone, thus emphasizing sounds from one direction. They are useful for relatively short-range recordings, such as songbirds recorded from thickets, or from low woodland areas where singing birds are not too high up in the canopy. Ordinary shotgun microphones are less successful at longer range recordings of distant singing birds. For distant birds, most audio recordists would favor a parabolic reflector microphone. Parabolic reflector microphones combine a conventional microphone with a bowl-shaped plastic or metal reflecting dish. This dish collects and focuses sound into the head of the microphone, giving the combination of reflector and microphone the ability to gather sounds from far away and to exclude virtually all extraneous sounds from the surrounding environment. Parabolic reflector microphones are the long-range telescopes of the audio world and are used in police surveillance, intelligence gathering, and military operations as well as in wildlife recording (Davis 1979, 1981; McKenzie 1989; Wickstrom 1988).

A parabolic reflector microphone and a "shotgun" directional microphone with a foam windsleeve fitted to reduce wind noise. The shotgun microphone is held with a pistol grip handle to reduce handling noises.

8 KEEPING FIELD NOTES

One of the most valuable habits you can develop is to write up your day's field notes, no matter where you are or how short your day in the field was. These notes will prove invaluable not only for future reference but also because in writing them you gradually accumulate a comprehensive overview of your birdwatching over long periods of time. Most experienced birdwatchers carry a notebook to record unusual or confusing observations for later study. Jot down on a simple pocket notebook or steno pad any features of a bird you may need help to identify later. If you are a careful observer, the many details you accumulate over a day's birding (plumage notes, behavior notes, habitat descriptions, etc.) will be much too numerous to commit to memory alone. With descriptive field notes before you, a careful identification should be possible long after your mystery bird has flown off and you have gone home to consult your library. A solid knowledge of avian topography (surface anatomy) is a must for recording field observations accurately. For example, you must understand which feather groups are visible in a bird's folded wing in order to describe accurately the wing coloration in a field report. Here are the most important features to record in a field description:

Shape and Color — Subtle nuances of shape are often important in the field identification of birds. Poor lighting conditions may make it difficult or impossible to see color patterns or surface detail on distant birds. The overall body shape or the shape of the wings, head, and tail may be your only solid clues to the species. Draw a simple sketch of the bird. Field sketches do not require great artistry, and in most cases some features of the bird will stand out that you can describe in even the crudest sketch. Annotate your drawing with specific notes on the color, pattern, and plumage details you saw. Don't forget to record the color of the fleshy parts of the bird's body: iris color, the color of bare skin near the eyes, bill color, and leg coloration. Try recording the characteristics of several common birds, and notice how much more detail you observe and remember when you make a careful written record of what you see in the field.

Size — Most people find it difficult to estimate the size of distant birds. Try to compare the bird to the size of plants or other nearby objects or relate it to a species you are very familiar with. For example, is the bird smaller than a Blue Jay or larger than a House Sparrow? In judging the size of distant flying birds with which there are no other species to compare, note the speed of the wingbeats and how much gusting winds affect the bird. Larger birds beat their wings more slowly and are less affected by the wind. These can be useful clues when, for example, you are trying to tell a Cooper's Hawk (*Accipiter cooperii*) from the very similar but smaller Sharp-shinned Hawk (*A. striatus*), or to separate a small Merlin (*Falco columbarius*) from the larger Peregrine Falcon (*F. peregrinus*) or Prairie Falcon (*F. mexicanus*).

Family — Did it look like a sparrow, a warbler, or a thrush?

Bill shape and color — This is often a key point in identifying songbirds.

Tail — Note the color patterns and overall shape of the tail. How long is the tail compared with the wingtips when the bird stands with folded wings? Is the end of the tail notched, straight, or rounded?

Wings — Note any markings (wingbars), the length of wings in relation to the tip of the tail, and any special features or patterns.

Legs — The length and color of the legs can be useful field marks.

Eyes — Note the color of the eyes and any eye ring, as well as the position of the eyes relative to the base of the bill.

Head markings or patterns — Important markings are often seen here, especially in sparrows and buntings.

Back pattern — Note whether it is streaked or plain. Does the color vary?

Underparts — Note any unusual markings.

Last, and certainly not least, record the behavior and physical condition of the bird, and make notes on the plant community or environment where you observed the bird. These additional observations are particularly important when reporting the appearance of rare or unusual birds in your area—it makes your record of their appearance that much more valuable to ornithologists and birdwatchers who study the birds of your area in the future.

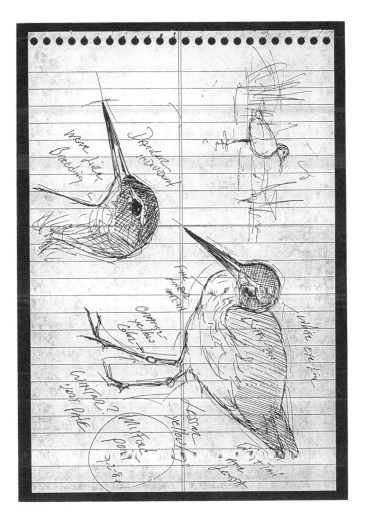

9 BANDING AND HANDLING WILD BIRDS

When studying wild birds, ornithologists often mark them with devices that make it easier to identify individuals of a species. The most common such technique is **banding** (called "ringing" in Britain and Europe), in which a numbered aluminum band is placed on one of the bird's legs. Data on birds in North America are then sent to the U.S. Fish and Wildlife Service's Bird Banding Laboratory, which keeps track of the band numbers and will notify the bander if anyone ever recaptures the banded bird or finds the bird dead and reports its band number to the laboratory. Ornithologists use banding and other forms of marking to study the lives, breeding success, and longevity of individual birds and to gain data on the age structure and population dynamics of bird species. Banding also allows ornithologists to study the migration of birds in much greater detail than would be possible otherwise; the recovery of banded birds far from the area in which they were banded can reveal a wealth of data on the geographic pathways and seasonal timing of annual migrations in a species.

Many thousands of birds must usually be banded to obtain enough recovery data for statistical analysis, particularly for studies of migrating songbirds in which the birds are captured on route, away from their breeding territories. Although more than 40 million birds have been banded since the U.S. Fish and Wildlife Service began banding programs in the 1920s, less than 5 percent of banded birds are recaptured or recovered. Even this low number is misleading, because most band recoveries are made by licensed hunters who return bands and other tags only from such game birds as pheasants, doves, geese, swans, and ducks. The recovery rates for non-game bird species are much lower, but long-term banding studies done on stable populations of breeding birds such as seabird colonies or heron rookeries can yield a high rate of recovery and a great deal of data on both the overall population and the lives and behavior of individual birds.

In 1804, John James Audubon performed what were probably the earliest bird "banding" experiments in North America. While studying phoebes and wood-pewees near his Pennsylvania home, Audubon tied silver threads around the birds' legs, so that he could tell whether the same individuals were returning year after year to nest (Chancellor 1978). The modern practice of attaching numbered metal bands to the legs of wild birds was originated in the late 1890s by a Danish schoolteacher, Hans Mortensen (Mortensen 1899–1922). Mortensen demonstrated the value of following individual birds throughout their lives, and the practice of bird banding quickly spread throughout Europe in the early 1900s. By 1902 scientific bird banding studies had begun in the United States, but the practice did not become well established until 1919, when the U.S. Fish and Wildlife Service assumed formal responsibility for supplying bands and keeping bird banding records (Hickey 1963).

Most birds are captured for banding with mist nets, which resemble extremely fine badminton netting. Mist nets are most often used near foliage, where the netting is nearly invisible against a backdrop of leaves. Birds moving through the foliage can not see the mist netting until it is too late to avoid hitting the net. The fine mesh of the net then entangles the bird

> **Note: It is a violation of U.S. federal law to handle any non-game wild bird** (dead or alive), or to handle any part of such a bird (including feathers and bones), unless you have a bird banding, salvage, or other special permit from the U.S. Fish and Wildlife Service or are under the direct supervision of a licensed bird bander or other licensed wildlife professional.

Handling Wild Birds

Holding a Small Bird for Photography

Most small songbirds should normally be held for examination or measurement as shown in the figure at the bottom of the page. However, this conventional method of handling birds is unsuitable for photographing captured birds because the hand covers too much of the bird's body. Photography should be the last step in the process of examination or banding, as there is always the possibility that the bird will escape while being handled.

Small to medium-sized birds may be held up for examination or photographs by gripping the very tops of the bird's legs with your fingertips, as shown here. Some birds will flap their wings when first held up in this position; don't try to restrain the bird's wings—just let it flap. Most birds quiet down quickly. Most birds will allow you to extend one wing gently to get a better view of wing colors and patterns. If the bird struggles violently or your grip on the legs slips, **LET THE BIRD GO.** If you attempt to hold onto the bird, you risk injuring its legs or wings.

Small and Medium-sized Songbirds

Passerines and other small birds should be held by laying the bird across the palm of your hand, with the bird's head and neck lying between your index and middle fingers. Wrap your fingers in a loose "cage" around the bird to keep it from struggling or escaping. Never directly squeeze the body or neck of any small bird—even mild pressure can interfere with the bird's breathing or cause fatal internal injuries. You can use your thumb and ring finger to hold and extend the bird's wings or legs for examination, measurement, or banding. To release the bird, hold your hand out with the palm up and spread out your fingers. Do not toss the bird into the air if it hesitates to fly; give the bird a few seconds to realize that it is free to leave before prodding it with your fingers.

When handling any wild animal it is good practice to wear some eye protection. Normal eyeglasses will do if they have safety glass lenses. Otherwise use high-impact "sports" sunglasses or plastic safety glasses available from hardware stores.

9 BANDING AND HANDLING WILD BIRDS, *continued*

just enough to hold it until the bander can recover the bird from the net. Various sorts of wire ground traps baited with food are also used to attract and capture birds, particularly songbirds that gather near feeders in winter. Wire traps may also be used to capture ground-nesting birds—bottomless traps are placed over the birds' nest to capture the breeding adults. Many long-term banding studies are done by banding nestlings before they leave the nest. This strategy is particularly effective in colonial breeding species but is also widely used to study raptors, bluebirds, and many other species where individual nests must be located and visited. More complex and specialized trapping and snaring methods must be used to capture migrating raptors, shorebirds, and waterfowl for banding (U.S. Fish and Wildlife Service 1980).

> Note: It is a violation of U.S. federal law to handle any non-game wild bird (dead or alive), or to handle any part of such a bird (including feathers and bones), unless you have a bird banding, salvage, or other special permit from the U.S. Fish and Wildlife Service or are under the direct supervision of a licensed bird bander or other licensed wildlife professional.

Small Raptors

The key to handling raptors is to control their legs and feet. Falcons and some owls may attempt to bite your hands, but the feet are a raptor's primary defensive weapons, and even a small hawk or owl can seriously injure your hands unless you are quickly able to locate and immobilize the legs. You should wear some form of eye protection at all times when handling any wild animal, but particularly when handling large birds.

To pick up a small raptor from a mist net, first be sure you know exactly how the bird is hanging and where its feet are. Hold the bird around its shoulders with one hand, with the bird's neck *loosely* between your index and middle fingers. If possible, turn the bird so that it cannot see your movements, then quickly slip your other hand around the *base* of the bird's legs. Use both hands if necessary to hold the bird's body and legs until you can place it in a cage or are otherwise able to release it. Never attempt to handle a large hawk or owl without expert, licensed supervision.

Form and Function

Skull ossification in the passerines

When skulling a songbird, the bird's crown feathers are soaked with water and parted along the midline of the skull to expose a patch of skin over the frontal bones. Bird skin is normally thin and almost transparent, so it is relatively easy to assess the degree of skull ossification.

It can be difficult to determine the age of birds banded during fall migration. In many species the drab first-winter plumage of immature birds hatched that summer is nearly identical to the winter, or basic, plumage of the adults. But in most songbirds the frontal bones of the skull are not completely ossified until six months or more after hatching. Bird banders use this delayed ossification as a tool to separate older birds in fall plumage from immature birds. Skulling does not harm the birds and takes just minutes to perform (Wood 1969).

In most immature passerines the frontal bones consist of a single translucent layer of bone overlying the brain. The immature bone appears dark pink as seen through the skin. Over the first four to six months of life the frontal bone develops into a double-layered calcified sheet of bone, with tiny columns of bone separating the inner and outer layers. When seen from above (as the bander sees them), these columns within the pneumatic air space of the mature frontal bone look like tiny white spots. Mature, fully calcified frontal bones are an opaque pearly whitish color speckled with hundreds of tiny white dots. The area of mature calcified bone gradually grows back from the eye sockets and forward from the base of the skull, enclosing and eliminating the dark immature areas of bone (see the figures below and left). By early winter most immature passerines have skulls that are fully ossified, but during their first fall migration the immature songbirds will show a mixed pattern of unossified and fully calcified frontal bone (Baird 1963; Miller 1946; Norris 1961; U.S. Fish and Wildlife Service 1980).

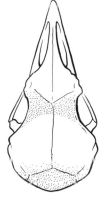

Immature

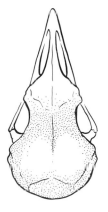

Intermediate

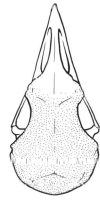

Fully ossified

Figures after Baird 1963; Miller 1946; Norris 1961; U.S. Fish and Wildlife Service 1980.

The Banding Process

Bird bands from the U.S. Fish and Wildlife Service (USFWS) are made of aluminum and marked with a serial number and the legend "Avise Bird Band; Write Washington DC USA." Special banding pliers are used to crimp the bands around the lower leg of the bird. Bands for most smaller birds simply wrap loosely around the leg like a collar. For raptors, which can easily remove regular bands with their strong hooked beaks, bands with a locking flange are used. Banders obtain their bands from the USFWS Banding Laboratory free of charge and annually report back to the laboratory the serial numbers and species data on birds they have banded the previous year. To obtain a bird banding permit, beginners usually apprentice themselves to an established bird bander or ornithologist. After becoming familiar with banding procedures and the identification and aging of every common local species, the apprentice bander may apply for independent status with a "master permit" to band birds.

To report their activities accurately banders must determine the bird's age and sex and collect other useful information on its physical condition. Although some species of birds can be aged and sexed with a quick visual assessment of plumage colors and patterns, most species require the bander to check fine details of the feathering and the colors of such soft tissues as the iris, and even to determine the degree of skull ossification to

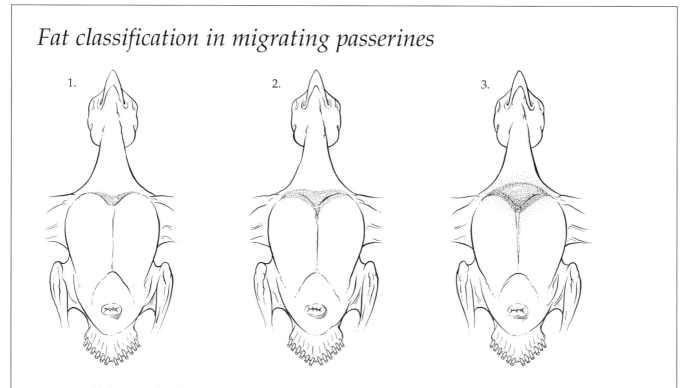

Fat classification in migrating passerines

1.　　　　　　2.　　　　　　3.

Migrating birds accumulate deposits of subcutaneous fat to fuel their journey. These fat deposits can easily be seen by blowing back the feathers of the breast and looking at the V-shaped notch between the clavicle bones (the furcula, or wishbone), just anterior to the breast muscles. Ornithologists and bird banders often use the amount of fat present in the interclavicular (furcular) area to gauge the condition of a bird, and record their findings using the four-step visual scale outlined here. The classes are: 0 – No fat visible; 1 – Some fat visible in interclavicular region; 2 – Interclavicular region nearly filled with fat; 3 – Interclavicular region completely filled with a bulging pad of fat and fat deposits visible elsewhere on the body as well (King and Farner 1965).

Various sizes of USFWS aluminum bird bands, shown at actual size. The smaller bands are for songbirds, the larger bands for birds with substantial legs, such as geese or cormorants. The special band at the bottom can be locked and is used to band larger hawks and owls.

		Typical birds in band size range
Size 0	○	Warblers, kinglets
Size 1	○	Barn Swallow, Swamp Sparrow
Size 1B	○	Grey Catbird, Fox Sparrow
Size 1A	○	Nuthatch, Veery
Size 2	○	Blue Jay, Hairy Woodpecker
Size 3	○	Male Sharp-shinned Hawk, flickers
Size 3A	○	Female Sharp-shinned Hawk

Size 8
Canada Goose

Size 7B Lock-on
Barred Owl

be sure the bird is properly identified and aged. The Bird Banding Laboratory provides all licensed banders with a manual that lists age and sex characteristics for most species in North America. Many species (hawks and sparrows, for example) can be sexed by measuring the wing chord—this measurement is a standard part of most banders' routines. In autumn many passerines can be aged by checking the degree of skull ossification (see opposite). Banders may check the overall physical condition of birds, including body mass, the degree of body fat in migrating birds, and deformities or gross injuries, as well as look for such parasites as mallophagan mites or hopoboscid flies. Many banders keep a camera handy to record unusual species, deformities, or injuries.

Several kinds of markers can be used to identify birds in a study population. The most common way to mark individual birds in behavioral studies is with colored plastic leg bands. These are particularly useful in studies on colonial breeding birds or in studies that follow other small resident populations of birds. Colored leg bands allow scientists to identify particular birds through binoculars or a spotting telescope, thus avoiding the need to recapture birds to follow their behaviors and movements on a daily or even hourly basis. Colored dyes are often used to mark birds in migration research, in which the identification of members of a group is more important than identifying specific individuals. Prominently dyed birds are easy to spot over long distances, and the highly visible dyes enable researchers to follow birds over great distances in migration. Larger birds such as geese and eagles are sometimes marked with plastic tags on the leading edge of their wings. Wing tags are easily visible and are especially useful if individual birds must be identified while in flight. Many studies of ducks and geese employ a numbered plastic bill saddle that sits atop the beak without harming the bird. Geese are frequently marked with numbered plastic neck collars.

Reporting Band Recoveries
If you find a dead or injured bird with a USFWS band or see a wild bird with any other sort of wing or leg tag or dye marking, please report it to the USFWS banding office. Before removing bands or other markers from a dead bird first note the wing or leg on which the marker is placed and the sequence of colored bands on the leg—this information is often crucial to the bander. Record the exact location and time the bird was recovered, the age and sex of the bird if you can determine it, and the cause of death if it is known. If you see a dye-marked bird while birding, write down the date, time, and exact location where you saw the bird, as well as any information on the activity of the bird (for example, feeding, roosting, or direction of flight). Send this information to the Bird Banding Laboratory at this address:

Department of the Interior
U.S. Fish and Wildlife Service
Bird Banding Laboratory
Laurel, MD 20811

10 MEASURING AND WEIGHING BIRDS

Bird banders rely heavily on wing and tail measurements to determine the age or sex of many species, and in these measurements a difference of one or two millimeters may be critical. Measuring birds requires careful, consistent procedures and practice to ensure accuracy. Multiple measurements taken from specimens ideally should vary by not more than 1 percent (USFWS 1980). The most commonly used measurements of live birds are described here:

Wing chord

The length between the leading edge of the wrist joint at the bend of the wing and the tip of the longest primary feather. This measurement is best taken with a stiff metal ruler that has an upright stop secured at the "0" point of the ruler. Butt the bend of the bird's wing against the stop on the ruler, then gently lower the wing onto the ruler until the tip of the longest primary just touches the ruler's surface. Do not flatten the wing against the ruler, which will change the curvature of the wing and distort the measurement.

Bill length

The precise method used in measuring should always be noted when recording bill measurements.

Bill from base—The standard bill length is the distance between the bill tip and the point at which the bill meets the frontal bones of the skull. Usually this point is covered by feathers and must be exposed to identify the true base of the bill.

Bill from feathers—Exposed bill length measures the exposed culmen of the bill, from the bill tip to the point at which it disappears under the forehead feathers.

Bill from nostril—The distance from the distal (outermost) edge of the nostril to the bill tip.

Bill from cere—In birds with strongly curved bills with a horny cere at the base, such as hawks, owls, and parrots, the bill is measured from the distal edge of the cere to the tip of the bill.

Bill depth

Measured with calipers or dividers, the maximum distance from the top edge of the bill (the culmen) to the bottom edge of the bill (the gonys). Usually taken at the base of the bill.

Tarsus length

Measured with calipers or dividers, the distance from the rear middle of the intertarsal joint to the distal edge of the last complete scale before the toes begin (this corresponds to the distal, or lowermost, end of the tarsal bone).

Tail length

Measured with a very thin plastic or metal ruler. Slide the ruler up under the upper tail covert feathers until the ruler edge hits the base of the feathers and stops. Hold the ruler against the tail feathers and read the length of the longest tail feather (rectrix).

Light spring-loaded scales like this 300-gram model are often used to weigh songbirds and other small birds. The bird is usually placed into a small plastic net bag, which is then clipped to the scale for weighing.

Bill from base

Bill from feathers

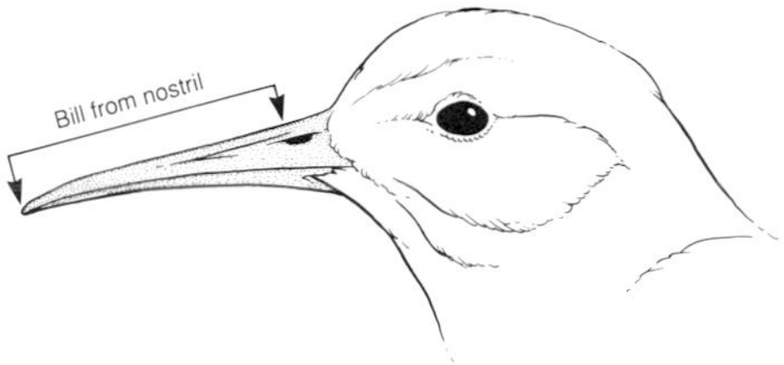

Bill from nostril

Bill from cere

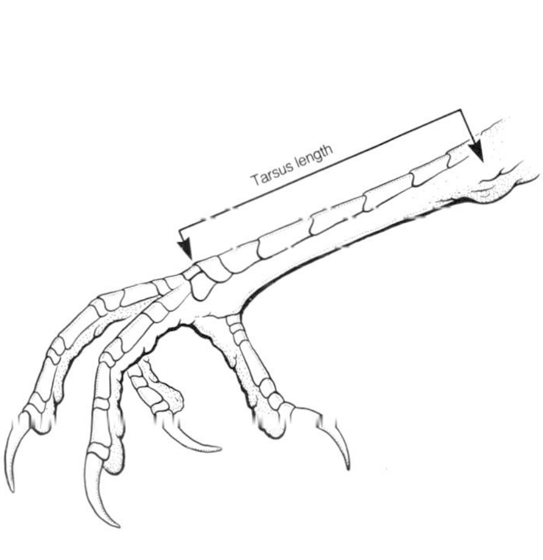

Tarsus length

Weighing Live Birds

Passerines and other small birds are often weighed in the field with small spring scales called Pesola scales, after the most popular brand of such scales. Typically the bird is slipped into a small net bag (like the kind in which onions are packed at the supermarket) and the bag is then clipped onto the Pesola scale. The mass in grams is measured, and then the mass of the net bag is subtracted to give the final mass of the bird. Larger birds may be immobilized by placing them into cardboard tubes, or metal tubes made from juice cans or coffee cans taped together. Tubes for this purpose should be not be tight around the bird's body but just snug enough to keep the bird from struggling or escaping from the tube while it is being weighed. Cut air holes at the head end of the tube to allow the passage of air around the bird's head. Pesola scales are only accurate in specific weight ranges, so most songbird banders have both a small 100-gram scale and a larger 300- or 500-gram scale for bigger birds like Blue Jays, flickers, or small hawks. Some banders prefer laboratory triple-beam balances, which are accurate over a wide range of weights from the smallest songbirds to hawks and other medium-sized birds. Triple-beam balances are much more costly than Pesola spring scales but are often more convenient for large banding operations that process hundreds of birds on a busy day.

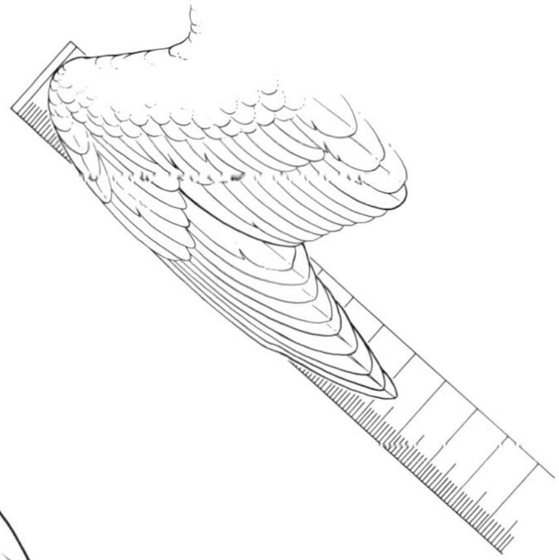

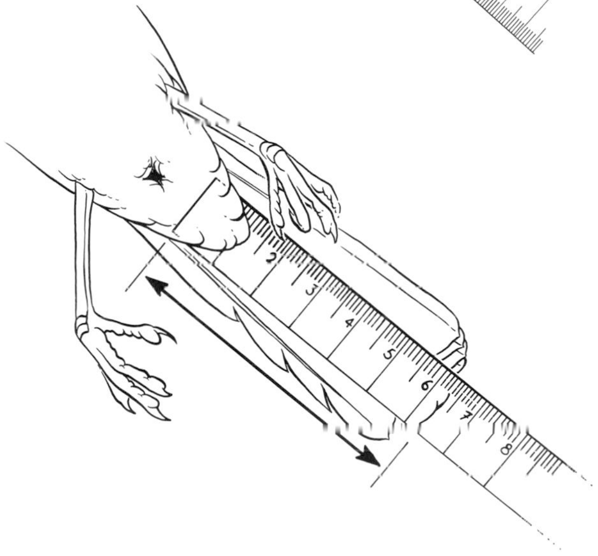

11 PREPARING STUDY SKINS

The dried, preserved remains of a bird's skin and feathers are called "study skins" or sometimes simply "skins." Preserved bird skins are a major reference in ornithology, and most museums of natural history and university biology departments have collections of study skins for research and teaching. Every field biologist should be familiar with the basic procedures for preserving a bird specimen, as outlined below. With proper preparation and careful storage, a bird's skin and plumage may last as a scientific resource for hundreds of years (Hall 1962; Grantz 1969).

Fresh Specimens

Always try to work with fresh specimens—birds that have been dead no more than a few hours or that have been properly refrigerated or frozen to preserve the skin tissues. Never assume that a bird killed along the roadside is fresh. Most birds decay rapidly after death (especially in warm weather), so always examine the bird's body for signs of decomposition. Foul smells from the specimen are the most obvious sign of decomposition, but there are also more subtle clues you can use to evaluate whether a specimen is worth skinning. A study skin of a bird specimen is primarily composed of the preserved feathers and skin of the bird, so there is no point in preparing the specimen if the skin tissues are not in reasonable condition. To evaluate a specimen, check the skin surrounding the bird's vent (cloaca). The lower belly around the vent is one of the first areas to deteriorate or "slip"—the layers of skin begin to separate and the skin no longer holds the feather shafts in place. Lay the bird on its back and spread the feathers in the lower belly region. If the feathers fall out easily, the skin has begun to slip. If the skin in the region is brownish and becomes watery when rubbed, it is probably past the point of making a good preparation. If this is your first time preparing a study skin, or if you have not done many, the best thing to do is to work with another specimen.

Frozen Birds

Birds that have been frozen should be allowed to thaw to a point of flexibility before attempting to prepare a skin. Larger birds like ducks may need to thaw overnight, but small songbirds may defrost in just an hour. In either case, be attentive to the thawing process. As soon as the bird is flexible the skinning process should begin. The specimen should never be allowed to sit at room temperature for long periods or the skin will begin to rot and the internal organs will begin to decompose. Special care is needed if the skin to be worked with has been kept in a freezer for an extended time. This is usually indicated by deeply sunken eyes and "freezer burn," or dehydration of the exposed skin. To try to save a freezer-burned specimen, soak the bird in water until it thaws and absorbs enough water so that the skin of the head region moves around when rubbed with your fingers. This indicates that the dried adherent skin is now workable and that the specimen can possibly be saved. The feathers can then be dried either by working corn meal into the plumage as an absorbing agent or by drying the plumage with a regular household blow drier. Dry the feathers until they are fluffy again before proceeding with the skinning.

To store a bird specimen in a freezer for an extended period, always wrap

Tools and Materials Needed

Scalpel
Scissors
Tweezers or forceps
Toothbrush
Sewing needles, both curved and straight
Dissecting tray or newspaper
Blow drier (if the specimen needs to be dried)
Styrofoam or wood pinning board
Dissection pins or heavy sewing pins
White corn meal (do not use yellow corn meal, which is too greasy)
Fine black sewing thread
Roll of white surgical cotton
Wooden applicator sticks (for small songbirds)
Stiff card-stock paper for the specimen information tag

Note: It is a violation of U.S. federal
law to handle any non-game wild bird
(dead or alive), or to handle any part of
such a bird (including feathers and
bones), unless you have a bird banding,
salvage, or other special permit from the
U.S. Fish and Wildlife Service or are
under the direct supervision of a
licensed bird bander or other licensed
wildlife professional.

the bird completely in plastic household wrap. Use several layers to be sure
the enclosure is airtight. This will help prevent dehydration of the bird's
tissues. You should record basic data on how, when, and where the bird
was collected, along with the color of any soft, exposed tissues such as iris,
bill, eye ring, lores, legs, and feet. Ideally you might refer to a naturalist's
color guide (such as Smithe 1975) for references and standard names of
colors. Enclose your notes on the specimen within the plastic wrap, so that
you or whoever prepares the skin will have the information handy when
the tag for the skin is prepared.

Preparing a Study Skin of a Songbird Specimen

Lay the bird on its back and measure the length
of its body from bill tip to the tip of the tail for
future reference when stuffing the bird. If your
specimen is very fresh, note the color of the iris,
any fleshy eye ring, and the lores, as well as the
colors of the bill, legs, feet, and any other
visible soft tissue. These areas of soft tissue
darken rapidly after drying, so future users of
the skin will have only your notes to guide
them. You may also wish to sketch the overall
shape of the bird for use in accurately shaping
the skin once stuffed.

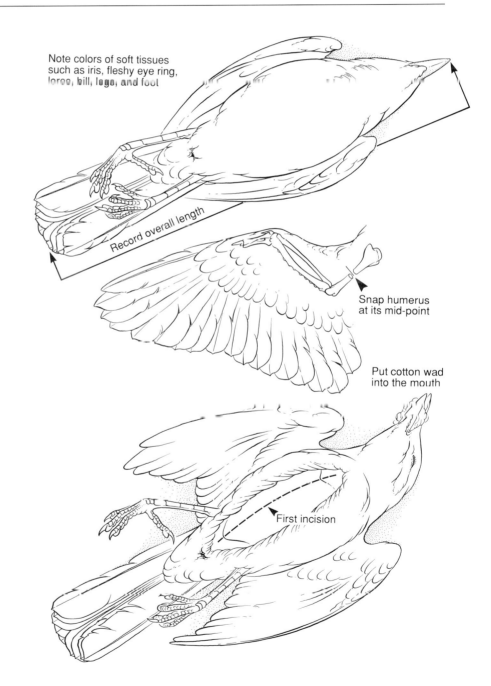

Note colors of soft tissues
such as iris, fleshy eye ring,
lores, bill, legs, and foot

Record overall length

Snap humerus
at its mid-point

Put cotton wad
into the mouth

First incision

Flex the bird's body and limbs to loosen skin
attachments and free any muscles that are stiff
from freezing or rigor mortis. Extend the wings
and break the humerus at the midpoint of the
bone shaft in both wings. This will make it
easier to free the wings from the shoulder
muscles later in the skinning process. Insert a
small ball of cotton into the bird's mouth to
prevent body fluids from leaking out and
soaking the head plumage.

Lay the specimen on its back and spread the
feathers of the belly and underbelly. The
breastbone and large pectoral muscles will be
apparent, as will the unsupported lower
abdomen. Make an incision with sharp pointed
scissors or scalpel into the upper breast area
and cut backwards alongside the breastbone to
the end of that bone. Try not to cut into the
peritoneal membrane and muscles around the
abdominal viscera, because this will cause
body fluids to weep onto the skin and you will
have to wash the skin later. If the peritoneum is
punctured, immediately blot the area with a
large quantity of white corn meal. White corn
meal is needed throughout the skinning
process to absorb any blood or fluids that may
leak from the body. The corn meal can later be
wiped, dusted, or blown off the specimen much
more easily than blood can be washed off the
plumage.

Preparing a Study Skin, continued

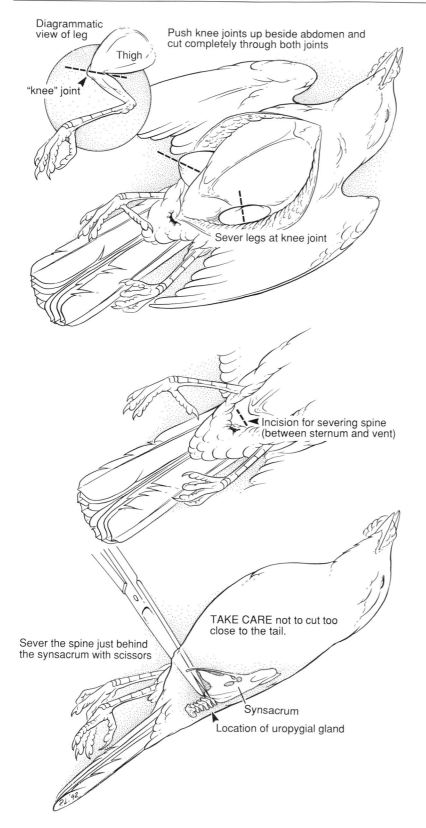

Diagrammatic view of leg

Thigh

"knee" joint

Push knee joints up beside abdomen and cut completely through both joints

Sever legs at knee joint

◄ Incision for severing spine (between sternum and vent)

Sever the spine just behind the synsacrum with scissors

TAKE CARE not to cut too close to the tail.

Synsacrum

Location of uropygial gland

Grasp the skin of the chest region with your fingertips, and separate the skin from the abdominal viscera. Work this separation process along the body wall laterally until you can see the knee joint of the leg, which will protrude up along the side of the body. Flex the knee upward by pushing on the leg from the outside until the knee joint arches high enough to allow a pair of scissors to pass under the joint. Insert the scissors and snip through the tibiotarsus of the leg. Push the leg bone up and scrape it clean, then carefully pull on the leg from the outside until it slips back into place. Repeat this procedure for the opposite leg.

Peel the skin from around the abdomen on both sides and work your way back toward the base of the tail. You will now cut into the abdominal peritoneum for the first time. Be prepared to apply a good deal of corn meal. Hold the bird in your hand with your fingertips along the line of the vertebral column. Puncture the abdominal cavity just anterior to the cloaca and insert the tip of a pair of scissors into and through the abdominal cavity. Push through until you can feel the scissor tips against the fingertips of your opposite hand, which you should position along the base of the vertebral column. Back the scissors off slightly to prevent cutting through the skin of the back, and then snip through the vertebral column to sever it. **Do not cut too close to the tail.** If you cut too close to the tail base, the whole tail will simply fall off. Now separate the skin from the abdominal viscera, pelvis, and thorax. Use as much corn meal as necessary to keep any blood or fluids off the plumage.

Invert the skin at the tail region and clean off excess muscle from around the tail vertebra. Be careful not to clip the base of the tail feathers, which are now visible on either side of the caudal vertebrae and pygostyle. Peel the skin away from the bird's rump area to expose the uropygial gland (a heart-shaped swelling) at the dorsal base of the tail feathers. Puncture the two lobes of the gland and squeeze out the oily material inside, blotting the area with corn meal to absorb the secretions. If the uropygial oil is not removed, oils will weep onto the skin, discolor the plumage, and weaken the skin that holds the tail feathers in place. After cleaning the gland area, peel the skin back along the body up to the level of the wing roots.

Now you are ready to clip the wing muscles to separate the wings from the body. Stretch the wings out to the side and pull the skin toward the body until you can see or feel under the wing at the humerus. If you remembered to break the humerus earlier in the skinning process it should be easy to snip through the muscles around the base of the humerus and free the wing skin from the shoulder. Do this on both sides of the body. Peel the skin back as far as you can along the wing bones and remove as much wing muscle as possible. Blot the area with corn meal.

Once you have completed this cleaning, you can finish peeling the skin up to the base of the skull. At this point you should have in your hand what looks like a miniature roasting chicken, with the skin hanging free of the body except for its attachment to the head.

Although the skin of the neck seems small, it should stretch over the head. Be sure to **remove the cotton ball** you earlier placed in the bird's mouth. Depending on the size of your specimen's head, use one of the two variations on the next procedure given below:

1. In small songbirds, grasp the skin and slowly work it up over the head with the thumbs. If needed, press in on the skull. This flex will allow the skin to slide over the large rear portion of the skull.

2. In larger birds, tendinous adhesions between skull and skin may be cut with a scalpel or scissors. Carefully peel the skin back up to the ear region. As you approach the ear openings, grasp the skin with your fingernails and pull it free. This will break the skin away from the ear openings, usually without tearing the skin. If you do tear the skin, come back later and sew up the torn skin before stuffing the bird. Peel the skin away from the skull until you reach the eyeballs, which will appear as dark globes under the translucent skin. Carefully pull the skin away from the eyeballs and skull, so that you can clearly see the edges of the eyelids (the eye ring). Snip or cut close to the skin, and try to keep the eye ring intact. Do not puncture the eyeball, or it may leak over the head plumage. Repeat the procedure to free the skin around the opposite eyeball.

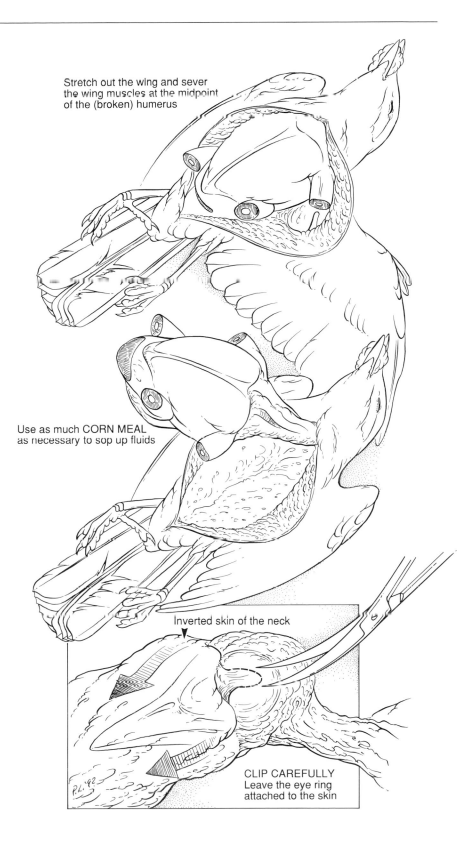

Stretch out the wing and sever the wing muscles at the midpoint of the (broken) humerus

Use as much CORN MEAL as necessary to sop up fluids

Inverted skin of the neck

CLIP CAREFULLY
Leave the eye ring attached to the skin

Preparing a Study Skin, continued

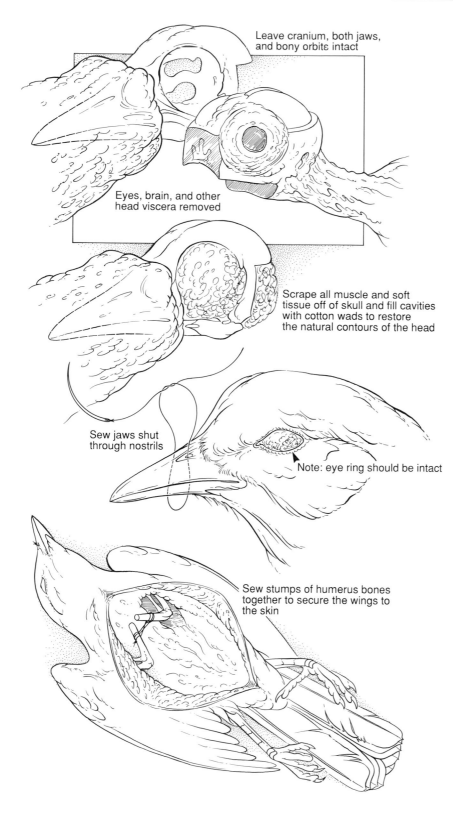

Leave cranium, both jaws, and bony orbits intact

Eyes, brain, and other head viscera removed

Scrape all muscle and soft tissue off of skull and fill cavities with cotton wads to restore the natural contours of the head

Sew jaws shut through nostrils

Note: eye ring should be intact

Sew stumps of humerus bones together to secure the wings to the skin

You are now ready to separate the body from the skin. To do this you will need to make four cuts on the skull: (1) between the lower mandibles, (2) at the base of the cranium, and (3 & 4) from the mandible cut to the cranial cut, on the right and left sides of the skull.

The object of these cuts is to leave the bones of the beak and upper skull attached to the skin to preserve the natural shape of the bird's head. The lower and posterior portions of the skull and neck are left attached to the body viscera.

After making the four final cuts into the head, pull the body and neck viscera away from the upper skull bones and skin. The skin and plumage should now be completely separate from the internal organs. Scoop out the brain tissue from the cranial cavity. Use tweezers to remove the eyeballs from the eye sockets. Scrape any remaining muscles and soft tissue off the skull and jaw bones, and use corn meal to dry the area out. The less soft tissue that remains on the skin or bones, the more durable the study skin will be.

Dust off the skin as much as possible with an old paintbrush or toothbrush. Make two small wads of cotton about the size of the bird's eyeballs and insert them into the eye sockets. The cotton balls should not be too large—just big enough to keep the eye area from indenting when the skin is pulled back over the skull. Now you are ready to turn the skin "right-side out" again to stuff it with cotton.

Lay the skin flat and align the bill, which cannot be seen but can be felt, with the tube of the neck skin. Carefully push the head back through the neck skin. Don't puncture the thin neck skin with the sharp bill point. Grasp the bill as it reappears and pull the skin right-side out. You should now have a complete skin with cotton in the eye sockets and the top of the cranium in place. Place your finger inside the skin and up into the neck tube, dust the skin off, and smooth down the neck plumage with a toothbrush.

Grasp the stumps of the wing bones and pull them toward the center of the body. Tie these together loosely to hold them in place. Some ornithologists omit the step of tying the humerus bones and simply place the wings into proper alignment with the body and allow them to dry in place. But if such a specimen were to be roughly handled, the wings might fall off the study skin, and extensive repairs would be necessary to restore it.

Sharpen a wooden applicator, and then build a body of cotton on it by taking thin layers of cotton and spinning them around the applicator. Be sure to wrap the layers of cotton in one direction. Use the bird's viscera as a guide to how bulky the cotton body should be, and try to duplicate the original contours of the neck, thorax, and abdomen in the cotton body. Make sure the cotton body "blank" is the same overall length as the viscera you removed from the skin. The most common faults with study skins are in the preparation of the neck area. Be sure not to overstuff the neck or make it too long. Most birds have very thin necks that are normally held in a compressed S position, so a straight, overstuffed neck on a study skin will always look unnatural and will distort the overall length of the bird.

Thread a curved needle with a short piece of fine black thread. Pass it through the nostril openings and sew the bill closed. Carefully slip the spun cotton body blank into the skin, and guide the tip of the cotton blank up through the neck tube. The forward tip of the cotton body should push all the way to the base of the bill. Sewing the bill shut prevents the tip of the cotton body from poking through the mouth.

Once the cotton body blank is settled within the cavity, grasp the sides of the incision and pull them closed. Sew the incision shut using a continuous running stitch, starting at the anterior end of the incision and running back toward the tail. When you reach the posterior end of the incision, sew a few extra stitches to secure the stitching and then cut the thread as close to the skin as possible. Fluff the belly feathers and fold them over the incision to hide the seam.

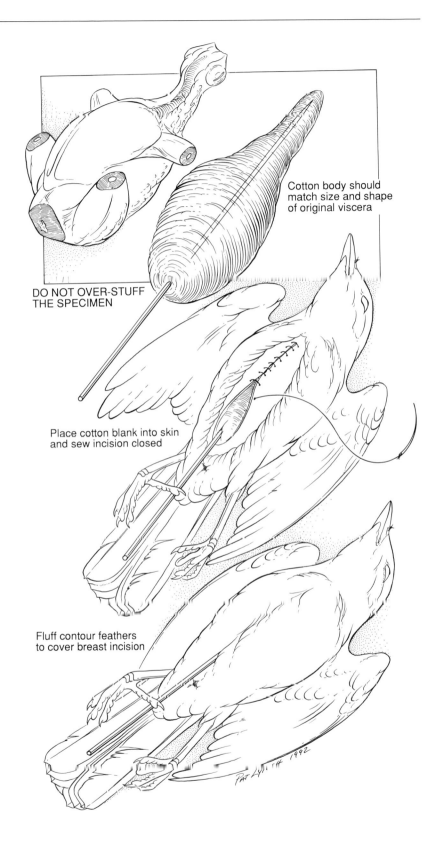

Cotton body should match size and shape of original viscera

DO NOT OVER-STUFF THE SPECIMEN

Place cotton blank into skin and sew incision closed

Fluff contour feathers to cover breast incision

Preparing a Study Skin, continued

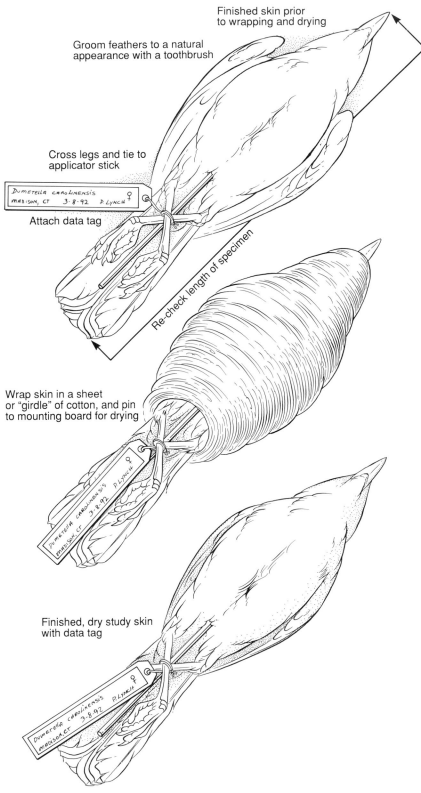

Finished skin prior
to wrapping and drying

Groom feathers to a natural
appearance with a toothbrush

Cross legs and tie to
applicator stick

Dumetella carolinensis
Madison, CT   3-8-92   P. Lynch ♀

Attach data tag

Re-check length of specimen

Wrap skin in a sheet
or "girdle" of cotton, and pin
to mounting board for drying

Finished, dry study skin
with data tag

The study skin is now ready for final shaping before drying. Place your fingers under the wings and work the skin up over the back to make the dorsal plumage look as natural as possible, smoothing out any wrinkles or kinks that may have developed. Use an old toothbrush to smooth and groom all the feathers and restore them to their normal positions. Make sure the cotton in the eyes looks full; with tweezers you can add small tufts of cotton to fill out the eye sockets. Place the bird on its back and fluff the underbelly feathers to hide the incision. Fold the wings against the body in a natural-looking position. You might fluff a few of the breast feathers over the forward edges of the wings (as seen in most living birds), to give the wings a more natural appearance. Check the current overall length of the skin against the measurements you took before you began the skinning process. If the skin has been stretched, try working the skin over the cotton blank with your fingers to compress it and restore a more natural overall length.

The bird is now ready for drying. Make a very thin rectangle of cotton from the roll. This will act as a girdle to hold things in place. Place the bird on its back. Cross the legs over each other and tie them together where they cross with a few loops of thread. Wrap the thin rectangle of cotton completely around the bird's body and wings. This will keep the folded wings in place and preserve the natural contours of the body. Then place the bird on its back on a styrofoam or soft wooden board and anchor it with pins crossed over the bill and on the wing sides. Spread the tail feathers a little and pin them in place. Use strips of card stock and pins to secure the tail feathers in position for drying. Dry the specimen for two to three days in a warm, dry location. When dried, the study skin will remain in the same shape for decades if not mishandled.

Data tag

To determine the sex of your specimen, place
the dissected body on its back. With a pair of
scissors, cut along the ribs of the bird's left side
and flex open the body cavity. Embedded in the
back will be the three-lobed kidneys. At the
forward end of the kidneys there will either be
the paired testes of the male or the granular-
looking single ovary of the female. Use the
drawings below for reference, or see chapter 10,
The Urogenital and Endocrine Systems, for
more information on the sexual organs of birds.
Measure the length of the testes or ovaries and
the length of the oviduct, if visible, and record
this information (in millimeters) on the
specimen tag.

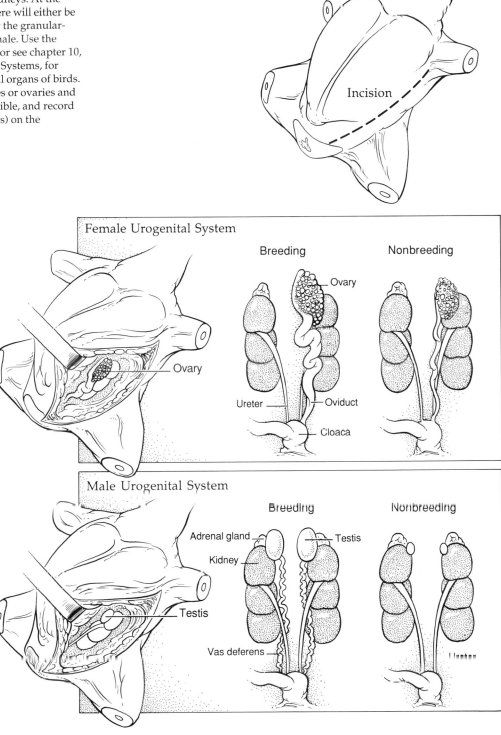

Incision

Female Urogenital System

Breeding   Nonbreeding

Ovary

Ovary

Ureter   Oviduct

Cloaca

Male Urogenital System

Breeding   Nonbreeding

Adrenal gland   Testis

Kidney

Testis

Vas deferens   Ureter

In the Field | FIELD ORNITHOLOGY PROJECTS

1. Chose a bird species that is common in your area. Set a schedule of viewing times, and during those periods record as much data as you can on the bird's habits. Vary the time of day that you observe the birds to cover the early morning, mid-morning, afternoon, and evening. After a month of such observations compile a description of the field marks, behavior, and natural history of the bird, entirely from your own observations. Then compare what you were able to discover about the bird with descriptions written by John James Audubon or Arthur Bent in their published works on the natural history of North American birds.

2. During the spring (March–May) or fall (September–November) migration seasons save the daily weather charts from your local newspaper. In spring, note how bird movements differ on cool mornings with little wind compared with warm mornings with a strong southerly breeze. In fall, compare cool rainy days with southerly winds with days when cold fronts bring strong northwest winds. Record your observations, and match them on heavy migration days with the weather maps from your newspaper.

3. Nocturnal bird migrations are spectacular, though hard to observe and evaluate. During spring or fall migration choose a still night in an area with little background noise and listen for the sound of migrating birds overhead. Note the times you are able to hear birds and whether you hear many birds calling or just a few. Are any of the calls identifiable? Listen to a tape of migrating songbirds, such as *Nocturnal Flight Calls of Migrating Thrushes*, by Bill Evans (P.O. Box 46, Mecklenburg, N.Y., 14863). Can you identify any of the thrushes as they pass overhead? Read the two articles listed below. Set up a spotting telescope on a clear night with a moon at least three-quarters full and look for birds as they pass in front of the moon. Birds close to you will usually pass too quickly to identify, but if the birds are far enough away it is occasionally possible to identify particular species by their flight silhouettes. Compare your night migration observations with those listed in the two papers, and try to estimate how many birds may have passed your observation point that night.

4. Choose a specific plant community or a location with several plant communities, and keep a daily log of the bird species that occur there. This will not only sharpen your birding skills but will give you insights into the interrelationships of the bird and plant communities of the area. Note the community preferences of the birds you observe and how specific species use particular plants for food, shelter, nesting, and perches for hunting insects.

References

Lowery, G. 1951. A quantitative study of nocturnal migration in birds. *Univ. Kans. Publ. Nat. Hist.* 3(2): 361–472.

Lowery, G., and R. Newman. 1955. Direct studies of nocturnal migration. In A. Wolfson, ed., *Recent studies in avian biology.* Urbana: University of Illinois Press, 238–74.

Field Techniques   *Chapter Worksheet*

1. What is the best type of microphone to use when recording wild bird songs?

2. What are the relative advantages and disadvantages of Porro prism binoculars versus roof prism binoculars?

3. Why do birdwatchers look forward to cold fronts in the fall months? What weather conditions are best for spring migrations of songbirds?

4. Why do most migratory birds favor inland routes when moving northward in the spring and coastal routes when traveling south in the fall?

Chapter Worksheet, continued

5. Why are peninsulas often such good birdwatching sites? What factors might influence the attractiveness of a peninsula to migrating birds?

6. What precautions should you take before attempting to handle a small hawk or falcon? What should you watch for when handling a Cardinal or Evening Grosbeak?

7. What data should be included on the paper label of a study skin that will be included in a formal ornithology collection?

References

Angel, H. 1982. *The book of nature photography*. New York: Alfred A. Knopf.

Bonney, J., and J. Baringer. 1988. Scanning for optics. *Living Bird Quarterly* 7(4):18–21.

Connor, J. 1988. *The complete birder*. Boston: Houghton Mifflin.

Davis, T. 1981. The microphone comes first. *Birding* 13(5): 161–63.

Dorst, J. 1962. *The migration of birds*. Boston: Houghton Mifflin.

Dunne, P. 1985. Binoculars for birders. *Living Bird Quarterly* 4(1):16–21.

Ehrlich, P. R., D. S. Dobkin, and D. Wheye. 1988. *The birder's handbook*. New York: Simon and Schuster.

Gooders, J. 1990. In S. Weidensaul, ed., *The practical ornithologist*. New York: Simon and Schuster.

Grantz, G. J. 1969. *Home book of taxidermy and tanning*. Harrisburg, Pa.: Stackpole Books.

Hall, E. R. 1962. Collecting and preparing study specimens of vertebrates. *Univ. Kans. Mus. Nat. Hist. Misc. Publ.* 30:1–46.

Hickey, J. J. 1963. *A guide to bird watching*. Garden City, N.Y.: Doubleday. Reprint, 1975; New York: Dover.

Kaufman, K. 1990. *Advanced birding*. Boston: Houghton Mifflin.

Lincoln, F. C. 1950. *Migration of birds*. Cir. 16. Washington, D.C.: U.S. Fish and Wildlife Service.

McKenzie, B. 1989. Recording equipment. *Birding* 21(5):250.

Mead, C. 1976. *Bird migration*. New York: Facts on File.

Meriwether, W. 1986. ABA area: bird sound recordings; some suggestions. *Birding* 18(1):30–33.

Pasquier, R. F. 1977. *Watching birds: an introduction to ornithology*. Boston: Houghton Mifflin.

Smithe, F. B. 1975. *Naturalist's color guide*. New York: American Museum of Natural History.

U.S. Fish and Wildlife Service. 1980. *North American bird banding techniques*. Washington, D.C.: Department of the Interior, U.S. Fish and Wildlife Service.

Wickstrom, D. 1988. Tools of the trade: bird recording equipment. *Birding* 20(4): 262–66.

Appendix:
Classification of Birds

A condensation of the Sibley and Monroe (1990) classification of the Class Aves, under review by the Check-list Committee of the American Ornithologists' Union (A.O.U.). Group affinities are based on DNA-DNA hybridization data (Sibley and Ahlquist 1985, 1990).

	Distribution
CLASS AVES	
Subclass Neornithes (2,057 genera/9,672 species)	
Infraclass Eoaves	
Parvclass Ratitae	
ORDER STRUTHIONIFORMES (5 genera/10 species)	
Suborder Struthioni	
Infraorder Struthionides	
Family Struthionidae (1/1) – Ostrich	Africa
Infraorder Rheides	
Family Rheidae (1/2) – Rheas	South America
Suborder Casuarii	
Family Casuariidae (2/4)	Australia, New Guinea
Tribe Casuariini – Cassowaries	
Tribe Dromaiini – Emu	
Family Apterygidae (1/3) – Kiwis	New Zealand
ORDER TINAMIFORMES (9 genera/47 species)	
Family Tinamidae (9/47) – Tinamous	Central and South America
Infraclass Neoaves (2,043 genera/9,615 species)	
Parvclass Galloanserae	
Superorder Gallomorphae	
ORDER CRACIFORMES (17 genera/69 species)	
Suborder Craci	
Family Cracidae (11/50) – Guans, chachalacas, currassows	Central and South America
Suborder Megapodii	
Family Megapodiidae (6/19) – Megapodes	Australian
ORDER GALLIFORMES (58 genera/214 species)	
Parvorder Phasianida	
Superfamily Phasianoidea	
Family Phasianidae (45/177) – Pheasants, grouse, turkeys, partridges	Primarily Old World
Superfamily Numidoidea	
Family Numididae (4/6) – Guineafowls	Africa, Madagascar
Parvorder Odontophorida	
Family Odontophoridae (4/6) – New World quails	Americas
Superorder Anserimorphae	

ORDER ANSERIFORMES (48 genera/161 species)
  Infraorder Anhimides
    Superfamily Anhimoidea
      Family Anhimidae (2/3) – Screamers   South America
    Superfamily Anseranatoidea
      Family Anseranatidae (1/1) – Magpie goose   Australasian
  Infraorder Anserides
      Family Dendrocygnidae (2/9) – Whistling-ducks   Pantropical
      Family Anatidae (43/148)   Worldwide
        Subfamily Oxyurinae – Stiff-tailed ducks
        Subfamily Stictonettinae – *Stictonetta*
        Subfamily Cygninae – Swans
        Subfamily Anatinae – Typical ducks, geese
Parvclass Turnicae
  ORDER TURNICIFORMES (2 genera/17 species)
      Family Turnicidae – Buttonquails   Old World, Australia
Parvclass Picae
  ORDER PICIFORMES (51 genera/355 species)
  Infraorder Picides
      Family Indicatoridae (4/17) – Honeyguides Africa, Southern Asia
      Family Picidae (28/215) – Woodpeckers, wrynecks, piculets Worldwide
  Infraorder Ramphastides
    Superfamily Megalaimoidea
      Family Megalaimoidae (3/26) – Asian barbets Asia
    Superfamily Lybioidea
      Family Lybiidae (7/42) – African barbets Africa
    Superfamily Ramphastoidea
      Family Rhamphastidae (9/55) South America
        Subfamily Capitoninae – New World barbets
        Subfamily Ramphastinae – Toucans
Parvclass Coraciae
  Superorder Galbulimorphae
  ORDER GALBULIFORMES (15 genera/51 species)
  Infraorder Galbulides
      Family Galbulidae (5/18) – Jacamars Central and South America
  Infraorder Bucconides
      Family Bucconidae (10/33) – Puffbirds Central and South America
  Superorder Bucerotimorphae
  ORDER BUCEROTIFORMES (9 genera/56 species)
      Family Bucerotidae (8/54) – Typical hornbills Africa, Southern Asia
      Family Bucorvidae (1/2) – Ground-hornbills Africa
  ORDER UPUPIFORMES (3 genera/10 species)
  Infraorder Upupides
      Family Upupidae (1/2) – Hoopoes Eurasia, Africa
  Infraorder Phoeniculides
      Family Phoeniculidae (1/5) – Woodhoopoes Central Africa
      Family Rhinopomastidae (1/3) – Scimitarbills Central Africa
  Superorder Coraciimorphae
  ORDER TROGONIFORMES (6 genera/39 species)

Family Trogonidae (6/39)	Pantropical
Subfamily Apaloderminae – African trogons	
Subfamily Trogoninae	
Tribe Trogonini – New World trogons	
Tribe Harpactini – Asian trogons	

ORDER CORACIIFORMES (34 genera /152 species)

Suborder Coracii

Superfamily Coracioidea

Family Coraciidae (2/12) – Typical rollers	Old World tropics
Family Brachypteraciidae (3/5) – Ground-rollers	Madagascar

Superfamily Leptosomoidea

Family Leptosomidae (1/1) – Cuckoo-roller	Madagascar, Comores

Suborder Alcedini

Infraorder Alcedinides

Parvorder Momotida

Family Momotidae (6/9) – Motmots	Central and South America

Parvorder Todida

Family Todidae (1/5) – Todies	Greater Antilles

Parvorder Alcedinida

Family Alcedinidae (3/24) – Alcedinid kingfishers	Worldwide

Parvorder Cerylida

Superfamily Dacelonoidea

Family Dacelonidae (12/61) – Dacelonid kingfishers	Australasia

Superfamily Ceryloidea

Family Cerylidae (3/9) – Cerylid kingfishers	Worldwide

Infraorder Meropides

Family Meropidae (3/26) – Bee-eaters	Africa and Eurasia, Australia

Parvclass Coliae

ORDER COLIIFORMES (2 genera/6 species)

Family Coliidae (2/6)	Central and Southern Africa
Subfamily Coliinae – Typical mousebirds	
Subfamily Urocoliinae – Long-tailed mousebirds	

Parvclass Passerae

Superorder Cuculimorphae

ORDER CUCULIFORMES (30 genera/143 species)

Infraorder Cuculides

Parvorder Cuculida

Superfamily Cuculoidea

Family Cuculidae (17/79) – Old World cuckoos	Africa and Eurasia

Superfamily Centropodoidea

Family Centropodidae (1/30) – Coucals	Africa, Australasia

Superfamily Coccyzida

Family Coccyzidae (4/18) – American cuckoos	New World

Infraorder Crotophagides

Parvorder Opisthocomida

Family Opisthocomidae (1/1) – Hoatzin	Central and South America

Parvorder Crotophagida

Family Crotophagidae (2/4)	Tropical New World
Tribe Crotophagini – Anis	

Tribe Guirini – Guira cuckoo
    Parvorder Neomorphida
        Family Neomorphidae (5/11) – Roadrunners, ground-cuckoos   North America, Asia
Superorder Psittacimorphae
  ORDER PSITTACIFORMES (80 genera/358 species)
        Family Psittacidae (5/11) – Parrots and allies   Worldwide
Superorder Apodimorphae
  ORDER APODIFORMES (19 genera/103 species)
        Family Apodidae (18/99) – Typical swifts   Worldwide
        Family Hemiprocnidae (1/4) – Crested swifts   Southeast Asia, Pacific islands
  ORDER TROCHILIFORMES (109 genera/319 species)
        Family Trochilidae (109/319)   New World
          Subfamily Phaethornithinae – Hermits
          Subfamily Trochilinae – Typical hummingbirds
Superorder Strigimorphae
  ORDER MUSOPHAGIFORMES (5 genera/23 species)
        Family Musophagidae (5/23)   Africa
          Subfamily Musophaginae – Turacos
          Subfamily Criniferinae – Plaintain-eaters
  ORDER STRIGIFORMES (45 genera/291 species)
    Suborder Strigi
      Parvorder Tytonida
        Family Tytonidae (2/17) – Barn owl, grass owls   Worldwide
      Parvorder Strigida
        Family Strigidae (23/161) – Typical owls   Worldwide
    Suborder Aegotheli
        Family Aegothelidae (1/8) – Owlet-nightjars   Australia
    Suborder Caprimulgi
      Infraorder Podargides
        Family Podargidae (1/3) – Australian frogmouths
        Family Batrachostomidae (1/11) – Asian frogmouths   Southeast Asia
      Infraorder Caprimulgides
      Parvorder Steatornithida
        Superfamily Steatornithoidea
        Family Steatornithidae (1/1) – Oilbird   Tropical South America
        Superfamily Nyctibioidea
        Family Nyctibiidae (1/7) – Potoos   Central and South America
      Parvorder Caprimulgida
        Superfamily Eurostopodoidea
        Family Eurostopodidae (1/7) – Eared-nightjars   Australasia
        Superfamily Caprimulgoidea
        Family Caprimulgidae (14/76)   Worldwide
          Subfamily Chordeilinae – Nighthawks
          Subfamily Caprimulginae – Nightjars
Superorder Passerimorphae
  ORDER COLUMBIFORMES (42 genera/313 species)
        Family Raphidae (2/3) – Dodos, solitaires (extinct)   Macarene Islands
        Family Columbidae (40/310) – Pigeons, doves   Worldwide
  ORDER GRUIFORMES (53 genera/196 species)

Suborder Grui
  Infraorder Eurypygides
      Family Eurypygidae (1/1) – Sunbittern   Central and South America
  Infraorder Otidides
      Family Otididae (6/25) – Bustards   Africa, Eurasia
  Infraorder Gruides
   Parvorder Gruida
    Superfamily Gruoidea
      Family Gruidae (2/15)   Northern Hemisphere,
        Subfamily Balearicinae – Crowned-cranes   Africa, Australia
        Subfamily Gruinae – Typical cranes
      Family Heliornithidae (4/4)   Central and South America
        Tribe Aramini – Limpkin
        Tribe Heliornithini – Sungrebes
    Superfamily Psophioidea
      Family Psophiidae (1/3) – Trumpeters   Tropical South America
   Parvorder Cariamida
      Family Cariamidae (2/2) – Seriemas   South America
      Family Rhynochetidae (1/1) – Kagu   New Caledonia
  Suborder Ralli
      Family Rallidae (34/142) – Rails, gallinules, coots   Worldwide
  Suborder Mesitornithi
      Family Mesitornithidae (2/3) – Mesites, monias, roatelos   Madagascar

ORDER CICONIIFORMES (254 genera/1,027 species)

  Suborder Charadrii
  Infraorder Pteroclides
      Family Pteroclidae (2/16) – Sandgrouse   Deserts of Africa and
  Infraorder Charadriides   Eurasia
   Parvorder Scolopacida
    Superfamily Scolopacoidea
      Family Thinocoridae (2/4) – Seedsnipes   South America
      Family Pedionomidae (1/1) – Plains-wanderer   Australia
      Family Scolopacidae (21/88)   Worldwide
        Subfamily Scolopacinae – Woodcocks, snipes
        Subfamily Tringinae – Sandpipers, curlews, phalaropes
    Superfamily Jacanoidea
      Family Rostratulidae (1/2) – Paintedsnipe   Pantropical, Australia
      Family Jacanidae (6/8) – Jacanas, lily-trotters   Pantropical, Australia
   Parvorder Charadriida
    Superfamily Chionidoidea
      Family Chionididae (1/2) – Sheathbills   Antarctic Ocean, islands
    Superfamily Charadrioidea
      Family Burhinidae (1/9) – Thick-knees   Pantropical, Australia
      Family Charadriidae (16/89)   Worldwide
        Subfamily Recurvirostrinae
          Tribe Haematopodini – Oystercatchers
          Tribe Recurvirostrini – Avocets, stilts
        Subfamily Charadriinae – Plovers, lapwings
    Superfamily Laroidea

        Family Glareolidae (6/18)   Africa, S. Europe, Asia,
            Subfamily Dromadinae – Crab-plover   Australia
            Subfamily Glareolinae – Pratincoles, coursers
        Family Laridae (28/129)   Worldwide
            Subfamily Larinae
               Tribe Stercorariini – Skuas, jaegers
               Tribe Rynchopini – Skimmers
               Tribe Larini – Gulls
               Tribe Sternini – Terns
            Subfamily Alcinae – Auks, murres, puffins
Suborder Ciconii
 Infraorder Falconides
  Parvorder Accipitrida
        Family Accipitridae (65/240)   Worldwide
            Subfamily Pandioninae – Osprey
            Subfamily Accipitrinae – Hawks, eagles, accipiters, kites
        Family Sagittariidae (1/1) – Secretarybird   Africa
  Parvorder Falconida
        Family Falconidae (10/63) – Falcons, caracaras   Worldwide
 Infraorder Ciconiides
  Parvorder Podicipedida
        Family Podicipedidae (6/21) – Grebes   Worldwide
  Parvorder Phaethontida
        Family Phaethontidae (1/3) – Tropicbirds   Pantropical, pelagic
  Parvorder Sulida
   Superfamily Suloidea
        Family Sulidae (3/9) – Boobies, gannets   Widespread, pelagic
        Family Anhingidae (1/4) – Anhingas, darters   Pantropical
   Superfamily Phalacrocoracoidea
        Family Phalacrocoracidae (1/38) – Cormorants, shags Worldwide
  Parvorder Ciconiida
   Superfamily Ardeoidea
        Family Ardeidae (20/65) – Herons, egrets, bitterns   Worldwide
   Superfamily Scopoidea
        Family Scopidae (1/1) – Hamerkop (hammerhead)   Africa, Madagascar
   Superfamily Phoenicopteroidea
        Family Phoenicopteridae (1/5) – Flamingos   Isolated areas of tropics
   Superfamily Threskiornithoidea
        Family Threskiornithidae (14/34) – Ibises, spoonbills Worldwide
   Superfamily Pelecanoidea
        Family Pelecanidae (2/9)   Widespread in temperate
            Subfamily Balaenicipitinae – Shoebill   and tropical regions
            Subfamily Pelecaninae – Pelicans
   Superfamily Ciconioidea
        Family Ciconiidae (11/26)   Worldwide
            Subfamily Cathartinae – New World vultures
            Subfamily Ciconiinae – Storks
   Superfamily Procellarioidea
        Family Fregatidae (1/5) – Frigatebirds   Pantropical pelagic

Family Spheniscidae (6/17) – Penguins .... S. Hemisphere pelagic
Family Gaviidae (1/5)   Loons .... Holarctic
Family Procellariidae (24/115) .... Worldwide pelagic
  Subfamily Procellariinae – Petrels, shearwaters, diving-petrels
  Subfamily Diomedeinae – Albatrosses
  Subfamily Hydrobatinae – Storm-petrels

ORDER PASSERIFORMES (1,161 genera / 5,712 species)
  Suborder Tyranni (Suboscines)
    Infraorder Acanthisittides
      Family Acanthisittidae (2/4) – New Zealand wrens .... New Zealand
    Infraorder Eurylaimides
      Superfamily Pittoidea
        Family Pittidae (1/31) – Pittas .... Africa, Australasia
      Superfamily Eurylaimoidea
        Family Eurylaimidae (8/14) – Broadbills .... Old World tropics
        Family Philepittidae (2/4) – Asities (asitys) .... Madagascar
    Infraorder/Family *Incertae sedis* (*Sapayoa*)
    Infraorder Tyrannides
      Parvorder Tyrannida
        Family Tyrannidae (146/537) .... North and South America
          Subfamily Pipromorphinae – Mionectine flycatchers, *Corythopis*
          Subfamily Tyranninae – Tyrant flycatchers
          Subfamily Tityrinae
            Tribe Schiffornithini – *Schiffornis*
            Tribe Tityrini – Tityras, becards
          Subfamily Cotinginae – Cotingas, plantcutters, sharpbill
          Subfamily Piprinae – Manakins
      Parvorder Thamnophilida
        Family Thamnophilidae (45/188) – Typical antbirds .... Central and South America
      Parvorder Furnariida
        Superfamily Furnarioidea
          Family Funariidae (66/280) .... South America
            Subfamily Furnariinae – Ovenbirds
            Subfamily Dendrocolaptinae – Woodcreepers
        Superfamily Formicarioidea
          Family Formicariidae (7/56) – Ground antbirds .... South America
          Family Conopophagidae (1/8) – Gnateaters .... South America
          Family Rhinocryptidae (12/28) – Tapaculos .... Central and South America
  Suborder Passeri (Oscines)
    Parvorder Corvida
      Superfamily Menuroidea
        Family Climacteridae (2/7) – Australo-Papuan treecreepers .... Australia and New Guinea
        Family Menuridae (2/4) .... Australia
          Subfamily Menurinae – Lyrebirds
          Subfamily Atrichornithinae – Scrub-birds
        Family Ptilonorhynchidae (7/20) – Bowerbirds .... Australia and New Guinea
      Superfamily Meliphagoidea
        Family Maluridae (2/4) .... Australia, New Guinea,
          Subfamily Malurinae   and New Zealand
            Tribe Malurini – Fairywrens

Tribe Stipiturini – Emuwrens
Subfamily Amytornithinae – Grasswrens
Family Meliphagidae (42/182) – Honeyeaters, *Ephthianura, Ashbyia*   Australia and New Guinea
Family Pardalotidae (16/68)   Australia
Subfamily Pardalotinae – Pardalotes
Subfamily Dasyornithinae – Bristlebirds
Subfamily Acanthizinae
Tribe Sericornithini – Scrubwrens
Tribe Acanthizini – Thornbills, whitefaces
Superfamily Corvoidea
Family Eopsaltriidae (14/46) – Australo-Papuan robins, *Drymodes*   Australia and New Guinea
Family Irenidae (2/10) – Fairy-bluebirds, leafbirds   Southeast Asia
Family Orthonychidae (1/2) – Logrunner, chowchilla   Australia and New Guinea
Family Pomatostomidae (1/5) – Australo-Papuan babblers   Australia and New Guinea
Family Laniidae (3/30) – True shrikes   Holarctic, Africa
Family Vireonidae (4/51) – Vireos, greenlets, peppershrikes, shrike-vireos   North and South America
Family Corvidae (127/647)   Worldwide
Subfamily Cinclosomatinae – Quail-thrushes, whipbirds
Subfamily Corcoracinae – Australian chough, apostlebird
Subfamily Pachycephalinae
Tribe Neosittini – Sittellas
Tribe Mohouini – *Mohoua, Finschia*
Tribe Falcunculini – Shrike-tits, *Oreoica, Rhagologus*
Tribe Pachycephalini – Whistlers, shrike-thrushes
Subfamily Corvinae
Tribe Corvini – Crows, magpies, jays, nutcrackers
Tribe Paradisaeini – Birds-of-paradise, *Melampitta*
Tribe Artamini – Currawongs, woodswallows, *Peltops, Pityriasis*
Tribe Oriolini – Orioles, cuckooshrikes
Subfamily Dicrurinae
Tribe Rhipidurini – Fantails
Tribe Dicrurini – Drongos
Tribe Monarchini – Monarchs, magpie-larks
Subfamily Aegithininae – Ioras
Subfamily Malaconotinae
Tribe Malaconotini – Bush-shrikes
Tribe Vangini – Helmet-shrikes, vangas, *Batis, Platysteira*
Family Callaeatidae (3/3) – New Zealand wattlebirds   New Zealand
Parvorder *Incertae sedis* – Family Picathartidae – *Picathartes, Chaetops*
Parvorder Passerida
Superfamily Muscicapoidea
Family Bombycillidae (5/8)   Holarctic, Central America
Tribe Dulini – Palmchat
Tribe Ptilogonatini – Silky-flycatchers
Tribe Bombycillini – Waxwings
Family Cinclidae (1/5) – Dippers   Eurasia, New World
Family Muscicapidae (69/449)   Worldwide

Subfamily Turdinae – True thrushes, *Chlamydochaera, Brachypteryx, Alethe*

Subfamily Muscicapinae

    Tribe Muscicapini – Old World flycatchers

    Tribe Saxicolini – Chats

Family Sturnidae (38/148)   Worldwide

    Tribe Sturnini – Starlings, mynas

    Tribe Mimini – Mockingbirds, thrashers, catbirds

Superfamily Sylvioidea

Family Sittidae (2/25)   Holarctic, Australia

    Subfamily Sittinae – Nutchatches

    Subfamily Tichodrominae – Wallcreeper

Family Certhiidae (22/97)   Holarctic, Northern Africa

    Subfamily Certhiinae

        Tribe Certhiini – Northern creepers

        Tribe Salpornithini – African creeper

    Subfamily Troglodytinae – Wrens

    Subfamily Polioptilinae – Gnatcatchers, verdin, gnatwrens

Family Paridae (7/65)   Holarctic, Africa

    Subfamily Remizinae – Penduline-tits

    Subfamily Parinae – Titmice, chickadees

Family Aegithalidae (3/8) – Long-tailed tits, bushtits   Holarctic, East Indies

Family Hirundinidae (14/89)   Worldwide

    Subfamily Pseudochelidoninae – River-martins

    Subfamily Hirundininae – Swallows, martins

Family Regulidae (1/6) – Kinglets   Eurasia, Africa, Australia

Family Pycnonotidae (21/137) – Bulbuls, greenbuls   Africa, Southern Asia

Family Hypocoliidae (1/1) – *Hypocolius*   South America

Family Cisticolidae (14/119) – African warblers,   Africa
    Cisticola, Apalis, Prinia

Family Zosteropidae (13/96) – White-eyes   Africa, Southern Asia

Family Sylviidae (101/552)   Eurasia, Australasia

    Subfamily Acrocephalinae – Leaf-warblers

    Subfamily Megalurinae – Grass-warblers

    Subfamily Garrulacinae – Laughingthrushes

    Subfamily Sylviinae

        Tribe Timaliini – Babblers, *Rhabdornis*

        Tribe Chamaeini – Wrentit

        Tribe Sylviini – *Sylvia*

Superfamily Passeroidea

Family Alaudidae (17/91) – Larks   Worldwide

Family Nectariniidae (8/169)   Asia, Australasia

    Subfamily Promeropinae – Sugarbirds

    Subfamily Nectariniinae

        Tribe Dicaeini – Flowerpeckers

        Tribe Nectariniini – Sunbirds, spiderhunters

Family Melanocharitidae (3/10)   New Guinea

    Tribe Melanocharitini – *Melanocharis*

    Tribe Toxorhamphini – *Toxorhamphus, Oedistoma*

Family Paramythiidae (2/2) – *Paramythia, Oreocharis*   New Guinea

Family Passeridae (57/386)   Worldwide
    Subfamily Passerinae – Sparrows, rock-sparrows
    Subfamily Motacillinae – Wag-tails, pipits
    Subfamily Prunellinae – Accentors, dunnock
    Subfamily Ploceinae – Weavers
    Subfamily Estrildinae
      Tribe Estrildini – Estrildine finches
      Tribe Viduini – Whydahs
Family Fringillidae (240/993)   Worldwide
    Subfamily Peucedraminae – *Peucedramus*
    Subfamily Fringillinae
      Tribe Fringillini – Chaffinches, brambling
      Tribe Carduelini – Goldfinches, crossbills
      Tribe Drepanidini – Hawaiian honeycreepers
    Subfamily Emberizinae
      Tribe Emberizini – Buntings, longspurs, towhees
      Tribe Parulini – Wood warblers, *Zeledonia*
      Tribe Thraupini – Tanagers, Neotropical honeycreepers, seedeaters, flower-piercers
      Tribe Cardinalini – Cardinals
      Tribe Icterini – Troupials, New World blackbirds, meadowlarks

Bibliography

Akester, A. R. 1971. The blood vascular system. In D. J. Bell, ed. *Physiology and biochemistry of the domestic fowl.* 5 vols. New York: Academic Press, 2:783–839.

Akester, A. R. 1979. The autonomic nervous system. In A. S. King and J. McLelland, eds., *Form and function in birds.* 4 vols. New York: Academic Press, 1:381–441.

Amadon, D. 1950. The Hawaiian honeycreepers. *Bull. Amer. Mus. Nat. Hist.* 95:151-262.

American Ornithologists' Union. 1983. *Check-list of North American birds.* 6th ed. Lawrence, Kans.: American Ornithologists' Union.

Ames, P. 1971. The morphology of the syrinx in passerine birds. *Bull. Peabody Mus. Nat. Hist.* 37:1-194.

Ammann, G. A. 1937. Number of contour feathers of *Cygnus* and *Xanthocephalus. Auk* 54:201–2.

Anderson, A. 1991. Early bird threatens *Archaeopteryx*'s perch. *Science.*253(5017):35.

Angel, H. 1982. *The book of nature photography.* New York: Alfred A. Knopf.

Assenmacher, I. 1973. The peripheral endocrine glands. In D. S. Farner, J. R. King, and K. C. Parkes, eds., *Avian biology.* 8 vols. New York: Academic Press, 3:183–286.

Austin, O. L., Jr. 1961. *Birds of the world.* New York: Golden Press.

Austin, O. L., Jr., and A. Singer. 1985. *Families of birds.* New York: Golden Press.

B

Bagg, A., W. Gunn, D. Miller, J. Nichols, W. Smith, and F. Wolfarth. 1950. Barometric pressure patterns and spring bird migration. *Wilson Bull.* 62:5–19.

Bailey, R. E. 1952. The incubation patch of passerine birds. *Condor* 54:121–36.

Baird, J. 1963. On aging birds by skull ossification. *Ring* 37:253–55.

Baird, J., and I. Nisbet. 1959. Observations of diurnal migration in the Narragansett Bay areas of Rhode Island in fall 1958. *Bird-Banding* 30:171–81.

Bakker, R. T. 1975. The dinosaur renaissance. *Sci. Amer.* 232(4):58–78.

Bakker, R. T. 1986. *The dinosaur heresies.* New York: William Morrow.

Bakker, R. T., and P. M. Galton. 1974. Dinosaur monophyly and a new class of vertebrates. *Nature* 248:168–72.

Bang, B. G. 1960. Anatomical evidence of olfactory function in some species of birds. *Nature* 188:547-49.

Bang, B. G., and S. Cobb. 1968. The size of the olfactory lobe in 108 species of birds. *Auk* 85(1):55-61.

Bang, B. G., and B. M. Wenzel. 1985. Nasal cavity and olfaction system. In A. S. King and J. McLelland, eds., *Form and function in birds.* 4 vols. New York: Academic Press, 3:195–225.

Baumel, J. J. 1988. Functional morphology of the tail apparatus of the pigeon (*Columba livia*). *Advances in anatomy, embryology and cell biology* 110. Berlin: Springer-Verlag.

Baumel, J. J., A. S. King, A. M. Lucas, J. E. Breazille, and H. E. Evans, eds. 1979. *Nomina anatomica avium.* New York: Academic Press.

Beehler, B. 1989. The birds of paradise. *Sci. Amer.* 261(6):116–23.

Bellairs, A. d'A. 1964. Skeleton. In A. L. Thomson, ed., *A new dictionary of birds.* New York: McGraw-Hill.

Bellairs, A. d'A., and C. R. Jenkin. 1960. The skeleton of birds. In A. J. Marshall, ed., *Biology and comparative physiology of birds.* 2 vols. New York: Academic Press, 1:241–300.

Bellrose, F. C. 1967a. Orientation in waterfowl migration. In R. M. Storm, ed., *Animal orientation and navigation.* Corvallis: Oregon State University Press.

Bellrose, F. C. 1967b. Radar in orientation research. *Proc. XIV Int. Ornith. Cong.* 281–309.

Bellrose, F. C. 1968. Waterfowl migration corridors east of the Rocky Mountains in the United States. *Biol. Notes* 61. Urbana: Illinois Natural History Survey.

Bennett, G. F. 1961. On three species of hippoboscid flies (Diptera) on birds in Ontario. *Can. J. Zool.* 39:379–406.

Bennett, T. 1974. The peripheral and autonomic nervous systems. In D. S. Farner, J. R. King, and K. C. Parkes, eds., *Avian biology.* 8 vols. New York: Academic Press, 4:1–77.

Bent, A. C. 1946. Life histories of North American crows, jays, and titmice. *U.S. Natl. Mus. Bull.* 191:32–52.

Benton, M. J. 1979. Ectothermy and the success of dinosaurs. *Evolution* 33:983–97.

Berger, A. J. 1956. Anatomic variation and avian anatomy. *Condor* 58:433–41.

Berger, A. J. 1960. The musculature. In A. J. Marshall, ed., *Biology and comparative physiology of birds.* 2 vols. New York: Academic Press, 1:301–344.

Berger, A. J. 1961. *Bird study.* New York: John Wiley. Reprint, 1971; New York: Dover.

Berger, M., O. Roy, and J. Hart. 1970. The co-ordination between respiration and wing beats in birds. *Z. Vergl. Physiol.* 66:190-200.

Bergman, C. A. 1981. The glass of fashion. *Audubon* 83(6):74–80.

Bingham, V., K. Able, and P. Kerlinger. 1982. Wind drift, compensation, and the use of landmarks by nocturnal bird migrants. *Anim. Behav.* 30:49–53.

Blackburn, D. G., and H. E. Evans. 1986. Why are there no viviparous birds? *Amer. Nat.* 128:165–90.

Blake, C. H. 1956. Topography of a bird. *Bird Banding* 27:22-31.

Bock, W. J. 1964. Kinetics of the avian skull. *J. Morph.* 114:1–42.

Bock, W. J. 1974. The avian skeletomusculature system. In D. S. Farner, J. R. King, and K. C. Parkes, eds., *Avian biology.* 8 vols. New York: Academic Press, 4:119–257.

Bock, W. J. 1985. The arboreal theory of the origin of birds. In M. K. Hecht, J. H. Ostrom, G. Viohl, and P. Wellnhofer, eds., *The beginnings of birds.* Eichstatt: Freunde des Jura–Museums.

Bock, W. J. 1986. Species concepts, speciation, and macroevolution. In K. Iwatsuki, P. H. Raven, and W. J. Bock, eds., *Modern Aspects of Species.* Tokyo: University of Tokyo Press, 31–55.

Bock, W. J., and R. S. Hikida. 1969. Turgidity and the function of the hatching muscle. *Amer. Midl. Nat.* 81:99–106.

Bock, W. J., and W. DeW. Miller. 1959. The scansorial foot of woodpeckers, with comments on the evolution of perching and climbing feet in birds. Amer. Mus. Nat. Hist. *Novitates* 193.

Bonney, J., and J. Baringer. 1988. Scanning for optics. *Living Bird* 7(4):18–21.

Borror, D. J. 1975. Bird song. *Amer. Birds* 29:3–7.

Borror, D. J., and R. E. White. 1970. *A field guide to the insects.* Boston: Houghton Mifflin.

Boyd, E. M. 1951. The external parasites of birds; a review. *Wilson Bull.* 63:363–69.

Breazile, J. E., and H.-G. Hartwig. 1989. Central nervous system. In A. S. King and J. McLelland, eds., *Form and function in birds.* 4 vols. New York: Academic Press, 4:485–566.

Brodkorb, P. 1949. The number of feathers in some birds. *Q. J. Fla. Acad. Sci.* 12:1–5.

Brodkorb, P. 1971. Origin and evolution of birds. In D. S. Farner, J. R. King, and K. C. Parkes, eds., *Avian biology.* 8 vols. New York: Academic Press, 1:19–55.

Broom, R. 1913. On the South African pseudosuchian *Euparkeria* and allied genera. *Proc. Zool. Soc. London* 1913:619–33.

Brown, J. L. 1964. The integration of agonistic behavior in the Stellar's Jay, *Cyanocitta stelleri* (Gmelin). *Univ. Calif. Publ. Zool.* 60.

Brown, L. 1977. *Birds of prey: their biology and ecology.* New York: A & W.

Brown, L., and D. Amadon. 1968. *Eagles, hawks, and falcons of the world.* New York: McGraw-Hill.

Bruderer, B. 1972. Radar studies in spring migration in northern Switzerland. *Proc. XV Inter. Ornith. Cong.* 635.

Brush, A. H., and G. A. Clark, Jr., eds. 1983. *Perspectives in ornithology: essays presented for the centennial of the American Ornithologists' Union.* Cambridge: Cambridge University Press.

Bull, J. 1974. *Birds of New York State.* Garden City, N.Y.: Doubleday/Natural History Press.

Burton, J. A., ed. 1973. *Owls of the world: their evolution, structure, and ecology.* New York: A & W Visual Library.

Burton, R. 1990. *Bird flight.* New York: Facts on File.

C

Calder, W. 1968. Respiratory and heart rates of birds at rest. *Condor.* 70:358–65.

Campbell, B., and E. Lack, eds. 1985. *A dictionary of birds.* Vermillion, S.D.: Buteo Books.

Caple, G. R., R. T. Balda, and W. R. Willis. 1983. The physics of leaping animals and the evolution of flight. *Amer. Nat.* 121:455–67.

Carey, C., ed. 1980. Physiology of the avian egg. *Amer. Zool.* 20:325–484.

Carey, C. 1983. Structure and function of avian eggs. In R. F. Johnson, ed., *Current ornithology.* 5 vols. New York: Plenum Press, 1:69–103.

Carroll, R. 1988. *Vertebrate paleontology and evolution.* New York: W. H. Freeman.

Chamberlain, F. 1943. *Atlas of avian anatomy.* Mich. Agric. Exp. Sta. Mem. Bull. 5.

Chancellor, J. 1978. *Audubon: a biography.* New York: Viking Press.

Chatterjee, S. 1991. Cranial anatomy and relationships of a new Triassic bird from Texas. *Philos. Trans. Royal Soc. London, Series B,* 332(1265):277–342.

Clark, G. A., and J. B. de Cruz. 1989. Functional interpretation of protruding filoplumes in oscines. *Condor* 91:962–65.

Clark, W. S., and B. K. Wheeler. 1987. *A field guide to hawks of North America.* Boston: Houghton Mifflin.

Clay, T. 1964. Ectoparasites. In A. L. Thomson, ed., *A new dictionary of birds.* New York: McGraw-Hill.

Cody, M. L., ed. 1985. *Habitat selection in birds.* New York: Academic Press.

Collias, N., and E. Collias. 1984. *Nest building and bird behavior.* Princeton, N.J.: Princeton University Press.

Connor, J. 1988. *The complete birder.* Boston: Houghton Mifflin.

Conway, W. P. 1989. Long lenses for birds. *Birding* 21:14–17.

Corral, M. 1989. *The world of birds: a layman's guide to ornithology.* Chester, Conn.: Globe Pequot Press.

Cowen, R. 1989. Alimentary, my dear hoatzin: ruminations on a gutsy bird. *Science News* 136:269–70.

Cracraft, J. 1974. Continental drift, paleoclimatology, and the evolution and biogeography of birds. *J. Zool.* 169:455–545.

Craighead, J. J., and F. C. Craighead. 1956. *Hawks, owls, and wildlife.* Harrisburg, Pa.: Stackpole Books.

Cramp, S., and K. E. L. Simmons, eds. 1977–83. *Handbook of the birds of Europe, the Middle East, and North Africa: the birds of the western Palearctic.* 3 vols. Oxford: Oxford University Press.

Curtis, E., and R. Miller. 1938. The sclerotic ring in North American birds. *Auk* 55:225–43.

D

Darwin, C. 1859. *On the origin of species by means of natural selection.* London: J. Murray. (Many more modern editions are available.)

Davis, T. 1978. Cassette recorder update. *Birding* 10(4):185–92.

Davis, T. 1979. Microphones and headsets for bird recording. *Birding* 11(5):240–43.

Davis, T. 1981. The microphone comes first. *Birding* 13(5):161–63.

Dennis, J. 1954. Meteorological analysis of occurrence of grounded migrants at Smith Point, Texas, April 17–May 17, 1951. *Wilson Bull.* 66:102–11.

DeQueiroz, K., and D. Good. 1988. The scleral ossicles of *Opisthocomus* and their phylogenetic significance. *Auk* 105:29–35.

Dorst, J. 1962. *The migration of birds.* Boston: Houghton Mifflin.

Dorst, J. 1974. *The life of birds.* 2 vols. New York: Columbia University Press.

Drent, R. H. 1970. Functional aspects of incubation in the Herring Gull. *Behav. Suppl.* 17:1–132.

Drent, R. H. 1975. Incubation. In D. S. Farner, J. R. King, and K. C. Parkes, eds. *Avian biology.* 8 vols. New York: Academic Press, 5:333–420.

Duncan, C. J. 1964. Smell. In A. L. Thomson, ed., *A new dictionary of birds.* New York: McGraw-Hill.

Duncan, R. 1990. Weather and birds: a gulf coast perspective. *Birding* 22(2):163–74.

Dunne, P. 1985. Binoculars for birders. *Living Bird* 4(1):16–21.

Dunne, P., D. Sibley, and C. Sutton. 1988. *Hawks in flight.* Boston: Houghton Mifflin.

Dwight, J., Jr. 1900. The sequence of plumages and moults of the passerine birds of New York. *Ann. N.Y. Acad. Sci.* 13:73–360.

E

Eastwood, E. 1967. *Radar ornithology.* London: Methuen.

Ehrlich, P. R., D. S. Dobkin, and D. Wheye. 1988. *The birder's handbook*. New York: Simon and Schuster.

Elder, W. 1954. The oil glands of birds. *Wilson Bull.* 66:6–31.

Epple, A., and M. H. Stetson, eds. 1980. *Avian endocrinology.* New York: Academic Press.

Espinasse, P. G. 1964. Feathers. In A. L. Thomson, ed., *A new dictionary of birds.* New York: McGraw-Hill.

F

Fabricius, E. 1959. What makes plumage waterproof? *Rep. Waterfowl Trust* 10:105–13.

Farner, D. S. 1960. Digestion and the digestive system. In A. J. Marshall, ed., *Biology and comparative physiology of birds.* 2 vols. New York: Academic Press, 2:411–68.

Farner, D. S., J. R. King, and K. C. Parkes, eds. 1971–85. *Avian biology.* 8 vols. New York: Academic Press.

Fedde, M. R. 1976. Respiration. In P. D. Sturkie, ed., *Avian physiology.* 3d ed. New York: Springer–Verlag.

Feduccia, A. 1980. *The age of birds.* Cambridge: Harvard University Press.

Feduccia, A. 1985. On why the dinosaur lacked feathers. In M. K. Hecht, J. H. Ostrom, G. Viohl, and P. Wellnhofer, eds., *The beginnings of birds.* Eichstatt: Freunde des Jura-Museums.

Feduccia, A., and H. B. Tordoff. 1979. Feathers of *Archaeopteryx*: asymmetric vanes indicate aerodynamic function. *Science* 203:1021–22.

Ficken, M. S., and R. W. Ficken. 1968. Courtship of Blue-winged and Golden-winged Warblers and their hybrids. *Wilson Bull.* 80:161-72.

Fisher, J., and R. T. Peterson. 1964. *The world of birds.* Garden City, N.Y.: Doubleday.

Forsyth, A. 1988. *The nature of birds.* Ontario: Camden House.

Fox, G. 1976. Eggshell quality: its ecological and physiological significance in a DDE-contaminated Common Tern population. *Wilson Bull.* 88:459–77.

Francis, E. T. B. 1964a. The heart. In A. L. Thomson, ed., *A new dictionary of birds.* New York: McGraw-Hill.

Francis, E. T. B. 1964b. The vascular system. In A. L. Thomson, ed., *A new dictionary of birds.* New York: McGraw-Hill.

G

Gadow, H. F. 1892. On the classification of birds. *Proc. Zool. Soc. London* 1892:229–56.

Gadow, H. F. 1893. Systematischer Theil. In H. G. Bronn, ed., *Klassen und Ordnungen des Theirreichs.* Vol. 6, pt. 4: *Vogel II.* Leipzig: C. F. Winter.

Gauthier, J., and K. Padian. 1985. Phylogenetic, functional, and aerodynamic analyses of the origin of birds and their flight. In M. K. Hecht, J. H. Ostrom, G. Viohl, and P. Wellnhofer, eds., *The beginnings of birds.* Eichstatt: Freunde des Jura-Museums.

Gauthier, J. 1986. Saurichian monophyly and the origin of birds. In K. Padian, ed., *The origin of birds and the evolution of flight.* Mem. Calif. Acad. Sci. 8.

Gauthreaux, S. 1971. A radar and direct visual study of passerine spring migration in southern Louisiana. *Auk* 88:343–65.

George, J., and A. J. Berger. 1966. *Avian myology.* New York: Academic Press.

Getty, R., ed. 1975. *Sisson and Grossman's The anatomy of the domestic animals.* 5th ed. Vol. 2. Philadelphia: W. B. Saunders.

Gill, F. B. 1980. Historical aspects of hybridization between Blue-winged and Golden-winged Warblers. *Auk* 97:1–18.

Gill, F. D. 1990. *Ornithology.* New York: W. H. Freeman.

Gingerich, P. D. 1971. Skull of *Hesperornis* and early evolution of birds. *Nature* 243:70–73.

Goldby, F. 1964. Nervous system. In A. L. Thomson, ed., *A new dictionary of birds.* New York: McGraw–Hill.

Gooders, J. 1990. In S. Weidensaul, ed., *The practical ornithologist.* New York: Simon and Schuster.

Goodrich, E. S. 1930. *Studies on the structure and development of vertebrates.* London: Macmillan. Reprint, 1958; New York: Dover.

Gould, S. J. 1982. The telltale wishbone. In S. J. Gould, *The panda's thumb.* New York: W. W. Norton, 267–77. (Reprint of the original 1977 article, *Nat. Hist.* 86:26–38.)

Gould, S. J. 1991a. Not necessarily a wing. In S. J. Gould, *Bully for Brontosaurus.* New York: W. W. Norton, 139–51.

Gould, S. J. 1991b. Opus 200. *Nat. Hist.* 100(8):12–18.

Gould, S. J. 1991c. Of Kiwi eggs and the Liberty Bell. In *Bully for Brontosaurus*. New York: W. W. Norton, 109–23.

Gould, S. J., and N. Eldridge. 1977. Punctuated equilibria: the tempo and mode of evolution reconsidered. *Paleobiol.* 3:115–51.

Gower, C. 1936. The cause of blue color as found in the Bluebird and the Blue Jay. *Auk* 53:178–85.

Grahal, A., and S. D. Strahl. 1991. A bird with the guts to eat leaves. *Nat. Hist.* 100(8):48–54.

Grahal, A., S. D. Strahl, and R. Parra. 1989. Foregut fermentaion in the hoatzin, a neotropical leaf-eating bird. *Science.* 245:1236–38.

Grant, P. J. 1986. *Gulls: a guide to identification.* 2d ed. Vermillion, S.D.: Buteo Books.

Grant, P. R. 1986. *Ecology and evolution of Darwin's finches.* Princeton, N.J.: Princeton University Press.

Grantz, G. J. 1969. *Home book of taxidermy and tanning.* Harrisburg, Pa.: Stackpole Books.

Greenewalt, C. H. 1968. *Bird song: acoustics and physiology.* Washington, D.C.: Smithsonian Institution Press.

Greenewalt, C. H. 1969. How birds sing. *Sci. Amer.* 221(5):126-39.

Griffin, D. R. 1964. *Bird migration.* New York: Doubleday.

Groebbels, F. 1932. *Der vogel.* Vol. 1: *Atmungswelt and Nahrungswelt.* Berlin: Gebruder Borntraeger.

Guravich, D., and J. E. Brown. 1983. *The return of the Brown Pelican.* Baton Rouge: Louisiana State University Press.

H

Hall, E. R. 1962. Collecting and preparing study specimens of vertebrates. *Univ. Kans. Mus. Nat. Hist. Misc. Publ.* 30:1–46.

Hardy, J. W. 1961. Studies in behavior and physiology of certain New World jays. *Univ. Kans. Sci. Bull.* 42:13–149.

Harrison, C. J. O. 1976. Feathering and flight evolution in *Archaeopteryx. Nature* 263:762.

Harrison, J. H. 1964. Plumage. In A. L. Thomson, ed., *A new dictionary of birds.* New York: McGraw-Hill.

Harrison, P. 1983. *Seabirds: an identification guide.* Boston: Houghton Mifflin.

Hartman, F. A. 1955. Heart weight in birds. *Condor.* 57:221–38.

Hartman, F. A. 1961. Locomotor mechanism in birds. *Smithsonian Misc. Coll.* 143(1):1–99.

Hayman, P., J. Marchant, and T. Prater. 1986. *Shorebirds: an identification guide to the waders of the world.* Boston: Houghton Mifflin.

Hazelhoff, E. H. 1951. Structure and function of the lung of birds. *Poultry Sci.* 30:3–10.

Hecht, M. K., J. H. Ostrom, G. Viohl, and P. Wellnhofer, eds. 1985. *The beginnings of birds.* Eichstatt: Freunde des Jura-Museums.

Heilmann, G. 1927. *The origin of birds.* New York: D. Appleton. Reprint, 1972; New York: Dover.

Hickey, J. J. 1963. *A guide to bird watching.* Garden City, N.Y.: Doubleday. Reprint, 1975; New York: Dover.

Hinchliffe, J. R. 1977. The chondrogenic pattern in chick limb morphogenesis. In D. A. Ede, J. R. Hinchliffe, and M. Balls, eds., *Vertebrate limb and somite morphogenesis.* Cambridge: Cambridge University Press, 293–308.

Hinchliffe, J. R., and P. J. Griffiths. 1983. The prechondrogenic patterns in tetrapod limb development and their phylogenetic significance. In B. C. Goodwin, N. Holder, and C. G. Wylie, eds., *Development and evolution.* Cambridge: Cambridge University Press, 99–121.

Hinchliffe, J. R., and D. R. Johnson. 1980. *The development of the vertebrate limb.* Oxford: Clarendon Press.

Hoage, R., and L. Goldman. 1986. *Animal intelligence: insights into the animal mind.* Washington, D.C.: Smithsonian Institution Press.

Hodges, R. D. 1981. Endocrine glands. In A. S. King and J. McLelland, eds., *Form and function in birds.* 4 vols. London: Academic Press, 2:149–234.

Houde, P. 1987. Critical evaluation of DNA hybridization studies in avian systematics. *Auk* 104:17–32.

Howard, H. 1929. The avifauna of Emeryville shellmound. *Univ. Calif. Publ. Biol.* 32(2):301–94.

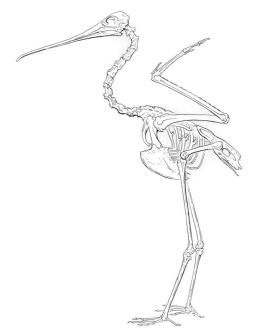

Howard, H. 1955. Fossil birds with special reference to the birds of Rancho LaBrea. *Los Angeles Co. Mus. Sci. Ser.* 17

Hudson, G. E. 1937. Studies on the muscles of the pelvic appendage in birds. *Amer. Midl. Nat.* 18:1-108.

Hudson, G. E., and P. J. Lanzilloti. 1955. Gross anatomy of the wing muscles in the family Corvidae. *Amer. Midl. Nat.* 53:1-44.

Hudson, G. E., and P. J. Lanzillotti. 1964. Muscles of the pectoral limb in galliform birds. *Amer. Midl. Nat.* 71:1-113.

Hudson, G. E., P. J. Lanzillotti, and G. D. Edwards. 1959. Muscles of the pelvic limb in galliform birds. *Amer. Midl. Nat.* 61:1-67.

Humphrey, P. S., and K. C. Parkes. 1959. An approach to the study of molts and plumages. *Auk* 76:1-31.

Huxley, J. S. 1939. A discussion on subspecies and varieties. *Proc. Linn. Soc. London* 151:105-6.

Huxley, T. H. 1867. On the classification of birds and on the taxonomic value of the modifications of certain cranial bones observable in that class. *Proc. Zool. Soc. London* 1867:415-72.

Huxley, T. H. 1868. On the animals which are most nearly intermediate between birds and reptiles. *Geol. Mag.* 5:357-65.

J

Jacob, J., and V. Ziswiler. 1982. The uropygial gland. In D. S. Farner, J. R. King, and K. C. Parkes, eds., *Avian biology.* 8 vols. New York: Academic Press, 6:199-324.

James, F. C. 1970. Geographic size variation in birds and its relationship to climate. *Ecology* 51:365-90.

James, F. C. 1983. Environmental component of morphological differentiation in birds. *Science* 221:184-86.

James, F. C. 1991. Complementary descriptive and experimental studies of clinal variation in birds. *Amer. Zool.* 31:694-706.

Jenkins, F., K. Dial, and G. Goslow, Jr. 1988. A cineradiographic analysis of bird flight: the wishbone in starlings is a spring. *Science* 241:1495-98.

Johnsgard, P. 1961. Tracheal anatomy of the Anatidae and its taxonomic significance. *Twelfth Ann. Rep. Wildfowl Trust,* 58-68.

Johnson, R. F., ed. 1983-88. *Current ornithology.* 5 vols. New York: Plenum Press.

Jones, D. R., and K. Johansen. 1972. The blood vascular system of birds. In D. S. Farner, J. R. King, and K. C. Parkes, eds., *Avian biology.* 8 vols. New York: Academic Press, 2:157-285.

K

Kaufman, K. 1990. *Advanced birding.* Boston: Houghton Mifflin.

King, A. S., and D. Z. King. 1979. Avian morphology: general principles. In A. S. King and J. McLelland, eds., *Form and function in birds.* 4 vols. New York: Academic Press, 1:1-38.

King, A. S., and J. McLelland, eds. 1979-89. *Form and function in birds.* 4 vols. New York: Academic Press.

King, G. M., and D. R. Custance. 1982. *Colour atlas of vertebrate anatomy.* Oxford: Blackwell Scientific.

King, J., and D. Farner. 1965. Studies of fat deposition in migratory birds. *Ann. N.Y. Acad. Sci.* 131:422-40.

Kinsky, F. 1971. The consistent presence of paired ovaries in the kiwi (*Apteryx*) with some discussion of this condition in other birds. *J. für Ornith.* 112:334-57

Kreithen, G., and C. Casler. 1979. Infrasound detection by the homing pigeon. *J. Comp. Phys.* 129:1-4.

Kroodsma, D. E., E. H. Miller, and H. Oullet, eds. 1982. *Acoustic communication in birds.* 2 vols. New York: Academic Press.

Kühne, R., and B. Lewis. 1985. External and middle ears. In A. S. King and J. McLelland, eds., *Form and function in birds.* 4 vols. New York: Academic Press, 3:227-71.

L

Lack, D. 1947. *Darwin's finches.* Cambridge: Cambridge University Press.

Lanyon, W. E. 1979. Hybrid sterility in meadowlarks. *Nature* 279:557-58.

Lasiewski, R. C. 1972. Respiratory function in birds. In D. S. Farner, J. R. King, and K. C. Parkes, eds., *Avian biology*. 8 vols. New York. Academic Press, 2.287–342.

Lasiewski, R. C., and W. A. Calder. 1971. A preliminary allometric analysis of respiratory variables in resting birds. *Resp. Physiol.* 11:152–66.

Laskey, A. R. 1958. Blue Jays at Nashville, Tennessee: movements, nesting, age. *Bird-Banding* 29:211–18.

Lawrence, L. de K. 1953. Nesting life and behaviour of the Red-eyed Vireo. *Can. Field-Nat.* 67:47–87.

Lewontin, R. C. 1978. Adaptation. *Sci. Amer.* 239(3):212–30.

Lincoln, F. C. 1935. *The waterfowl flyways of North America*. U.S. Dept. Agric. Cir. 342. Washington, D.C.: U.S. Fish and Wildlife Service.

Lincoln, F. C. 1939. *The migration of birds*. New York: Doubleday.

Lincoln, F. C. 1950. *Migration of birds*. Cir. 16. Washington, D.C.: U.S. Fish and Wildlife Service.

Lofts, B., and R. Murton. 1973. Reproduction in birds. In D. S. Farner, J. R. King, and K. C. Parkes, eds., *Avian biology*. 3:1–107.

Lowery, G. 1951. A quantitative study of nocturnal migration in birds. *Univ. Kans. Publ. Nat. Hist.* 3(2): 361–472.

Lowery, G., and R. Newman. 1955. Direct studies of nocturnal migration. In A. Wolfson, ed., *Recent studies in avian biology*. Urbana: University of Illinois Press.

Lucas, A. M., and P. R. Stettenheim. 1972. *Avian anatomy: integument*. Agric. Handb. 362:1–340. Washington, D.C.: U.S. Government Printing Office.

M

McClure, H. E. 1984. The occurrence of hippoboscid flies on some species of birds in southern California. *J. Field Ornith.* 55:230-40.

McKenzie, B. 1989. Recording equipment. *Birding* 21(5):250.

McLelland, J. 1979. Digestive system. In A. S. King and J. McLelland, eds., *Form and function in birds*. 4 vols. New York: Academic Press, 1:69–181.

Marsh, O. C. 1880. *Odontornithes: a monograph on the extinct toothed birds of North America*. Engineering Dept., U.S. Army, 18:1–201. Washington, D.C.: U.S. Government Printing Office.

Marshall, A. J., ed. 1960–61. *Biology and comparative physiology of birds*. 2 vols. New York: Academic Press.

Martin, G. R. 1985. Eye. In A. S. King and J. McLelland, eds., *Form and function in birds*. 4 vols. New York: Academic Press, 3:311–73.

Martin, L. D. 1983. The origin and early radiation of birds. In A. H. Brush and G. A. Clark, Jr., eds., *Perspectives in ornithology: essays presented for the centennial of the American Ornithologists' Union*. Cambridge: Cambridge University Press, 291–337.

Martin, L. D. 1985. The relationship of *Archaeopteryx* to other birds. In M. K. Hecht, J. H. Ostrom, G. Viohl, and P. Wellnhofer, eds., *The beginnings of birds*. Eichstatt: Freunde des Jura–Museums, 177–84.

Martin, L. D., J. D. Stewart, and K. N. Whetstone. 1980. The origin of birds: structure of the tarsus and teeth. *Auk* 97:86–93.

Matthiessen, P. 1967. In G. D. Stout, ed., *The shorebirds of North America*. New York: Viking Press.

Mayr, E. 1942. *Systematics and the origin of species*. New York: Columbia University Press.

Mayr, E. 1969. *Principles of systematic zoology*. New York: McGraw-Hill.

Mayr, E. 1970. *Populations, species, and evolution*. Cambridge: Belknap Press of Harvard University Press.

Mead, C. 1976. *Bird migration*. New York: Facts on File.

Meriwether, W. 1986. ABA area: bird sound recordings; some suggestions. *Birding* 18(1):30–33.

Miller, A. H. 1928. The molts of the Loggerhead Shrike *Lanius ludovicianus*. *Univ. Calif. Publ. Zool.* 30:393–417.

Miller, A. H. 1946. A method of determining the age of live passerine birds. *Bird-Banding* 37:33–35.

Miller, A. H., and R. C. Stebbins. 1964. *The lives of desert animals in Joshua Tree National Monument*. Berkeley: University of California Press.

Miller, M. R. 1975. Gut morphology of Mallards in relation to diet quantity. *J. Wildl. Mgmt.* 39:168–73.

Mitård, U. 1980. Arteriovenous anastomoses and vascularity in the feet of eiders and gulls (Aves). *Zoomorph.* 96:263–70.

Mitård, U. 1984. Blood vessels and the occurrence of arteriovenous anastomoses in cephalic heat loss areas of mallards, *Anas platyrhynchos* (Aves). *Zoomorph.* 104:323 35.

Mortensen, H. C. C. 1899–1922. *Studies in bird migration, being the collected papers of H. Chr. C. Mortensen*, ed. P. Jespersen and A. V. Tanning. Copenhagen: Dansk Orn. Foren. Tidsskr.

Mueller, H., and D. Berger. 1961. Weather and fall migration of hawks at Cedar Grove, Wisconsin. *Wilson Bull.* 73:171–92.

Murphy, R. C. 1936. *Oceanic birds of South America.* New York: Macmillan.

Myers, A. A., and P. S. Giller, eds. 1988. *Analytical biogeography: an integrated approach to the study of animal and plant distributions.* London: Chapman and Hall.

N

National Geographic Society. 1983. *Field guide to the birds of North America*, ed. S. L. Scott. Washington, D.C.: National Geographic Society.

Neal, H. V., and H. W. Rand. 1939. *Chordate anatomy.* Philadelphia: Blakiston's Son.

Necker, R. 1985. Receptors in the skin of the wing of the pigeon and their possible role in bird flight. In W. Nachtigall, ed., *Bird flight.* Stuttgart: Gustav Fisher, 433–44.

Nero, R. W. 1951. Pattern and rate of cranial ossification in the House Sparrow. *Wilson Bull.* 63(1):84–88.

Newton, A. 1896. *A dictionary of birds.* London: Adam and Charles Black.

Norberg, R. Å. 1977. Occurrence and independent evolution of bilateral asymmetry in owls and implications for owl taxonomy. *Phil. Trans. Royal Soc. London* B280:375–408.

Norberg, R. Å. 1978. Skull asymmetry, ear structure and function, and auditory localization in Tengmalm's Owl, *Aegolius funereus* (Linné). *Phil. Trans. Royal Soc. London* B282:325–410.

Norris, R. 1961. A modification of the Miller method of aging live passerine birds. *Bird-Banding* 32:55–57.

Norris, R., and F. Williamson. 1955. Variation in the relative heart size of certain passerines with increase in altitude. *Wilson Bull.* 67:78–83.

O

O'Brien, S. J., and E. Mayr. 1991. Bureaucratic mischief: recognizing endangered species and subspecies. *Science.* 251:1187–88.

O'Connor, R. 1984. *The growth and development of birds.* New York: John Wiley and Sons.

Olson, S. L. 1985. The fossil record of birds. In D. S. Farner, J. R. King, and K. C. Parkes, eds. *Avian biology.* 8 vols. New York: Academic Press, 8:79–238.

Olson, S. L., and A. Fedducia. 1979. Flight capability and the pectoral girdle of *Archaeopteryx. Nature* 278:247-48.

Ostrom, J. H. 1969. The osteology of *Deinonychus antirrhopus*, an unusual theropod from the lower Cretaceous of Montana. *Bull. Peabody Mus. Nat. Hist.* 30:1–165.

Ostrom, J. H. 1970. Archaeopteryx: notice of a "new" specimen. *Science* 170:537–38.

Ostrom, J. H. 1974. *Archaeopteryx* and the origin of flight. *Q. Rev. Biol.* 49:27–47.

Ostrom, J. H. 1975. The origin of birds. *Ann. Rev. Earth Planet Sci.* 3:55–77.

Ostrom, J. H. 1976a. Some hypothetical stages in the evolution of avian flight. In S. L. Olsen, ed., *Collected papers in avian paleontology. Smithsonian Contr. Paleobiol.* 27:1–21.

Ostrom, J. H. 1976b. *Archaeopteryx* and the origin of birds. *Biol. J. Linn. Soc.* 8:91–182.

Ostrom, J. H. 1978. The osteology of *Compsognathus longipes* Wagner. *Zitteliana* 4:73–118.

Ostrom, J. H. 1979. Bird flight: how did it begin? *Amer. Sci.* 67:46–56.

Ostrom, J. H. 1985. The meaning of *Archaeopteryx*. In M. K. Hecht, J. H. Ostrom, G. Viohl, and P. Wellnhofer, eds., *The beginnings of birds.* Eichstatt: Freunde des Jura-Museums, 161–76.

Ostrom, J. H. 1987. Reasurring the dinosaurs. *The Sciences* 27(3):56 63.

Owen, D. F. 1963. Polymorphism in the Screech Owl in eastern North America. *Wilson Bull.* 75:183–90.

P

Padian, K. 1985. The origins and aerodynamics of flight in extinct vertebrates. *Paleontology* 28(3):413–33.

Padian, K., ed. 1986. *The origins of birds and the evolution of flight*. Mem. Calif. Acad. Sci. 8.

Palmer, R. S., ed. 1962–1988. *Handbook of North American birds*. 5 vols. Vol. 1: *Loons through flamingos*. New Haven: Yale University Press.

Parkes, K. C. 1966. Speculations on the origin of feathers. *Living Bird* 5:77-86.

Pasquier, R. F. 1977. *Watching birds: an introduction to ornithology*. Boston: Houghton Mifflin.

Paul, G. S. 1988. *Predatory dinosaurs of the world*. New York: Simon and Schuster.

Payne, R. S. 1971. Acoustic location of prey by Barn Owls (*Tyto alba*). *J. Exp. Biol.* 54:535–73.

Peakall, D. 1970. DDT: effects on calcium metabolism and concentration of estradiol in the blood. *Science* 168:592-94.

Pearson, R. 1972. *The avian brain*. New York: Academic Press.

Pennycuick, C. J. 1975. Mechanics of flight. In D. S. Farner, J. R. King, and K. C. Parkes, eds., *Avian biology*. 8 vols. New York: Academic Press, 5:1–75.

Perrins, C. 1987. *New generation field guide to the birds of Britain and Europe*. Austin: University of Texas Press.

Peters, D. S., and W. Fr. Gutmann. 1985. Construction and functional preconditions for the transition to powered flight in vertebrates. In M. K. Hecht, J. H. Ostrom, G. Viohl, and P. Wellnhofer, eds., *The beginnings of birds*. Eichstatt: Freunde des Jura-Museums, 233–42.

Peters, H. S. 1936. A list of external parasites from birds of the eastern part of the United States. *Bird-Banding* 7:9-27.

Peterson, A. 1985. The locomotor adaptations of *Archaeopteryx*: glider or cursor? In M. K. Hecht, J. H. Ostrom, G. Viohl, and P. Wellnhofer, eds., *The beginnings of birds*. Eichstatt: Freunde des Jura-Museums, 99–103.

Peterson, R. T. 1963. *The birds*. New York: Time-Life Nature Library.

Peterson, R. T. 1980. *A field guide to the birds*. 4th ed. Boston: Houghton Mifflin.

Petrak, M. L., ed. 1969. *Diseases of cage and aviary birds*. Philadelphia: Lea and Febiger.

Pettingill, O. S., Jr. 1970. *Ornithology in laboratory and field*. 4th ed. Minneapolis: Burgess.

Pettingill, O. S., Jr. 1985. *Ornithology in laboratory and field*. 5th ed. New York: Academic Press.

Portmann, A., and W. Stingelin. 1961. The central nervous system. In A. J. Marshall, ed., *Biology and comparative physiology of birds*. 2 vols. New York: Academic Press, 2:1–36.

Portmann, A., and W. H. Stingelin. 1964. Development, embryonic. In A. L. Thomson, ed., *A new dictionary of birds*. New York: McGraw-Hill.

Pough, F. H., J. B. Heiser, and W. N. McFarland. 1989. *Vertebrate life*. 3d ed. New York: Macmillan.

Poulin, R. 1991. Group-living and infestation by ectoparasites in passerines. *Condor* 93(2):321–24.

Powell, F. L., and P. Scheid. 1989. Physiology of gas exchange in the avian respiratory system. In A. S. King and J. McLelland, eds., *Form and function in birds*. 4 vols. New York: Academic Press, 4:393–437.

Power, D. M., ed. 1989–91. *Current ornithology*. Vols. 6–8. New York: Plenum Press.

Prange, H. D., J. F. Anderson, and H. Rahn. 1979. Scaling of skeletal mass to body mass in birds and mammals. *Amer. Nat.* 113:103–22.

Prater, A. J., J. H. Marchant, and J. Vuorinen. 1977. *Guide to the identification and aging of Holarctic waders*. Tring: British Trust for Ornithology.

Preston, F. W. 1969. Shapes of birds' eggs: extant North American families. *Auk* 86:246–64.

Price, P. W. 1980. *Evolutionary biology of parasites*. Princeton, N.J.: Princeton University Press.

R

Rahn, H., A. Ar, and C. V. Pagnelli. 1979. How bird eggs breathe. *Sci. Amer.* 240(2):46–55.

Raikow, R. J. 1974. Species-specific foraging behavior in some Hawaiian Honeycreepers (*Loxops*). *Wilson Bull.* 86:471-74.

Raikow, R. J. 1976. The origin and evolution of the Hawaiian Honeycreepers (Drepanididae). *Living Bird*. 15th annual. Ithaca, N.Y.: Cornell University.

Raikow, R. J. 1985a. Locomotor system. In A. S. King and J. McLelland, eds., *Form and function in birds*. 4 vols. London: Academic Press, 3:57–147.

Raikow, R. J. 1985b. Problems in avian classification. In R. F. Johnson, ed., *Current ornithology*. Vol. 2. New York: Plenum Press, 187–212.

Raikow, R. J., L. Bicanovsksy, and A. Bledsoe. 1988. Forelimb joint mobility and the evolution of wing-propelled diving birds. *Auk* 105:446–51.

Rao, C., and P. C. Sereno. 1990. Early evolution of the avian skeleton: new evidence from the Lower Cretaceous of China. *Abs. J. Vert. Paleont.* 10(3):38A.

Rayner, J. M. V. 1988. Form and function in avian flight. In A. S. King and J. McLelland, eds., *Form and function in birds.* 4 vols. New York: Academic Press, 3:57–147.

Regal, P. J. 1975. The evolutionary origin of feathers. *Q. Rev. Biol.* 50:35-66.

Regal, P. J. 1985. Common sense and reconstructions of the biology of fossils: *Archaeopteryx* and feathers. In M. K. Hecht, J. H. Ostrom, G. Viohl, and P. Wellnhofer, eds., *The beginnings of birds.* Eichstatt: Freunde des Jura–Museums.

Remsen, J. V. 1977. On taking field notes. *Amer. Birds* 31(5):946-53.

Richardson, W. J. 1974. Spring migration over Puerto Rico and the western Atlantic: a radar study. *Ibis* 116:172–93.

Richardson, W. J. 1976. Autumn migration over Puerto Rico and the western Atlantic: a radar study. *Ibis* 118:309–32.

Richardson, W. J. 1978. Timing and amount of bird migration in relation to weather: a review. *Oikos* 30:224-72.

Ricklefs, R. E. 1979. *Ecology.* 2d ed. New York: Chiron Press.

Rising, J. D. 1983. The Great Plains hybrid zone. In R. F. Johnston, ed., *Current ornithology.* Vol. 1. New York: Plenum Press, 131-57.

Robbins, C. S., B. Bruun, and H. S. Zim. 1983. *Birds of North America.* 2d ed. New York: Golden Press.

Romanoff, A., and A. Romanoff. 1949. *The avian egg.* New York: John Wiley.

Romer, A. S. 1962. *The vertebrate body.* 3d ed. Philadelphia: W. B. Saunders.

Rüppell, G. 1977. *Bird flight.* New York: Van Nostrand Reinhold.

Russell, F. 1983. Winter: the enduring challenge. In R. M. Poole, ed., *The wonder of birds.* Washington, D.C.: National Geographic Society, 190–221.

S

Sacilotto, I. H. 1989. Finding the equipment that's right for you. *Birding* 21:10–13.

Salt, G., and E. Zeuthen. 1960. The respiratory system. In A. J. Marshall, ed., *Biology and comparative physiology of birds.* 2 vols. New York: Academic Press, 1:363–409.

Salt, G. W. 1964. Respiratory evaporation in birds. *Biol. Rev. Camb. Phil. Soc.* 39:113–36.

Scheid, P., and J. Piiper. 1971. Direct measurement of the pathway of respired air in duck lungs. *Resp. Physiol.* 11:308-14.

Scheid, P., and J. Piiper. 1989. Respiratory mechanics and air flow in birds. In A. S. King and J. McLelland, eds., *Form and function in birds.* 4 vols. New York: Academic Press, 4:369–91.

Schmidt-Nielsen, K. 1971. How birds breathe. *Sci. Amer.* 225(6):72-79.

Schmidt-Nielsen, K. 1983. *Animal physiology: adaptation and environment.* 3d ed. Cambridge: Cambridge University Press.

Schmidt-Nielsen, K., J. Kanwisher, R. C. Lasiewski, J. E. Cohn, and W. L. Bretz. 1969. Temperature regulation and respiration in the Ostrich. *Condor* 71:341–52.

Schwartzkopff, J. 1973. Mechanoreception. In D. S. Farner, J. R. King, and K. C. Parkes, eds., *Avian biology.* 8 vols. New York: Academic Press, 3:417–77.

Sheldon, W. P. 1967. *The book of the American woodcock.* Amherst: University of Massachusetts Press.

Shields, G. F., and K. M. Helm–Bychowski. 1988. Mitochondrial DNA of birds. In R. F. Johnson, ed., *Current ornithology.* Vol. 5. New York: Plenum Press, 273–95.

Short, L. L., Jr. 1963. Hybridization in wood warblers *Vermivora pinus* and *V. chrysoptera. Proc. XIII Int. Ornith. Cong.* 147–60.

Shulfeldt, R. 1909. Osteology of the birds. *Bull. N.Y. State Mus.* 130:5–381.

Sibley, C. G., and J. E. Ahlquist. 1983. Phylogeny and classification of birds based on the data of DNA–DNA hybridization. In R. F. Johnson, ed., *Current ornithology.* Vol. 1. New York: Plenum Press, 245–92.

Sibley, C. G., and J. E. Ahlquist. 1986. Reconstructing bird phylogeny by comparing DNAs. *Sci. Amer.* 254(2):82–92.

Sibley, C. G., and J. E. Ahlquist. 1990. *Phylogeny and classification of birds: a study in molecular evolution.* New Haven: Yale University Press.

Sibley, C. G., J. E. Ahlquist, and B. L. Monroe. 1988. A classification of the living birds of the world based on DNA-DNA hybridization studies. *Auk* 105:409–23.

Sibley, C. G., and B. L. Monroe. 1990. *Distribution and taxonomy of birds of the world.* New Haven: Yale University Press.

Sillman, A. J. 1973. Avian vision. In D. S. Farner, J. R. King, and K. C. Parkes, eds., *Avian biology.* 8 vols. New York: Academic Press, 3:349–87.

Simmons, K. E. L. 1964. Feather maintenance. In A. L. Thomson, ed., *A new dictionary of birds.* New York: McGraw-Hill.

Simons, J. R. 1960. The blood vascular system. In A. J. Marshall, ed., *Biology and comparative physiology of birds.* 2 vols. New York: Academic Press, 1:345–62.

Skutch, A. F. 1973. *The life of the hummingbird.* New York: Crown.

Skutch, A. F. 1976. *Parent birds and their young.* Austin: University of Texas Press.

Small, A. 1975. *The birds of California.* New York: Winchester Press.

Smith, N. G. 1968. The advantages of being parasitized. *Nature* 219:690–94.

Smithe, F. B. 1975. *Naturalist's color guide.* New York: American Museum of Natural History.

Sparks, J., and T. Soper. 1970. *Owls: their natural and unnatural history.* New York: Tapplinger.

Stager, K. 1964. The role of olfaction in food location by the Turkey Vulture (*Cathartes aura*). Los Angeles Co. Mus. *Contr. Sci.* 81.

Stettner, L., and K. A. Matyniak. 1968. The brain of birds. *Sci. Amer.* 218:64–76.

Stettner, L., and W. Schultz. 1967. Brain lesions in birds: effects on discrimination acquisition and reversal. *Science* 155:1689-92.

Stewart, R. E. 1953. A life history study of the Yellow-throat. *Wilson Bull.* 65:99–115.

Stiles, E. 1982. Expansion of Mockingbirds and *Multiflora* roses in the northeast U.S. and Canada. *Amer. Birds* 36:358–64.

Stokes, D. W. 1979. *A guide to the behavior of common birds.* Vol. 1. Boston: Little, Brown.

Stone, W. 1965. *Bird studies at Old Cape May.* New York: Dover.

Storer, R. W. 1971. Classification of birds. In D. S. Farner, J. R. King, and K. C. Parkes, eds., *Avian biology.* 8 vols. New York: Academic Press, 1:1–18.

Streseman, E. 1927–34. Aves. In W. Kukenthal and T. Krumbach, eds., *Handbuch der Zoologie. Sauropsida: Aves.* 7 vols. Berlin: W. de Gruyter, VIIB:729–853.

Sturkie, P. D. 1954. *Avian physiology.* Ithaca, N.Y.: Comstock.

Sturkie, P. D., ed. 1976. *Avian physiology.* 3d ed. New York: Springe-Verlag.

Sturkie, P. D., ed. 1986. *Avian physiology.* 4th ed. New York: Springer-Verlag.

Sturtevant, A. H. 1965. *A history of genetics.* New York: Harper and Row.

Sy, M. 1936. Funktionell-anatomische Untersuchungen am Vogelflügel. *J. für Ornith.* 84:199–296.

T

Tansley, K. 1964. Vision. In A. L. Thomson, ed., *A new dictionary of birds.* New York: McGraw-Hill.

Tarsitano, S., and M. K. Hecht. 1980. A reconsideration of the reptilian relationships of *Archaeopteryx. Zool. J. Linn. Soc.* 69:149–82.

Taylor, T. G. 1970. How an eggshell is made. *Sci. Amer.* 222(3):88–95.

Terres, J. K. 1980. *The Audubon Society encyclopedia of North American birds.* New York: Alfred A. Knopf.

Terres, J. K. 1987. *How birds fly.* New York: Harper and Row. (Reprint of *Flashing wings.* 1968. New York: Doubleday.)

Thomson, A. L., ed. 1964. *A new dictionary of birds.* New York: McGraw-Hill.

Thomson, J. A. 1923. *The biology of birds.* London: Macmillan.

Tinbergen, N. 1951. *The study of instinct.* Oxford: Clarendon Press.

Tinbergen, N. 1960. *The Herring Gull's world: a study of the social behavior of birds.* New York: Basic Books.

Tucker, V. 1969. The energetics of bird flight. *Sci. Amer.* 220(5):70-78.

U V

U.S. Fish and Wildlife Service. 1980. *North American bird banding techniques.* Washington, D.C.: Department of the Interior, U.S. Fish and Wildlife Service.

Van Tyne, J., and A. J. Berger. 1959. *Fundamentals of ornithology.* New York: John Wiley.

Van Tyne, J., and A. J. Berger. 1976. *Fundamentals of ornithology,* 2d ed. New York: John Wiley

and Sons.

Vazquez, R. J. 1992. Functional osteology of the avian wrist and the evolution of flapping flight. *J. Morph.* 211:259–68.

Voous, K. H. 1988. Owls of the northern hemisphere. Cambridge: MIT Press.

W

Walker, A. D. 1972. New light on the origin of birds and crocodiles. *Nature* 237:257–63.

Walker, A. D. 1974. Evolution, organic. In D. L. Lapedes, *Yearbook of science and technology.* New York: McGraw-Hill.

Walker, A. D. 1977. Evolution of the pelvis in dinosaurs and birds. In S. M. Andrews, R. S. Miles, and A. D. Walker, eds., *Problems in vertebrate evolution.* New York: Academic Press, 319–58.

Wallace, G. J. 1963. *An introduction to ornithology.* 2d ed. New York: Macmillan.

Wallace, G. J., and H. D. Mahan. 1975. *An introduction to ornithology.* 3d ed. New York: Macmillan.

Walls, G. 1942. *The vertebrate eye and its adaptive radiation.* Bloomfield Hills, Mich.: Cranbrook Institute of Science.

Walter, H. E., and L. P. Sayles. 1949. *Biology of the vertebrates.* 3d ed. New York: Macmillan.

Webster, D., and M. Webster. 1974. *Comparative vertebrate morphology.* New York: Academic Press.

Wellnhofer, P. 1974. Das fünfte skelettexemplar von *Archaeopteryx. Paleontographica* 147:169–216.

Welty, J. C., and L. F. Baptista. 1988. *The life of birds.* 4th ed. New York: W. B. Saunders.

Wenzel, B. M. 1973. Chemoreception. In D. S. Farner, J. R. King, and K. C. Parkes, eds., *Avian biology.* 8 vols. New York: Academic Press, 3:389–415.

Wetmore, A. 1936. The number of contour feathers in passeriform and related birds. *Auk* 53:159–69.

Wetmore, A. 1960. A classification for the birds of the world. *Smithsonian Misc. Coll.* 139(11):1-37.

Weymouth, R., R. Lasiewski, and A. J. Berger. 1964. The tongue apparatus in hummingbirds. *Acta Anat.* 58:252–70.

Whetstone, K. N., and L. D. Martin. 1979. New look at the origin of birds and crocodiles. *Nature* 279:234–36.

Whetstone, K. N., and L. D. Martin. 1981. Common ancestry of birds and crocodiles: a reply. *Nature* 289:98.

Wheeler, T. A., and W. Threlfall. 1986. Observations on the ectoparasites of some Newfoundland passerines (Aves: Passeriformes). *Can. J. Zool.* 64:630–36.

Wickstrom, D. 1988. Tools of the trade: bird recording equipment. *Birding* 20(4):262–66.

Wilbur, S. R, and J. A. Jackson, eds. 1983. *Vulture biology and management.* Berkeley: University of California Press.

Williams, T. C., J. M. Williams, L. C. Ireland, and J. M. Teal. 1977. Autumnal bird migration over the western North Atlantic Ocean. *Amer. Birds* 31:251–67.

Wilson, B., ed. 1980. *Birds: readings from* Scientific American. San Francisco: W. H. Freeman.

Wilson, E. O. 1972. Animal communication. *Sci. Amer.* 227(3):52-60.

Wilson, E. O. 1977. *Sociobiology: the new synthesis.* Cambridge: Belknap Press of Harvard University Press.

Wingstand, K. G., and O. Munk. 1965. The pecten oculi of the pigeon with particular regard to its function. *Biol. Skr. Dan. Videnskab.* 14:1–64.

Witschi, E. 1961. Sex and secondary sexual characters. In A. J. Marshall, ed., *Biology and comparative physiology of birds.* 2 vols. New York: Academic Press, 2:115–68.

Wood, M. 1969. *A bird-bander's guide to determination of age and sex of selected species.* College of Agriculture, Pennsylvania State University.

Worden, A. N. 1964. Alimentary system. In A. L. Thomson, ed., *A new dictionary of birds.* New York: McGraw-Hill.

Z

Zeigler, H. 1963. Effects of endbrain lesions on visual discrimination learning in pigeons. *J. Comp. Neur.* 120(2):161-94.

Zimmer, C. 1992. Ruffled Feathers. *Discover*. 13(5):44–54.

Zink, R. M., and J. V. Remsen, Jr. 1986. Evolutionary processes and patterns of geographical variation in birds. In R. F. Johnson, ed.*Current ornithology*. Vol. 4. New York: Plenum Press, 1-69.

Ziswiler, V., and D. S. Farner. 1972. Digestion and the digestive system. In D. S. Farner, J. R. King, and K. C. Parkes, eds., *Avian biology*. 8 vols. New York: Academic Press, 2:343–430.

Zuckerkandl, E., and L. Pauling. 1962. Molecular disease, evolution, and genetic heterogeneity. In M. Kasha and B. Pullman, eds., *Horizons in biochemistry*. New York: Academic Press, 189–225.

Zusi, R. L. 1987. A feeding adaptation of the jaw articulation in the new world jays (Corvidae). *Auk* 104:655–80.

Index